China Builds the Bomb

China Builds the Bomb

John Wilson Lewis and Xue Litai

Illustrated Edition

STANFORD UNIVERSITY PRESS
Stanford, California

Studies in International Security and Arms Control
John W. Lewis and William J. Perry, General Editors

Sponsored by the Center for International Security and Arms Control
Stanford University

Stanford University Press
Stanford, California
© 1988 by the Board of Trustees of the
Leland Stanford Junior University
Printed in the United States of America

Original printing 1988
Last figure below indicates year of this printing:
00   99   98   97   96   95   94   93   92   91

CIP data appear at the end of the book

FOR JACQUELYN AND QIPING

Preface

To the eye of history, many things once hidden are now visible. But, finally, when we glimpse some small part of important past events, we look for meaning, and it is truly said that "in every object there is inexhaustible meaning; the eye sees in it what the eye brings means of seeing." So it is with our quest to learn why and how China built the bomb.

That quest began more than a decade ago. As China and the United States started improving their relations in the early 1970s, we took the initial steps that eventually led to our coauthoring this book. For years, each of us had looked across the Pacific, albeit from opposite sides, to a place that could never quite be reached. Then, in what seemed a moment, reality replaced fantasy, and we reached out to a place that once had seemed so hidden, even threatening. And thus almost by accident we came to this joint effort on China's nuclear weapons program.

How well prepared we were for the task only others can judge and time tell. Our interests, long in developing, ranged from Chinese politics to arms control. The atomic cloud at Hiroshima had cast its spell over our careers, even as we lived under the shadow of past American and Chinese hostilities. In the first three postwar decades, the revolutions of man and warfare exacted an appalling toll, and in the attendant turmoil nations acted in ways that might have brought destruction to us all. The Chinese bomb was at once a legacy of those times and a new factor in future world affairs. Our study was more than an inquiry into Chinese history. It became a search for answers, however partial and tentative, to some of the great questions of our age.

Writing this book required that we try to grasp as fully as possible what happened—in politics, industrial technology, military thinking, high science, and national values. In our story of the Chinese bomb, we have dealt with each of these key aspects in examining the Politburo's decision in January 1955 to commence the bomb program—Project 02—and how that program progressed thereafter.

In Chapter 1, we note the many sources on which we have drawn. The published sources, all unclassified both in China and in the United States, provided an immense amount of data, and we have done our best to sort those sources out, reconcile the many discrepancies, and explain what they meant to the Chinese who participated and to others. Beijing clearly has been seeking to reveal the main historical outlines of its strategic weapons programs, and without these outlines we scarcely could have begun. How much the Chinese have elected to publish truly surprised us, and we expect much more to come.

As we delved into the subject, we recognized what a major task lay before us. How could we, social scientists, presume to sift through and make intelligent sense of data on uranium prospecting or the chemical makeup of a neutron initiator? What competence did we possess to evaluate the many technical sources with gaps and even outright contradictions? In this endeavor, we especially profited by being close to scientists, historians, and former governmental officials at Stanford University's Center for International Security and Arms Control. Need a scientific assessment, a lecture on uranium processing or postwar U.S. foreign policy, or just some common sense? Ask Barton Bernstein, Sidney Drell, Jack Evernden, Philip Farley, Alexander George, David Holloway, Gerald Johnson, Robert Mozley, Wolfgang Panofsky, Theodore Postol, Rudolph Sher, or Lynn Sykes (Columbia University). All of these colleagues gave liberally of their time and knowledge to discuss the history of other bomb programs, to explain a problem of physics or nuclear testing, to listen to a just-forming idea, or to read and comment on drafts.

Our special thanks go to Sid, our friend and special colleague at the Center, for his ceaseless encouragement and for so willingly agreeing to write the Foreword to this book.

We also faced the nightmare that comes from deciding on what must seem trivial matters. All things Chinese are cited and spelled in many different ways. No system is perfect, and the several "standard" orthographies all posed special problems. With some misgivings and an exception for the names of people and places under the jurisdiction of Taiwan, we have used standard pinyin and, for Chinese place-names and boundaries, the pinyin gazetteer in *Zhonghua Renmin Gongheguo*

Fen Sheng Dituji (Collection of Provincial Maps of the People's Republic of China; Beijing, 1983).

The result of our initial labors left an open field for gifted editors, mapmakers, proofreaders, manuscript preparers, and other heroines and heroes. We owe much to a great many. Here we especially wish to thank Barbara Mnookin of Stanford University Press and Miriam DeJongh of the International Strategic Institute at Stanford (ISIS) for their editorial assistance, which included but far surpassed putting the manuscript into shape. Rosemary Hamerton-Kelly, who thought she had other jobs at ISIS, helped on editing, proofreading, and managing the computer. So, too, did Justina Chau, whom we especially thank for digging out materials, typing, and preparing the maps for the cartographer. Anca Ruhlen, the ISIS librarian, made the unending search for references and special scientific materials a real adventure, and Alison Brysk provided important assistance in the early stages of our research.

The generous help that supported our research and writing came from many different sources. We wish to thank the Carnegie Corporation of New York, the William and Flora Hewlett Foundation, the Columbia Foundation, the General Service Foundation, and the John D. and Catherine T. MacArthur Foundation. A special word of appreciation also goes to Gerry Bowman, Marjorie Kiewit, Helen Morales, and Nancy Okimoto, each of whom helped in making this book a reality. We must add the obvious truth, of course, that we alone bear the responsibility for this volume, its faults as well as its ideas and judgments.

While completing this book, we incurred many debts. We have tried to acknowledge them even if we cannot truly repay them. For our wives who so wonderfully have sustained us during these many months, a special word of thanks: we dedicate this book to them.

John W. Lewis
Xue Litai

Contents

Foreword by Sidney D. Drell   xvii

1. China's Quest for Security   1
2. American Power and Chinese Strategy, 1953-1955   11
3. The Strategic Decision and Its Consequences   35
4. The Uranium Challenge   73
5. The Production of Fissionable Material   104
6. The Design and Manufacture of the Bomb   137
7. The Final Countdown   170
8. Strategic Doctrines and the Hydrogen Bomb   190
9. Chinese Lessons and the Global Nuclear Experience   219

   Photo sections follow pp. 108 and 196

   Appendixes
   A. Statement of the Government of the People's Republic of
      China, October 16, 1964   241
   B. China's Nuclear Weapons Tests, 1964-1978   244
   C. Key Figures in China's Nuclear Weapons Program,
      1954-1967   246

   Notes   253
   References Cited   293
   Index   313

Maps, Tables, and Figures

MAPS

People's Republic of China   xiv
1. East China   23
2. Southeast China   79
3. Northwest China   110
4. Xinjiang   173

TABLES

1. Characteristics of the Important Uranium-Mining
   Techniques Used in China   82
2. Principal Defects of the Uranium-Mining Techniques
   Used in China   83
3. Comparative Yields of the Uranium-Mining Techniques
   Used in China, 1966-1980   84

FIGURES

1. Organization of the Chinese Nuclear Weapons Program,
   1959-1964   56
2. Organization of the Second Ministry of Machine
   Building, 1959-1964   58

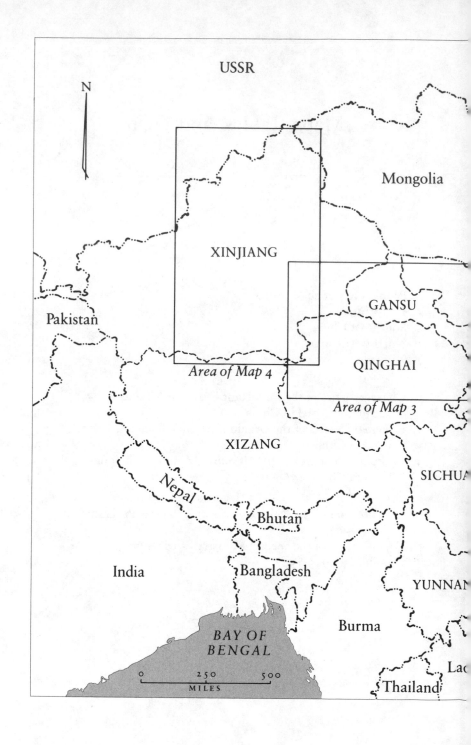

Foreword

Atomic bombs have been a major factor in international diplomacy since the first explosions over Hiroshima and Nagasaki in August 1945. The world community has wrestled ever since with the problem of reducing the dangers that these weapons of mass destruction pose to the very survival of our civilization. These efforts—and the deadly dangers of nuclear conflict—have been discussed extensively in a large body of writings. However, aside from extensive studies of the U.S. and British bomb programs, our knowledge of how other nations—and in particular China and the Soviet Union—went about building the bomb and starting a nuclear arsenal is relatively limited.

It is the important—and fascinating—contribution of this book that we now have the first in-depth information on China's development of the atom bomb—the hows, whys, and wherefores of the high-technology achievement of a society with a primitive economic and technical base. Drawing on new material and the authors' deep understanding of China itself, this account will interest not just scholars, but all who have wondered how such a sophisticated technological/military feat could have been accomplished by a poverty-stricken nation with limited industrial and scientific resources—a feat the more amazing for being accomplished amid the enormous political turmoil of the Great Leap Forward and the "three hard years."

In developing their history of the Chinese program to acquire nuclear weapons (a similar study of the Chinese missile program will follow), John Lewis and Xue Litai have had access to a great deal of information previously held secret by the Chinese government. They carefully piece together a picture of how the organization was forged, of its leaders and their management methods. They describe the relations between scientists and workers, the military and civilians, the government and the project, and—before the rupture in 1960—the People's Republic and the Soviet Union.

They argue that Chinese fears of the threat of U.S. nuclear blackmail, impelled by our own fears of a Communist takeover in Asia in the aftermath of the Korean War and the crisis in the Taiwan Strait in the early 1950s, made the Chinese leaders feel that nuclear weapons were essential to their security, and that "the leadership's commitment and faith in the nation's scientific prowess or potential over the long run played an important role" in achieving their goal.

They also chronicle the policy changes during China's emergence as a modern nuclear power. Mao originally formulated the revolutionary struggle in terms of "people's war" and in 1946 derided the bomb as a "paper tiger"; but with the development of modern nuclear bombs and missiles, he recast the struggle into one with a military-technical emphasis that relied on assured nuclear retaliation to ensure deterrence. Like so many other nations, China accommodated its policy to technology, rather than the other way around.

In between the analyses of the political and strategic issues lies the heart of this work: a description of how the Chinese created the means to "master the uranium challenge"—mining the materials, building the factories—all culminating at 3:00 P.M. on October 16, 1964, when their test team successfully exploded an atom bomb with a yield of roughly 20 kilotons. Curiously enough, it was an enriched uranium bomb implosively detonated. Chinese bomb builders, with a supply of fuel from their gaseous-diffusion plant for enriched uranium, had mastered the advanced technique of implosive detonation—originally designed for a bomb fueled by plutonium—and applied it to a bomb fueled by enriched uranium. This was an enormously impressive achievement. The project leader, Nie Rongzhen, working with the top layer of government leaders while forging a coherent and efficient scientific organization, beat his own schedule in spite of food shortages, political upheavals, and the Soviet defection.

It took less than ten years from the date of the commitment to build a bomb in 1955—before it was even known whether there were any concentrations of uranium in China—to the first detonation. The PRC detonated its first boosted nuclear weapons less than two years later, followed, the next year, by a large multistage thermonuclear bomb. By then the country's missile force was also growing. The authors have given a detailed and fascinating chronicle of this major achievement, and one that succeeds in conveying its full drama.

Sidney D. Drell

China Builds the Bomb

ONE

China's Quest for Security

On October 16, 1964, China detonated its first nuclear weapon in the Xinjiang desert near the oasis of Huangyanggou, 150 kilometers northwest of the Lop Nur marshes. The test marked a historic feat for a leadership then celebrating its fifteenth anniversary in power. The official statement announcing the explosion called the event a "major achievement."[1] China, the statement continued, has become a nuclear weapons state after a decade of struggle to strengthen its defenses and to "oppose the U.S. imperialist policy of nuclear blackmail and nuclear threats."

Two weeks before the mid-October test, Secretary of State Dean Rusk had publicly announced that a Chinese detonation was imminent and, while downplaying its effect on the U.S. "military posture and our own nuclear weapons program," he deplored such "atmospheric testing in the face of serious efforts made by almost all other nations to protect the atmosphere."[2] President Lyndon B. Johnson, right after the test, said that its military significance should not be overestimated and stressed America's superior nuclear strength and its defense commitments in Asia.[3] He characterized the Chinese nuclear weapons program as a tragedy for the Chinese people. China's "nuclear pretensions," he said, "are both expensive and cruel to its people."

The day after the explosion, Premier Zhou Enlai called for the "complete prohibition and thorough destruction of nuclear weapons"—a proposal that China had promulgated a year earlier at the time the United States, Britain, and the Soviet Union had signed the limited nuclear test ban treaty in Moscow. Though China had been "compelled" to conduct nuclear testing and develop nuclear weapons, the "Chinese Government solemnly [declared] that at no time and in no circumstances [would] China be the first to use nuclear weapons."[4] On October 22, *Renmin Ribao* (People's Daily) lashed back at Johnson: "The successful explosion of China's first atom bomb can be a tragedy only for U.S. imperialism, if it is a tragedy at all. . . . So long as U.S. imperialism possesses nuclear bombs, China must have them too."

China's nuclear capability grew steadily after 1964, and today the nation possesses a nuclear weapons inventory greater than that of the French and British strategic forces combined.[5] The sheer magnitude of the undertaking to create a substantial nuclear force seems to justify the Chinese case that the primary purpose was to strengthen the nation's defenses to meet a serious security threat, and the assistance provided by the Soviet Union in the initial phases of the nuclear program strongly suggests that the target was the United States. Both the cost and the character of the program make it unlikely that the purposes lay elsewhere.

Nevertheless, until the publication of three major histories in the 1980s, along with other revelations, no authoritative basis existed to interpret the causes and scope of Beijing's nuclear decision. In 1984, the senior director of the strategic program, Marshal Nie Rongzhen, published his memoirs, and the next year the Ministry of Nuclear Industry released an anecdotal account of the "secret course" to the 1964 detonation.[6] These two histories presaged the publishing in 1987 of an authoritative history of China's nuclear industry.[7] Such works make it possible, at least on a preliminary basis, to integrate the large number of articles that have been printed since 1979 on individual aspects of the nuclear weapons program. These works have thus lifted the veil of secrecy that long covered the program.

This volume, to be followed by a companion volume on the strategic missile program, constitutes a case history of the Chinese program to acquire nuclear weapons. In this sense, it is meant to complement the major histories of the American, British, German, and Soviet atomic weapons programs.[8] Though we shall attempt to compare aspects of the other nuclear programs in the conclusion, the principal focus will be on China itself.

The study of the Chinese nuclear program raises a number of questions, and these questions will guide our inquiry. They fall into three groups:

1. Under what conditions did China's leaders decide to undertake a nuclear weapons program, and how did the leadership's guidance of the program change over time?

2. How did Beijing organize the program and what role did the Soviet Union play? How affected was the program's structure and direction in the late 1950s by the sweeping policy changes at home and by Soviet involvement?

3. How did Beijing manage technological progress and innovation on such a scale in a poor society just recovering from major wars?

The third question introduces a theme to which we shall return throughout this study: how the poverty-stricken Chinese nation could devote huge resources and energies to building a weapon of mass destruction. We shall attempt to understand why officials just down from the hills of revolution would seek to obtain the world's most advanced weaponry for a land wracked by a century of turmoil.

The focus of our study, then, is on leadership, organization, and the management of research and development. In these terms, China's policy makers can be described in retrospect as having successfully accomplished a series of seven tasks:[9]

1. Define and evaluate the military threat or opportunity that would justify an all-out developmental effort

2. Assemble and condense sufficient information about nuclear weapons that senior leaders could understand and act on

3. Develop the high-level organizations and the subordinate infrastructure to supervise and implement the program both domestically and in cooperation with the Soviet Union

4. Assure through the use of experts that alternatives for each stage in the program would be examined thoroughly, and that the options chosen would provide both the authority to the responsible group and fallbacks should the group fail

5. Provide an environment that would encourage experimentation and appropriate risk taking in the critical subcomponents of the program

6. Develop mechanisms to coordinate the subcomponents and to align them in a workable sequence

7. Respond to the technological requirements and challenges with new ideas and doctrines that would instill purpose in the program and prepare the way for determining the eventual deployment of the nuclear arsenal

These seven tasks have helped organize the approach of this book. The chapters on the making of initial programmatic decisions, uranium mining and processing, uranium enrichment, bomb design, weapons testing, and strategic doctrines and deployments illuminate the operation of the Chinese political system and identify the key personalities in the accomplishment of these tasks. They reveal how the Chinese reached the decision to obtain atomic and hydrogen weapons and pursued the program's implementation. Threats to national security may have caused the urgent need for nuclear weapons, and immutable scientific realities and technological capabilities may have set the initial outer limits of the program's speed, scope, and autonomy; but the decision that China's security required the acquisition of nuclear weapons stemmed from values and perspectives shared by Beijing's revolutionary commanders. While most of the leaders of the strategic weapons program probably accepted those inherited values and perspectives in the beginning, the program's seven essential tasks gradually superimposed new perspectives and, ultimately, new values on the old.

Yet unyielding realities lay as much in human resolve as in predestined tasks or in an atom of uranium. Out of the latent genius of Chinese youth, a handful of older officials, scientists, and technicians would mold an effective cadre of men and women. These men and women, in their twenties and thirties, then committed their lives and talent to building the bomb. Without their commitment and inventiveness, the program would have faltered, perhaps failed. The special success of the program's leadership, organization, and technological management was that it was able to tap a reservoir of dedication that would outlast, indeed overcome, the recurrent follies of leadership, organization, and management in the 1950s and 1960s.

A rising tide of Chinese nationalism explains some of this dedication. We have been told by Chinese specialists that many of the young bomb builders saw the Korean Armistice signed in 1953 as the first equal treaty ever signed between modern China and the West. They experienced a new pride as their country fought the most powerful nation in the world to a standstill and felt embittered that American weapons had taken such a toll on their brothers in the Korean fighting. Many said, "Never again." For them, the Korean War had spawned an angry determination to make a difference to the strengthening of the New China.

Beyond this, we would argue, the lure of science and engineering as much as nationalism shaped the mind-set of those who met the nuclear challenge. In order to illustrate this mind-set, we shall introduce a number of Chinese scientists and technicians and examine their contributions to the nuclear weapons program. Their stories illuminate the excitement and joy they brought to the bomb project and highlight the extraordinary range of talent the Chinese nation could mobilize to solve so many mysteries.

One person in particular is worth singling out at the outset: Mao Zedong, China's supreme leader in the period under study. For Mao, values and other concepts originated in international and domestic experiences—or, as he called them, "practice."[10] Conceptual knowledge depends on social practice, Mao wrote in 1937, and social practice could take many forms: production, class struggle, political activity, and scientific and artistic pursuits. Such practice, as guided by the Party, determines valid knowledge and differentiates between right and wrong. Party-managed practice, Mao predicted, would imbue the Party's cadres and members with the confidence of having true knowledge and of being right.

In June 1949, Mao declared that the Party's revolutionary practice or experiences had taught China to "unite as one with the international revolutionary forces," meaning the Soviet Union.[11] Though there is some evidence to suggest that Mao and his lieutenants had once entertained the possibility of improving relations with the United States,[12] by mid-1949 Mao had embraced a set of values that drew a sharp line between "reactionaries and revolutionaries." He advocated a policy of not showing "the slightest timidity before a wild beast," including the leadership of the United States. From decades of practice in warfare, he had reached a conclusion: "Either kill the tiger or be eaten by him—one or the other." Only after the "internal and external reactionaries" had been beaten would China "be able to do business and establish diplomatic relations with all foreign countries on the basis of equality, mutual benefit and mutual respect for territorial integrity and sovereignty."[13] Revolutionary practice had set Mao on a likely collision course with the United States, though Mao considered that course manageable and, ultimately, of advantage to China.

Thus, by the time of the establishment of the People's Republic of China on October 1, 1949, Mao's worldview—a combination of Communist ideology, harsh memory, and tested values—evinced a marked bias in favor of the Soviet Union and against the United States. His view, or what he called "knowledge," provided a framework for his leadership's determination of national policies and confined the selection of its strategies to proven approaches and principles. As the Chinese Communist elite began to contemplate future foreign policy positions after 1949, nothing fundamental was to occur that shook Mao's worldview or forced him seriously to reconsider his policy assumptions. If anything, these assumptions were reinforced.

After the Second World War and prior to the Communist victory in 1949, Mao applied his worldview to global strategy. He had always regarded China's revolution as a force leading toward an international turning point, toward world communism. At the close of the war, he viewed the "forces of world reaction" as preparing for another war. Yet these forces faced a situation in which the pro-revolutionary or "democratic forces of the people of the world have surpassed the reactionary forces." Because they were "forging ahead," the democratic forces could overcome the danger of war and impose compromises on the "forces of world reaction" through peaceful negotiation.[14] This was the challenge–and opportunity–of Mao's revolution.

In August 1946, in a talk with the American left-wing journalist Anna Louise Strong, Mao expanded on this theme. After asserting that to "start a war, the U.S. reactionaries must first attack the American people," the Chinese revolutionary leader spelled out his principal thesis: "The United States and the Soviet Union are separated by a vast zone which includes many capitalist, colonial, and semi-colonial countries in Europe, Asia and Africa. Before the U.S. reactionaries have subjugated these countries, an attack on the Soviet Union is out of the question." He also said that Washington was "now trumpeting so loudly about a U.S.-Soviet war" because, "under the cover of anti-Soviet slogans," it was attempting to turn "all the countries which are the targets of U.S. external expansion into U.S. dependencies." Among its targets for military bases, Mao singled out "that part of China under Kuomintang rule." He concluded that, though the Chinese revolutionaries faced many difficulties, "the day will come when these reactionaries are defeated and we are victorious."[15]

Strategically, Mao warned against overestimating the strength of American imperialism and dubbed the atomic bomb "a paper tiger which the U.S. reactionaries use to scare people."[16] He acknowledged that "with regard to each part, each specific struggle," the Chinese "must never take the enemy lightly," but that in respect to the strategic whole they should "dare to win victory."[17] In June 1949, on the eve of establishing the People's Republic of China, Mao calculated that the experiences of the Communist Party "have taught us to lean to one side," and that leaning to the side of the Soviet Union would be necessary for victory. That victory would require "various forms of help from the international revolutionary forces," the Soviet Union.[18]

For Mao in the late 1940s, the strategic calculus was clear. The struggle against imperialism could be intensified and need not be intimidated by the American nuclear threat. As the Chinese leader reasoned, the Soviet Union and the United States would continue to compete for the lands that lay between them and would not directly fight one another. In such a competition, American nuclear supremacy had little relevance. The "vast zone" between the United States and the Soviet Union, Mao said, had become the battleground, and China as one of the targets in that zone could take actions to achieve victory in the struggle between the forces of peace and war. The anti-war forces, the Chinese revolutionaries, could hold their own and eventually defeat the American-led coalition if they sided wholeheartedly with the Soviet Union. For its part, the Soviet state would be spared direct conflict with the United States if Moscow steadfastly supported the "people's conflict" in the vast zone.

In China's case, the Sino-Soviet alliance of February 14, 1950, provided the joint contract to pursue that struggle.[19] Yet the United States, as the Chinese were concluding the treaty of alliance, had intelligence that all was not well between the two Communist powers. A secret message from Secretary of State Dean Acheson to the U.S. embassy in Paris revealed that Mao had "been highly dissatisfied with attempted exactions on China," and that Premier Zhou Enlai might "resign rather than accede to their [Soviet] demands as presented."[20] Nevertheless, by April 1950, Assistant Secretary of State Dean Rusk reported to his Secretary, Dean Acheson, that Soviet military jets had begun to appear on the China mainland, and that "highly effective Russian military assistance to the Chinese Communists [might] prompt further strong recommendations . . . for countervailing aid to the National Government [on Taiwan]."[21] The next month, Rusk told Acheson that the "loss of China to Communists" had marked "a shift in the balance of power in favor of Soviet Russia."[22]

The first test of Mao's strategic view and the alliance came in the Korean War, which China entered in October 1950.[23] At the outset, Mao's optimism seemed to be borne out. The war did lead to the expansion of Soviet assistance to China, as well as North Korea, and a strengthening of the alliance. By mid-1951, U.S. intelligence estimated that there were "10,000 Soviet military advisers throughout China."[24] Soviet assistance had made substantial improvements to the Chinese air force and navy and was expected to increase sharply.

The war also tested China's resolve. When the Chinese entered the Korean conflict, North Korea's armies had virtually collapsed and the

U.N. troops were closing in on the Sino-Korean border. On October 4, Mao called an urgent meeting of the Central Committee to discuss sending Chinese troops to rescue the North Koreans. After hearing "the list of disadvantages" involved in doing so, he said: "You have reasons for your arguments. But at any rate, once another nation is in crisis, we'd feel bad if we stood idly by." [25]

Present at the meeting was Peng Dehuai, soon to be the commander and political commissar of the "Chinese People's Volunteers" (CPV), and he saw Mao's comment as combining "internationalism with patriotism." [26] As Peng wrestled with the pros and cons of joining the war, he concluded that the "U.S. occupation of Korea, separated from China by only a river, would threaten Northeast China." [27] If this happened, he thought, the United States "could find a pretext at any time to launch a war of aggression against China." He found this prospect intolerable and argued that "without going into a test of strength with U.S. imperialism, it would be difficult for us to build socialism." On October 8, 1950, Mao ordered the CPV "to support the Korean people's war of liberation and to resist the attacks of U.S. imperialism," or as the Chinese press later put it, "to keep the wolf from the door." [28]

According to U.S. intelligence estimates, the Korean War resulted "in the deployment of major portions of Communist China's best military forces in Korea and/or Manchuria." [29] By June 1951, the CPV forces in Korea, largely as a result of General James Van Fleet's counteroffensive in May, had suffered an estimated 577,000 casualties, including 73,000 non-battle casualties, and had surrendered 16,500 prisoners of war. Moreover, they had suffered heavy losses of materiel and were "becoming increasingly dependent upon the USSR for logistic support." [30] Later, the Chinese were to allege that most of the loans from the Soviet Union were "used up in the war to resist U.S. aggression and aid Korea." [31] By the end of the war, in July 1953, the conflict had taken a fierce toll, including mounting burdens imposed on those who remained at home.[32] Consistent with Mao's thesis on conducting the global struggle in the vast zone between the United States and the Soviet Union, Chinese commentators later declared: "We ourselves preferred to shoulder the heavy sacrifices necessary and stood in the first line of defense of the socialist camp so that the Soviet Union might stay in the second line." [33] In recalling those sacrifices some ten years later, the Chinese denounced the Soviet claim that Beijing had ever favored a head-on clash between the United States and the Soviet Union.

Of special relevance to this study, the Korean War brought home to the Chinese high command the need for technological modernization. As the political scientist Alexander George writes, "Drawing upon large quantities of Soviet military equipment and military advisers the PLA [People's Liberation Army] commander in Korea, P'eng Te-huai [Peng Dehuai], began modernizing the PLA even while the war continued in a desultory but painful fashion on the bleak Korean terrain."[34] In general, the trauma of the 1950-51 losses forced the PLA to rely ever more heavily on advanced technology and professionalism, even while its leaders clung to the language of fighting a people's war with a people's army. As part of the change, the PLA emphasized fixed defensive fortifications and brought in heavy weapons, including tanks and artillery. The Chinese equipped and trained an air force with 1,800 aircraft, including 1,000 jet fighters by mid-1952, in what the military historian William Whitson has called a process of "modernization under fire."[35]

In his interviews with Chinese captured in Korea, Professor George has documented the political side of this change within the CPV, a large number of whose officers later had senior assignments in China's strategic programs. He notes that Chinese combat cadres in Korea became disillusioned with Mao Zedong's man-over-weapons fighting doctrine as American firepower took its toll.[36] Under the leadership of Peng Dehuai, says Whitson, CPV combat veterans had witnessed the "sacrifice [of] many tens of thousands of the PLA's best troops," and Peng "must have quickly communicated a fierce professional pragmatism to his subordinates."[37] By the end of the war, the expensive lessons of Korea had been used in strengthening and Sovietizing 25 armies, which between their "Russian mentors and American enemies" had received a prolonged education in modern professional warfare. Many CPV officers had attended the new Military Academy in Nanjing, where General Liu Bocheng passed on the lessons of Korea to the Chinese military elite.[38] Experience in Korea had established a new baseline of knowledge.

As Beijing leaned closer to Moscow during the Korean conflict, Washington increasingly focused on ways to exploit China's weaknesses. It first explored possible uses of Chinese Nationalist forces against the mainland, and then, more broadly, sought to capitalize on China's vulnerabilities.[39] The great uncertainty in American intelligence estimates, however, hinged on the degree of Soviet willingness to rescue China if American military pressure on Beijing increased. The former ambassador to the Soviet Union, George Kennan, worried that "if we continue to advance into North Korea without making vigorous efforts to achieve a cease-fire, I fear they [the Kremlin leaders] will see no alternative but to intervene themselves."[40]

Thus, when the war ended in 1953, the senior security specialists in both the People's Liberation Army and the Eisenhower administration shared a deep frustration about the war's lessons. For their part, the Chinese knew firsthand the devastating might of modern arms and the high cost and probable military irrelevance of earlier revolutionary doctrines. Both the requirement for and the upper boundary on the modernization of the Chinese army were being created largely by the level of foreign technology: what was arrayed against them and what they could hope to obtain. The war introduced Mao's China to advanced armaments and techniques and, as we shall see, to the threat of nuclear attack. To survive in the modern world, China would have to have modern arms.

TWO

American Power and Chinese Strategy, 1953-1955

Chinese leaders hold that they reached the decision to inaugurate the nuclear weapons program under duress. The Chinese Politburo had tentatively initiated the nation's first five-year plan in 1953, but then the Party leaders delayed formal approval for two years because of "objective difficulties."[1] Chinese specialists argue that, in addition to the country's internal problems, the government had to postpone the effort to modernize China because, during the years of the Korean War (1950-53) and the Taiwan Strait crisis (1954-55), the United States was actively seeking to unseat Communist rule in Beijing and restore Chiang Kai-shek to power on the mainland.[2]

That relations between the United States and China had deteriorated to the point of armed conflict is beyond dispute. Why this happened is less clear. To what degree did Beijing provoke American hostility? What forms did that hostility take? Was it simply reactive to unceasing Chinese aggressiveness? Did the United States threaten China with nuclear weapons? If so, was the threat designed to deter Chinese actions or could Mao Zedong legitimately have perceived U.S. nuclear threats as unrelated to such actions and directed instead toward the ultimate destruction of his fledgling revolutionary regime?

This chapter does not seek complete answers to these questions. Rather, it confines itself to those aspects of the questions that come, finally, to bear on January 1955, when the Chinese decided to obtain their own nuclear arsenal. Using both American and Chinese sources, we look briefly at three major events: the ending of the Korean War in 1953, hostilities at the time of and shortly after the 1954 Geneva Conference on Korea and Indochina, and the crisis in the Taiwan Strait in late 1954 and early 1955. As we shall note, U.S. officials believed these events proved that revolutionary expansionism on the Maoist model, if not halted, would sweep through Asia and imperil vital American and allied interests there. They rejected the idea of inevitable Communist victory on the Asian continent; strength and will could thwart it. In short, Washington believed that American military power could deter Communist aggression and eventually compel its retreat.

The Chinese saw these events through a very different strategic lens. They had fought the American-led U.N. forces to a standstill in Korea, and their brothers in Indochina were moving to defeat American-backed French colonial armies in that warring land. These would be only partial victories, however, for Beijing understood that the United States would not suffer such wounds lightly and would fight back. The Americans would increasingly focus on China as the real enemy and would find pretexts to strike directly at the Chinese mainland. The clash to come in the Taiwan Strait thus represented more than a contest for the offshore islands remaining in Nationalist hands or even for Taiwan itself. It constituted a direct test of strength between the People's Republic of China (PRC) and the United States.

Although Mao proclaimed the irrelevance of nuclear weapons to that test, he knew otherwise. The Sino-Soviet alliance of 1950 provided China a nuclear umbrella, and Mao welcomed its protection. He believed, nonetheless, that the United States, which would not risk a nuclear confrontation over marginal territories such as Korea or Indochina, might well do so if it came to a showdown with Mao's forces. And, should nuclear weapons be used, would the Soviet Union risk its national survival for its Chinese ally? In later years, during the negotiations leading to the Treaty on the Non-Proliferation of Nuclear Weapons (signed on July 1, 1968), diplomats of the great powers, in accordance with their U.N. Charter responsibilities, pledged to assist nonnuclear weapons states faced with the prospect of nuclear attack and to counter the threat or use of nuclear weapons against those states.[3] The Soviet Union had implied such a pledge in its treaty of alliance with China in 1950. If Beijing received atomic ultimatums in the Korean War and thereafter, we have a test case of how a nuclear guarantor acted at the moment of truth.

In this chapter, we treat the evolution of China's strategic uncertainties and the ways Mao's leadership chose to respond. We do not pretend to deal with all the issues involved in U.S.-China relations in the years 1953-55. We do believe that some knowledge of the events surrounding the ending of the Korean War and the Indochina and Taiwan Strait crises is essential to understanding the timing of the Chinese decision to build the bomb and to grasping the intensity of national feeling surrounding that decision. We argue that the Chinese felt them-selves alone in a tightening vise. We believe they feared an inward-curving sequence of threats; they could not predict where nuclear blackmail, as they perceived it, might lead. Nuclear weapons backed up Washington's threats and exaggerated their potential impact. They also structured China's responses and provoked its defiant anger and the decision to undertake the costly nuclear weapons program.

Eisenhower, China, and the End of the Korean War

The ordeal of Korea wrenched and tested all postwar conceptions of American relations with Communist states. It proved a searing cru-cible for Washington's policies toward the regime of Mao Zedong and for the Eisenhower presidency. During the election campaign of 1952, Dwight Eisenhower underscored his preoccupation with the war and his presidential credentials with a pledge to go to Korea and bring the truce negotiations to a rapid and successful conclusion. With two weeks of the campaign to go, Eisenhower's announced intention to "determine for myself what the conditions [are] in that unhappy coun-try" helped assure his election victory and impelled events that were to alter the course of Chinese defense policies.[4]

Following the Korean leg of his December tour, the president-elect used the return journey to ponder his options for ending the war and to meet several Cabinet-designees, including John Foster Dulles and Admiral Arthur W. Radford, the leading candidate to chair the Joint Chiefs of Staff. The trip thus mixed two important themes that were to govern the first months of the new administration: peace in Asia and global defense.

The first signal of how Eisenhower proposed to end the war came soon after his meetings with military commanders in Korea. These officers reportedly recommended that the U.N. forces "should con-sider the use of small atomic bombs and artillery shells. . . . A new offensive should be accompanied by a blockade of the Chinese Com-munist mainland and permission to attack the enemy's Manchurian base."[5] As Eisenhower was on the way home, his so-called high com-mand leaked word of a "new strategic plan for the conduct of the Korean War that, it believes, will exert so much pressure on the Com-munist forces that the Soviet Union will agree to an armistice."[6] Upon his return from Korea, Eisenhower voiced confidence that "taking measures that would 'induce' Communist aggressors to want peace also" would accelerate "a satisfactory solution" to the war.[7]

The president-elect explicitly warned the Chinese of his intention to escalate the war if the armistice negotiations remained stalemated. He publicly hinted at the possible use of nuclear weapons against Beijing, although his actions proved far more indecisive than he would later recall in his memoirs and to colleagues.[8] Looking beyond Korea to the war's impact on Soviet and Chinese expansionism, the American leader had come to believe that negotiating an acceptable truce in Korea would require a combined strategy of warnings and blandishments, and, somewhat hesitatingly, he weighed several possible combinations of the two.[9] As he later characterized his decision, the policy course chosen was "to let the Communist authorities understand that, in the absence of satisfactory progress, we intended to move decisively without inhibition in our use of weapons, and would no longer be responsible for confining hostilities to the Korean Peninsula."[10]

What this meant to his closest administration advisers was not all that obvious, though White House Assistant Sherman Adams, when he later asked Eisenhower what he had intended, was told that "it was indeed the threat of atomic attack that eventually did bring the Korean War to an end on July 26, 1953." In his memoirs, Adams says that during the previous spring the United States had moved atomic missiles into Okinawa, and that in May Secretary of State Dulles had informed Indian Prime Minister Nehru that the United States "could not be held responsible for failing [sic] to use atomic weapons if a truce could not be arranged."[11] Nehru later denied that he had grasped the intent of the Dulles communication, and in any case had not transmitted any atomic threat to Beijing.[12] The American president later claimed that the atomic threat had reached China through other channels, such as the armistice talks at Panmunjom, and forced the Chinese to compromise.

What the president and Sherman Adams assumed, of course, is that the Chinese not only knew of the threat, but were prompted to yield in response, as Eisenhower asserts to be the case. Chinese who were in positions of authority in Korea at the time point out that soon after Eisenhower's inaugural address the official Chinese news agency denounced U.S. plans to "resort to the use of atomic weapons."[13] Moreover, Chinese present at the Panmunjom talks and in the Ministry of Foreign Affairs state that they knew of Eisenhower's nuclear threat and had intelligence on the U.N. command's proposals for expanding the war, including a major landing of troops at the narrowest section of North Korea.[14]

The central issue stalemating the Korean negotiations was the U.N. command's insistence on the voluntary repatriation of prisoners-of-war (POWs).[15] In late 1952, the North Korean-Chinese side had rejected this demand as a violation of the Geneva convention on POWs, and there the matter stood when the American threats were communicated. The Chinese considered these threats to be signs that the "American side had not yet prepared for the conclusion of an armistice, and that they wanted to try again for a military solution." The prevailing Chinese view, we have been told, was that the Communist forces should exercise great caution in the face of American nuclear threats, but that American and world public opinion would make it quite improbable that the United States would make good on those threats.[16]

The Chinese appraisal of the possible American actions in Korea generated a dual response.[17] On the one hand, Beijing resolved to stand firm in the face of the American "blackmail." "Under the circumstances," the Chinese say, "there was no point in the [North] Korean-Chinese side's making concessions because any such concession would be perceived by the other side as a sign of weakness." Emulating the Americans, the Chinese were arguing that the "only way to turn the clock back to negotiation was to prepare for the coming battle." As one Chinese remembers the moment: "Our slogan was: everything for the front; if you want men, we will send men; if you want materiel, we will send materiel." Thereupon, the Chinese People's Volunteers launched an urgent campaign to construct fortifications, including, "in the frontline battlefield, Anti-Atom shelters . . . built deep in the middle of the mountains." And then too there were the deliberate leaks: "We purposefully let the spies of the other side . . . get some intelligence of the preparations we were waging."[18]

On the other hand, the Chinese were not averse to negotiating, as they demonstrated when the U.N. commanding general wrote what Eisenhower has called "a routine letter" to the Communist high command in February 1953 to inquire whether it would be willing to exchange sick and wounded POWs.[19] When they received the letter in Panmunjom, the Chinese there "were quite amused to think that the American side was playing but another trick." Since they were aware of Washington's threats of a nuclear attack and of a landing in the northern part of the peninsula, they did not consider the proposal a serious one. However, when the message reached Beijing, the response was different. Leaders in the Chinese capital "saw it as a sign of American intention to compromise." They ordered replies drawn up to facilitate the exchange of sick and wounded POWs and to resume the negotiations.[20]

Although the war was to drag on for several more months, the negotiating process quickly resumed and moved forward despite major South Korean opposition to a diplomatic settlement. In the final analysis, probably for a variety of reasons that included Stalin's death and the changing of the guard in Moscow, the Chinese accepted most of the American demands. They bowed to Washington's insistence on voluntary prisoner repatriation. Nevertheless, the perceptions of how the compromise was worked out are totally at odds. From the viewpoint of the Chinese, their delaying tactics and preparations in the face of nuclear threats had blunted the attempted American blackmail; from the U.S. viewpoint, atomic diplomacy had forced the Chinese back to the peace table.[21] To this day, each side believes it coerced the other into compromise.

The New Look and China Policy

Chinese officials, who at senior levels were kept fully informed of U.S. press reports, could not help knowing in early 1953 that a thoroughgoing shift of American defense policy toward greater reliance on nuclear weapons was under way.[22] They could not know the full details, to be sure—these have only recently been declassified[23]—but the general outlines and some vital details were well documented in the Western media. What was happening in these months was the formulation of the so-called New Look policy, and, given the urgency of Asian-Pacific conflicts, the Eisenhower administration closely linked the New Look to strategies for that region.[24]

A fundamental focus of the policy was how to conduct the long-term struggle with the Soviet Union and its allies. In a press conference in April 1953, the president rejected the view that Soviet strategic power would catch up with and offset that of the United States by the mid-1950s. Contrary to what in later years might have been termed a window-of-vulnerability view of the Soviet-American balance, Eisenhower looked to an enduring struggle against the Communist powers with no particular time of critical vulnerability for the United States. He reportedly said: "We reject the idea that we must build up to a maximum attainable strength for some specific date. . . . Defense is not a matter of maximum strength for a single date. It is a matter of adequate protection to be projected as far into the future as the actions and apparent purposes of others may compel us."[25] As the president spelled out the nature of that defense over the first year of his administration, he let it be understood that the military establishment, when reformed, would have to cost less as well as fulfill more effectively its multiple missions.

From Western news media and probably from intelligence sources, the Chinese had a fairly full picture of the New Look in U.S. defense policy by early 1954.[26] The concerns of PRC leaders about Washington's increased dependence on nuclear weapons and its readiness to use them against China appear to have been well founded. The administration's highly classified "Basic National Security Policy" (October 30, 1953), usually referred to by its number, NSC 162/2, called for a strong military posture based on nuclear striking power, a state of U.S. and allied troop readiness permitting forces to move rapidly to meet aggression and hold Western assets, and a mobilization base capable of ensuring victory in the event of nuclear conflict. The document stated that Indochina and Taiwan (Formosa) were of "such strategic importance that an attack on them probably would compel the United States to react with military force."[27] The key sentence was, "In the event of hostilities [with the Soviet Union or China], the United States will consider nuclear weapons to be as available for use as other munitions."[28]

In a follow-up top secret presidential memorandum (November 11, 1953), Eisenhower stated: "It was agreed that the dependence we are placing on new weapons would justify completely some reduction in conventional forces."[29] Earlier, in a National Security Council (NSC) meeting in August, Secretary of State Dulles, when asked about the course of action leading to the new defense policies, said he presumed that preserving security in areas such as Europe and Asia would mean greater dependence on air power and atomic weapons.[30] At the same meeting Admiral Arthur Radford confirmed that "in the event of a major conflict the United States would use atomic weapons both in the tactical and strategic realm."[31] The admiral, among other administration officials, reported to the nation and the international community on the new security policy in December. In public, he stressed the increased reliance on atomic weapons, which had "virtually achieved conventional status in our Armed Forces."[32]

By the end of Eisenhower's first year in office, the New Look had, for all to see, become synonymous with heavy reliance on atomic power, and the president formally presented its outlines to the nation in his State of the Union address on January 7, 1954.[33] Soon after, when Secretary Dulles publicly expounded on the New Look's focus on "massive retaliatory power," the press dubbed the new strategic doctrine "massive retaliation."[34]

Throughout the first year of the Eisenhower presidency, the Chinese had followed closely the public disclosures on the New Look, and the January statements brought no surprises. In response, Beijing predictably denounced the military tone of the president's address to the nation, with its stress on continuing support for Korea and Taiwan: "Its main points are . . . to build up atomic weapons [and] to develop new weapons . . . in order to intimidate and to maintain a tense international situation." China also detailed the elements of atomic diplomacy found in Dulles's speech and subsequent American thermonuclear weapons testing in the Pacific.[35]

In fact, President Eisenhower had begun to implement his New Look ideas with changes in the deployments of nuclear weapons within months after he assumed office.[36] In June 1953, he ordered steps taken to shift the nuclear arsenal from virtually total civilian control to military custody. Many complete atomic weapons were transferred to naval and air commands, and "by 1961 less than 10 percent of the stockpile remained in civilian control."[37] According to declassified military documents cited by the historian David Rosenberg, during the Eisenhower administration "tactical nuclear weapons were increasingly integrated into calculations of force requirements and effectiveness."[38] The Chinese closely monitored hints in the media of these changes in the American nuclear arsenal and force planning and reacted ever more seriously to their implications for China's security.[39]

While the principal target of these moves was the Soviet Union, the United States made no secret of its strong hostility toward China.[40] In considering the possibility of renewed fighting in Korea during the early part of 1954, for example, the commander of the Strategic Air Command, General Curtis LeMay, said: "There are no suitable strategic air targets in Korea. However, I would drop a few bombs in proper places in China, Manchuria and Southeastern Russia. In those 'poker games,' such as Korea and Indo-China, we . . . have never raised the ante—we have always just called the bet. We ought to try raising sometime."[41] Short of war, the policy of the United States toward China was "to reduce [its] relative power position."[42] Dulles, among others, had provided the justification for such thinking. "The Chinese Communist regime," he said, "has been consistently and viciously hostile to the United States."[43]

The basic White House document in this respect was entitled "U.S. Policy Towards Communist China" (NSC 166/1, November 6, 1953). It saw China as having formidable capabilities and prescribed a strategy of weakening its strength and of using all feasible means, covert and overt, to impair Sino-Soviet relations. The document deemed Taiwan to be "a considerable asset to the U.S. position in the Far East" and stated that the "military forces of the Nationalists [on Taiwan] constitute the only readily available strategic reserve in the Far East." In the event of all-out conflict with China, it concluded, U.S. power, "employing all available weapons, could impose decisive damage on the Chinese Communist air force and its facilities," though this "might absorb a considerable proportion of the U.S. atomic stockpile."[44]

The Geneva Conference and Moves Toward a U.S.-Taiwan Defense Pact

Nuclear diplomacy set the mood for the coming crisis in Indochina. In January and February 1954, the foreign ministers of the United States, Great Britain, France, and the Soviet Union met in Berlin to plan a conference to effect a "political settlement of the Korean question" as well as to discuss the problem of "restoring peace in Indochina." They agreed to invite the representatives of the People's Republic of China to participate in this conference, to be held in Geneva from April 26 to July 21, 1954.[45]

Almost from the outset, it became clear at Geneva that there would be no progress on Korea. Top-secret messages between Dulles and State Department officials quickly let Washington know that its bitter dispute with South Korean President Syngman Rhee over a compromise settlement had in no way been resolved.[46] The Korean president had made it plain as early as February that he wanted no part of a compromise. In what the State Department characterized as a highly offensive letter to Eisenhower, Rhee described the United States as weak "in the face of Communist intransigence." Although this letter, seen only in draft, was withdrawn, Rhee did transmit another one to Eisenhower on March 11 dealing with the long-simmering issue of Korean reunification.[47] In his reply, Eisenhower rejected Rhee's idea of South Korea alone achieving the unification of Korea by military means. Such action, he said, would expose the South Korean "armed forces to disastrous defeat and possible destruction."[48]

What concerns us in this study is that the Chinese, along with the rest of the world, knew that the American commitment to a rapid political settlement on the peninsula had become openly embroiled in controversy.[49] What Beijing could learn about this primarily came from the vocal Syngman Rhee, who had publicly demanded a renewal of the Korean fighting. Presumably, they did not know that Rhee was secretly calling for an attack by the United States on China itself. As the U.S. ambassador in Seoul reported to Washington: "Rhee asks, if we [the United States] are sincere about helping Indochina, why do we not bomb factories and supply lines on the Chinese mainland from advanced air bases in Korea."[50]

As we have seen, the Chinese had become highly sensitized to American nuclear threats at the end of the Korean War and would have placed the Korean president's public demands for a renewal of the Korean War in the context of the U.S. administration's official statements on Korean policy and the New Look. In December 1953, for example, Washington had announced that it would withdraw two divisions from the peninsula, but news reports added that the United States "would compensate for a planned reduction in ground forces in Korea by the increased fire power made possible by atomic cannon."[51] Secretary Dulles, echoing Admiral Radford's December call for readying "tremendous vast retaliatory and counter-offensive blows," addressed the issue of instant and massive retaliation in his January speech to the Council on Foreign Relations;[52] in March 1954, the United States resumed testing hydrogen bombs in the Pacific.

For many in the administration, China constituted both an enduring target of the New Look and the immediate focal point for the containment of Communist expansionism. These U.S. officials sought to supplement a military buildup in Asia with political steps to incorporate Taiwan into the American defense network. On a visit to Taipei in December 1953, Vice-President Richard Nixon conferred with Taiwan's leader, Chiang Kai-shek, who spoke of retaking the mainland and of having made all the necessary preparations to do so.[53] Two months later, Assistant Secretary of State Walter S. Robertson said that the United States and Taiwan must maintain a military threat against the mainland.[54] As a legal underpinning for such a military posture toward Beijing, many American leaders at this point advocated the conclusion of a defense treaty with Taiwan.

According to Dulles's later testimony, Taiwan first proposed a mutual defense treaty on December 1, 1953, though informal soundings had begun earlier.[55] One of its principal U.S. proponents was the ambassador to Taiwan, Karl Lott Rankin. Considering himself far more than the president's representative, Rankin had long advocated weighing the pros and cons of substantial military assistance to Taipei to enable Chiang's forces to "fight back to the mainland." He notes in his memoirs that the "question of one or more multilateral pacts of mutual defense in the Pacific area was being raised with increasing frequency" in 1954, and that these pacts offered states such as the Republic of China "the best hope for peace in the Western Pacific."[56] Particularly impressed by what he deemed the political importance of defense accords, the U.S. ambassador began by calling for a regional Pacific pact that would include Taiwan. Beijing became aware of the pressures for such a pact as early as March.[57]

But other, more urgent matters burdened official Washington in the first half of the year. Despite repeated recommendations for negotiations on either a Pacific-wide or a Taiwan-specific security treaty,[58] American concerns shifted to Southeast Asia in the spring of 1954. The mood was one of crisis, spurred in part by the collapse of the French garrison at Dienbienphu in Indochina on May 7 and in part by the fear of overt Chinese intervention in Southeast Asia.[59] The idea of a pact confined to that embattled region seemed to take precedence, and Rankin and Chiang became concerned that Washington might "lose interest in a more substantial military pact" with Taiwan.[60]

From the end of 1953 through the following spring, Taipei repeatedly prodded Washington to open talks on a mutual defense pact, and in February it submitted a draft for American review.[61] This was followed by an intense lobbying campaign to convince visiting senior American officials to back negotiations on the treaty. Two targets of this campaign were Secretary of Defense Charles E. Wilson, who visited Taiwan in May, and General James Van Fleet, who was there in June.[62] Wilson's visit coincided with the growing impasse in the Korean phase of the Geneva Conference and the collapse of the French bastion at Dienbienphu just before the Indochina phase. These events led Rankin to redouble his efforts. On a home visit soon after, he seems to have played a pivotal role, spending "three extremely busy weeks" promoting a mutual security treaty. He states that talks in Washington "with literally dozens of senators and representatives were on my schedule, in addition to calls on most of the principal officers of the Departments of State and Defense." [63] Steadily, the advocates of the treaty were making headway. Only the higher priorities accorded the Geneva Conference and the prompt negotiation of a Southeast Asian pact temporarily postponed the decision to undertake negotiations with Taipei.

As the prospects for a treaty with Taiwan faded momentarily, Washington put even greater emphasis on the military side of its China policy. Admiral Radford recalls in his memoirs that at this time "the central philosophy" of the Joint Chiefs of Staff (JCS) was that "the real solution to Far Eastern difficulties lay in the neutralization of Communist China." The "JCS strategic concept and plan of operations," he says, "called for destroying effective communist forces and their means of support in Indochina, as well as reducing Chinese communist capability for further aggression. . . . Should this not suffice to assure victory, the attack against China would have to be stepped up. It might require a highly selective atomic offensive, in addition to attacks with other weapons systems." The JCS further recommended that there should be "an appropriate degree of mobilization to provide for the greater risk of general war."[64] Such a war might result in the employment of atomic weapons by both sides, and to prepare for this contingency, the United States discussed at the presidential level and with allied military representatives the possible use of nuclear weapons against the Vietminh in Indochina and against China, should it intervene.[65]

As the French position in Indochina deteriorated and the Geneva Conference proceeded to its conclusion in July, Dulles's response was to fashion a "grand strategy" (Radford's words) of united action. By this time, Dulles embraced plans for a regional security arrangement that would reduce the power and influence of the Soviet Union in the Far East, "primarily through the containment and curtailment of Communist China's relative position of power." While most of the talk in the summer referred to the Southeast Asia Collective Defense Treaty (which was signed in Manila on September 8, 1954), the matter of a defense pact with Taiwan was getting increased attention within the Department of State.

In the end, the mutual defense pact was not signed until December 2, 1954. But what is important to this study is that Beijing had been long aware that it was under deliberation—which is to say, China's concern in this regard predated its subsequent campaign for the "liberation of Taiwan" and the crisis in the Taiwan Strait.

The Chinese, not without cause, linked the talk of a defense pact to ever more aggressive moves by Taiwan. They believed that the United States endorsed, perhaps even initiated, these moves. Throughout the first half of 1954, the Western press had commented on the maneuvers concerning a U.S.-Taiwan defense treaty,[66] and just prior to the conclusion of the Geneva Conference on July 21, the Nationalist government became emboldened to increase its attacks on the mainland. From the Chinese point of view, the merger and activation of Nationalist and American military interests required a response. It was Beijing's reaction in turn that intensified the Taiwan Strait crisis in September.

The Taiwan Strait Crisis

American leaders of the era remember the crisis over the offshore islands of Quemoy and Matsu as having begun as an unprovoked aggression on the part of Beijing (see Map 1).[67] Eisenhower recalls first

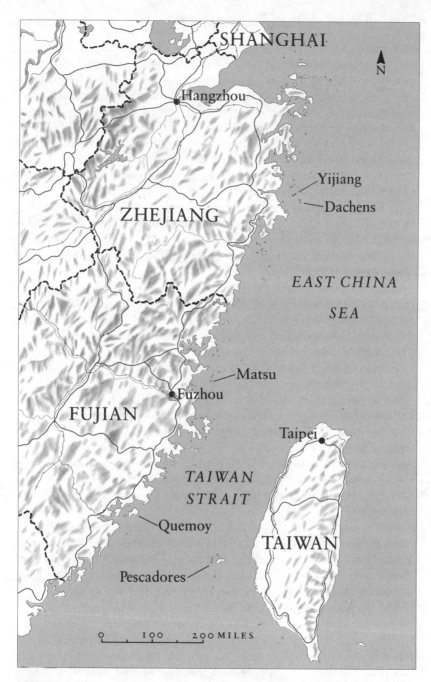

Map 1. East China

hearing the news "at about seven o'clock on the evening of September 3, 1954, [when] the Army Signal Corps at the summer White House in Denver brought me a message . . . that at one forty-five that morning, . . . the Chinese Communists had begun heavy artillery shelling of Quemoy Island."[68] Two American soldiers died in the bombardment, and the Pacific Fleet, alive with rumors of an impending Chinese invasion of Quemoy, was put on alert.[69] While Eisenhower states that the Chinese shelling "did not come as a complete surprise," he does convey a sense of urgency about an event that was "seemingly [to] carry the country to the edge of war."[70]

Secretary Dulles, in testimony to the Senate Foreign Relations Committee on February 1, 1955, clearly regarded the Chinese shelling as a surprise attack. According to him, "Relative quiet had prevailed off the China coast from 1949 until on September 3, 1954, the Chinese Communists opened heavy artillery fire on Quemoy Island." Dulles explained to the committee that, in his view, the bombardment had been timed to coincide with the opening of the Manila Conference, and "when it was known that the United States intended to negotiate this Treaty of Mutual Defense with the Republic of China, the Chinese Communists stepped up their anti-American activities."[71]

Dulles did not recall that the Taiwan Strait had in fact been far from quiet in the months prior to the opening volley on Quemoy. From the Chinese perspective, the United States during the Korean War had backed, at least tacitly, Taiwan's actions to blockade and harass the Chinese coast. In November 1953, a National Security Council statement of policy objectives toward Taiwan advocated the "increased effectiveness of the Chinese National armed forces . . . for raids against the Communist mainland and seaborne commerce with Communist China, and for such offensive operations as may be in the U.S. interest."[72]

In October 1953 and again in May 1954, Nationalist warships hijacked China-bound Polish merchant vessels in international waters off Taiwan, and Beijing accused the U.S. navy of providing assistance.[73] A month after the second hijacking, a Taiwanese warship seized a Soviet oil tanker making for a Chinese port, and the Soviet government lodged an official protest with Washington.[74] As the number of incidents against ships of both PRC and foreign registry mounted, the Chinese press lodged its own protests. On July 16, *Renmin Ribao* strongly editorialized against Nationalist "piracies" and Taiwan's American "backstage supporters." During the past several years, it alleged, "the United States and the Kuomintang bandit gang hijacked in the sea area near Taiwan more than seventy merchant ships that belonged to China or those foreign countries that were trading with China."[75]

In the summer, the United States added to the atmosphere of confrontation. It sent two nuclear-capable carrier battle groups into the East China Sea, and as a way to test Chinese defenses, carrier aircraft and destroyers patrolled virtually up to the international-waters boundary close to the Chinese coast. In the week following the signing of the Indochina armistice on July 21, tensions in East Asia mounted rapidly as Nationalist aircraft increased their overflights of PRC territory, and in the confusion Beijing's fighters downed a British airliner. The Chinese apologized for the mistake and offered to pay reparations, but added that Chinese planes would fire on any Nationalist ships or planes violating PRC airspace. During the rescue operations following the downing of the British aircraft, moreover, U.S. carrier-based fighters shot down two Chinese patrol fighters and strafed Chinese gunboats. The Chinese claimed the American fighters also attacked two Polish merchant ships.[76]

Meanwhile, to highlight American support for Taiwan, the United States stationed military advisers on several offshore islands held by Chiang Kai-shek's forces, and, as we have seen, the campaign to conclude a U.S.-Taiwan defense treaty was stepped up during the highly publicized visits of the Secretary of Defense and General Van Fleet in May and June. It was at this point that China's press, in two editorials, issued the call "We Must Liberate Taiwan" and made it the "Glorious Task of the People's Liberation Army" to do so.[77]

At his press conference on August 3, Dulles declared that the United States would apply force to prevent the Communist conquest of Taiwan and would pursue the negotiation of a military treaty with Taiwan.[78] The following week, the Nationalist navy reportedly engaged and sank eight PRC gunboats in a "furious sea battle" off the Chinese coast, and Taiwan's vice-president stated that war against the People's Republic would preclude the "final arrival of an all-destructive atomic war."[79] Zhou Enlai responded on August 11 with a widely distributed governmental report calling for the "liberation" of Taiwan,[80] even as the military limited its preparations to planned actions against islands far to the north, off Zhejiang Province.* President Eisenhower in turn

* By early August, the Central Military Commission had ordered its forces "not to provoke the Americans," and had substituted the shelling of Quemoy for a previous plan calling for the systematic liberation of the many offshore islands. The commission established the East Zhejiang Front-Line Command in this same month, and, on August

stated that the U.S. Seventh Fleet would go into action to break up any Chinese attempt to invade Taiwan.[81] Clearly, the Dulles characterization to the contrary, relative quiet had hardly prevailed off the China coast prior to September 3.

The initial shelling of Quemoy came virtually on the eve of the opening of the Manila Conference, called to conclude a Southeast Asia Collective Defense Treaty. The meeting ended on September 8 with the signing of the pact among Australia, France, New Zealand, Pakistan, the Philippines, Thailand, the United States, and the United Kingdom. The same day, Beijing's Government Administrative Council approved an order to begin national military conscription, for it judged the Southeast Asia pact to be the first post-Geneva move by Washington to forge a ring of alliances around China. The next move, it understood, would center on Taiwan.

In the midst of this crisis, from September 29 to October 12, 1954, Nikita Khrushchev, first secretary of the Central Committee of the Soviet Communist Party, made his first trip to the People's Republic of China. Some of the Soviet leader's deliberations with the Chinese at this time concentrated on defense matters, primarily from the perspective of how to augment Soviet security by assisting China. Khrushchev writes: "We agreed to send military experts, artillery, machine guns, and other weapons in order to strengthen China and thus strengthen the socialist camp."[82] A reading of the final joint declaration of the Mao-Khrushchev negotiations verifies that the Taiwan crisis was on the agenda, but that the Soviet government made no commitments to rescue the Chinese in the event the Americans retaliated.[83] Indeed, Khrushchev makes no reference to the Taiwan crisis in his memoirs; he vividly recollects Chinese demands for expensive military and economic aid and his own prediction of inevitable conflict with his Chinese allies.[84] The Chinese remember the Khrushchev visit as isolating China in the eye of a storm, though unquestionably a storm intensified by Beijing itself.

Tensions persisted during the following months as Washington and Taipei advanced toward the mutual security treaty. Government edicts issued from Beijing repeatedly affirmed the call for Taiwan's "liberation" and, in addition to the continued shelling of the offshore islands,

---

31, convened a conference to devise a new battle plan. According to Zhang Aiping, the head of this command, the plan called for an attack on the Dachens, beginning with an invasion of the nearby island of Yijiang. Nie Fengzhi et al., *Sanjun Huige Zhan Donghai*, pp. 38-40.

the Chinese chose this period to sentence as "spies" several American fliers who had been shot down over China in the Korean War.[85] Each side was escalating its provocations against the other as the United States and Taiwan completed negotiations on the security pact that was to be signed on December 2, 1954.

The Formosa Resolution and Deterrence

We believe that the Chinese proclamations and actions demonstrated a weak, reactive response to American and Nationalist actions they could not prevent. Had the United States excluded the possibility of defending the offshore islands, of course, China might have chosen to seize them. Beijing did not expect this, though given what happened later with the decision by Taipei and Washington to evacuate the Dachen (Tachen) islands farther to the north, some Chinese leaders may have hoped for Taiwan's retreat.

In any case, China authorized only minimal military preparations for possible actions beyond the artillery bombardment of Quemoy and Matsu. Nevertheless, official U.S. statements about this bombardment tended to assume or assert that China, in a high tide of expansionism, intended to invade the offshore islands opposite Taiwan, particularly Quemoy and Matsu, and then Taiwan itself. These statements expressed the belief that the U.S.-Taiwan defense treaty and American actions would deter a planned aggression. Two questions arise: what did China do, and did the U.S. response serve as a deterrent to Beijing's invasion plans, if they existed?

Part of the answer turns on whether by "deterrent" one means the general military might of the United States or specific U.S. military actions.[86] Certainly American power constrained and influenced Chinese actions; the very U.S. military presence, which was augmented in the Western Pacific during the crisis, may be defined as a political-military deterrent. Yet if this is so in the Taiwan Strait crisis, why would any additional U.S. actions become necessary, and why would China not have taken steps to back up its probing use of shelling with an invasion buildup (as in the case of the Dachens), if the United States flinched? For what it is worth, the Chinese say they did not intend to invade the offshore islands of Quemoy and Matsu and took no action to this end.[87]

We tend to agree with scholars who have argued against the "deterred" hypothesis as an explanation of what the Chinese did and why. John Gittings, for one, has made the case that China in the mid-1950s was attempting "to break out of the diplomatic box into which it had been confined by the two great powers." Beijing's goal after the Geneva Conference, according to Gittings, was "how to defuse the American threat" and decrease international tensions. The Chinese timed the shelling on September 3, he would agree, to precede the opening of the Manila Conference, and the crisis escalated as the United States pursued its plans for the mutual defense pact with Taiwan. However, Gittings views the bombardment as coming *in medias res*. He concludes: "It is a standard myth of cold war historians that this Treaty was an American *response* to the Chinese-inspired Offshore Islands crisis of the preceding months. This is chronologically upside down." [88]

Prior to the Taiwan Strait crisis, Zhou Enlai had believed that, with the convening of the Geneva Conference, the time had arrived for negotiations, and this belief lay behind the Chinese pressures on Ho Chi Minh to accept a compromise solution (to settle for a division of Vietnam) at the conference. One Chinese has stated: "Zhou Enlai . . . was very upset in seeing his efforts to maintain peace zones in Asia dashed, and he repeatedly mentioned to foreign visitors that he was 'inexperienced' in believing that the U.S. would actually abide by the Geneva Accord." [89] U. Alexis Johnson, like several other American delegates to Geneva, commented at the time on the independent role China had played, though most believed Zhou's statements to the conference "revealed no hint of concession." [90] During or just after the conference, Zhou had visited India, Burma, East Germany, and Poland, and his clear aim, as he said in his final speech at Geneva, was the "further relaxation in the tensions in Asia and all over the world." [91] That aim perished within days of the issuance of the Final Declaration of the Conference on Indochina.

Zhou's speech to the Central People's Government Council on August 11 marked a clear turning point. The Chinese premier contrasted the American effort to establish "antagonistic military blocs" in an attempt to create new tension in various parts of Asia with China's proposal at Geneva for "the Asian countries [to] consult among themselves and make joint efforts to safeguard the peace and security of Asia by mutually assuming obligations." [92] He reviewed the history of Sino-American confrontation in the past months and the American attempts "to intervene militarily in China and threaten us with war from three fronts: Taiwan, Korea and Indo-China." With the armistice in Korea and the restoration of peace in Indochina, he said, the focal point of the struggle now centered on Taiwan. The premier foresaw no way that the struggle might end peacefully. It is "imperative," he said, "that the People's Republic of China liberate Taiwan and do away with the Chiang Kai-shek traitor gang." [93]

From the Chinese perspective, the American-Taiwanese provocations after the Korean Armistice required confrontation. Mao's strategy, rejecting as it did passivity and fearfulness, could not have embraced deterrence for China. In 1945, he wrote: "If they start fighting, we fight back, fight to win peace. Peace will not come unless we strike hard blows at the reactionaries." [94] His words were to find an echo in Zhou Enlai's address of August 11: "It is therefore necessary to shatter the designs of the aggressive circles of the United States [and] to administer defeat after defeat to their aggressive policy if the peace and security of Asia and the world are to be safeguarded." Zhou's target this time was clear: to reduce tensions and preserve the peace it was necessary to "liberate Taiwan." [95]

Such declarations, however, heralded a general shift to the realities of enduring conflict, not a specific timetable of conquest. As Mao had said, China aimed to bring the United States up short and to challenge its policy of alliances and its designs on Taiwan. That challenge was political, not military.

Interestingly, most knowledgeable American officials understood this and said so at the time. Leading members of the U.S. administration differed on the importance of the Chinese bombardment of Quemoy and Matsu, but even Ambassador Rankin reported: "Rightly or wrongly, I do not take the present fracas around Kinmen [Quemoy] very seriously from a purely military standpoint." He guessed that Beijing was "simply trying us out, taking advantage of the coincidence of the Manila Conference and the summer invasion season." [96] Army Chief of Staff Matthew Ridgway agreed. The two islands, he wrote a few years later, had "little value as offensive bases." Then he added these key sentences: "To my mind, what the Chinese were doing there did not justify a conclusion that they were planning an attack on Formosa. Their activities could just as well be defensive as offensive in nature. . . . There was no indication they were concentrating ground troops there, or organizing an invasion force of their own." [97] For his part, the president, in his press conference on August 17, 1954, had spoken of a buildup of Communist strength opposite Taiwan, a threat he believed was continuing in November and January, but even the chairman of the Joint Chiefs of Staff could cite only a general "buildup and declarations" to substantiate his belief that an invasion threat existed toward Taiwan. [98]

Two contemporary intelligence reports, now declassified, provide the most convincing evidence on what U.S. officials thought China intended to do. The first of these, a special intelligence estimate titled "The Situation with Respect to Certain Islands Off the Coast of Mainland China," is dated September 4, 1954, just after the bombardment began. In a section on the "current situation," its authors noted that the islands served as bases for Nationalist "intelligence activities, escape and evasion, and raids on coastal traffic and on mainland targets." They singled out the Dachen islands, however, as extending "U.S. early warning capability for Okinawa." More to the point, they reported "new troop and naval concentrations in the islands and along the coast" as a prelude to the artillery shelling. As to capabilities, the Chinese had "long had sufficient troops" and amphibious craft to overwhelm all but Quemoy and Matsu, yet even though Chinese armies had returned from Korea, there had been "no great increase in troop strength" in the coastal region opposite Taiwan. Despite these capabilities, "Nationalist blockade efforts [had] greatly hampered the movement of seaborne cargo from Europe to North China." The authors concluded that "Peiping [Beijing] presently estimates that an all-out effort to take the major Nationalist-occupied off-shore islands might well involve a substantial risk of war with the U.S." China could be expected to resort to "probing actions designed to test U.S. intentions," and any unilateral American guarantee to defend Taiwan would be "less of an affront" than a formal mutual defense pact. In the current circumstances, they believed, the Chinese "would continue to be deterred from an all-out attempt to seize the major islands by the prospect of U.S. counteraction."[99]

The authors of the second major intelligence estimate, "Probable Developments in Taiwan Through Mid-1956" (September 14, 1954), came to almost the same conclusions. They did not believe China would invade Taiwan or the Pescadores, because Beijing understood that such an attack would lead to war, and that "over the long run they can further their objectives with respect to the Chinese Nationalists by means not involving war." The writers concluded that the pressure for a U.S.-Taiwan treaty derived, not from any immediate military danger, but from Taiwan's view that "the US should give more concrete evidence of long-term support for [Nationalist] China." Further, the Taipei government wished to join a regional security system in East Asia because it "strongly fears that any system excluding Nationalist China would compete with Taiwan for US military aid." The document mentioned no immediate danger to Taiwan or the offshore islands. Rather: "The means employed [by China] include propaganda, diplomatic effort, threats, and military demonstrations." The belief that "the Chinese Communists will not invade Taiwan or the Pescadores" was based on the then firmly established ties between Washington and Taipei. Further committing the United States was deemed not only unnecessary, but provocative.[100]

At the highest policy levels, others joined Ridgway while the crisis was in progress in testifying before the Congress that they saw no preparations for a Communist strike against Taiwan or Quemoy and Matsu. But in fact the Chinese at this point were preparing to attack islands much farther to the north, in the Dachen group. On November 1, Chinese planes began bombing the Dachens and mobilizing for their invasion.[101] As the attacks continued, Eisenhower took the view that "these islands . . . were much less significant to the Nationalist cause than were Quemoy and Matsu, farther south." And when, in mid-January, "nearly four thousand Communist troops, with heavy air bombardment and an amphibious attack, overwhelmed one thousand Nationalist guerrillas and seized the island of Ichiang [Yijiang; now Yijiangshan Dao], just seven miles north of the Tachens," he concluded that neither Yijiang nor the Dachens were vital to the defense of Taiwan, and ordered the Seventh Fleet "to assist in the evacuation of the Tachens."[102] Chiang Kai-shek's government relinquished control of these islands on February 12, 1955. On that day, the U.S. leaders acquiesced in the only major military offensive planned by the Chinese in the entire crisis: they did not deter it.*

It was in the midst of the Dachen phase of the crisis that Washington and Taipei signed the mutual defense treaty.[103] Under such treaties, as a National Security Council working paper pointed out in October, an armed attack on Formosa and the Pescadores "would, in accordance with constitutional processes, involve the United States in war . . . in the case of Asiatic countries at least with Communist China or the Communist satellite committing the aggression." Nevertheless, the paper's drafters also stated that the United States "must keep open the possibility of negotiating with the USSR and Communist China," and made no mention of any immediate jeopardy to the offshore islands or Taiwan.[104] In November, as fears of the threat to Taiwan were being voiced in Washington, a top-secret State Department paper still

---

* After the attack on Yijiang on January 18, the East Zhejiang command ordered its troops to rest a week and then to prepare to invade the Dachens. Just as Eisenhower was deciding to evacuate the Dachens, the Central Military Commission ordered a temporary postponement of the attack in order not to provoke a larger conflict with the United States. Nie Fengzhi et al., pp. 51, 57.

suggested that "despite its bellicose talk, . . . Communist China may well be more interested in strengthening its economy and its international position than in early major military ventures that would involve war with the U.S."[105] But just as the treaty was being completed, the NSC was drafting a top-secret statement of policy that, while stressing China's internal problems, reiterated that China was "ostensibly committed to the conquest of Formosa." In a footnote, the writers concluded that the Chinese would almost certainly "increase probing actions against the Nationalist-held offshore islands and [would] probably try to seize them, if they believe this can be done without bringing on major hostilities with the U.S."[106] Signed in Washington on December 2, the treaty was transmitted to the Senate on January 6 and approved on February 9; it entered into force on March 3, 1955.

The inexorable progress toward the treaty masked a time of controversy in American strategic policy and spotlighted the limits of power. Across the Pacific, Beijing watched the debates in Washington and tried to calculate how the president would respond. In early January, the Chinese fixed on a speech by Admiral Radford in which he advocated preparations to use nuclear weapons if hostilities resumed in Korea.[107] According to a Chinese source, Radford added that "as for the use of atomic weapons in other places in Asia," this would "depend on the situation." The Chinese recalled that Radford had advocated the use of nuclear weapons in the Indochina conflict and assumed that he was now proposing to employ them against China itself.[108] On January 8, *Renmin Ribao* declared that Radford had advocated using atomic weapons against both North Korea and China.[109]

After the surrender of Yijiang a few days later, however, the impotency of the U.S. massive retaliation strategy was becoming ever more obvious to Chinese and Americans alike. As Eisenhower contemplated the Dachen evacuation, he settled on a scheme to offset the psychological consequences of that decision, giving no thought to an atomic response. He proposed a message that would simultaneously announce the planned evacuation and bolster Chinese Nationalist morale.[110] On January 24, he requested authorization to employ the armed forces of the United States "as he deems necessary for the specific purpose of securing and protecting Formosa [Taiwan] and the Pescadores against armed attack." Within a matter of days, the Congress passed the Formosa Resolution, granting Eisenhower's request and giving him the authority to employ armed force in the "securing and protection of such related positions and territories of that area now in friendly hands."[111] According to one source, this reference to "that area now in friendly hands" presaged "a secret commitment to the nationalists to defend Quemoy and Matsu." Recently declassified documents suggest that the Taipei authorities believed that such a commitment had been made, but that the U.S. president, as he put it on January 30, wanted to "be sure we are not hooked into any agreement whereby we would have to join in the defense of Quemoy or Matsu." [112] This is consistent with a view expressed by Dulles (and endorsed by Eisenhower) that the United States "should declare that we will assist in holding Quemoy and possibly the Matsus, as long as the Chinese Communists profess their intention to attack Formosa." [113]

The congressional hearings concerning the mutual defense treaty and the Formosa Resolution reinforce the conclusion that the administration did not believe an invasion of either Taiwan or Quemoy and Matsu was likely. Indeed, the administration did not present the case for the treaty or the authorization to use U.S. troops as a response to any imminent military threat. In his testimony on the Formosa Resolution (January 13), Secretary Dulles stated: "There are at present, as far as our intelligence is aware, no formation on the land of landing forces, or any accumulation of ships which would be required for landing operations." He called the situation around Quemoy one of "relative quiet," though he later argued that "by everything that they have said" Communist operations were designed to be a "buildup into an attack against Formosa." More to the point, the secretary argued, the Chinese had challenged the "position of the United States," and "the free nations of that area . . . are watching to see what if any reaction will come from this probing operation." [114]

The problem, Dulles said, was not one of geography "but a problem very largely of psychology," and for this reason he felt that the abandonment of the Dachen islands could not be left to stand by itself. If that was all that happened, he told Congress, "I think that in itself would be very bad. I think it can be handled if at the same time that they [the Chinese Nationalists] regroup their forces there is the added weight on the other side of the scales of a clear determination by the United States to use its Armed Forces. . . . I believe it is essential that we should throw something to offset that into the scales." Dulles placed the righting of the scales in a political perspective. China, as he evaluated its policies, had achieved major political gains in Korea and Indochina and was about to chalk up a military victory in the Dachens. He argued that the United States had to show resolve and even risk war with China to offset its political achievements. If the United States was not willing to take that risk, he said, "let's make that decision and we get out and we make our defenses in California." [115]

Admiral Radford in his turn acknowledged that there was some risk of war, which "might cause the Chinese Communists to call on their Russian allies for help under the Sino-Russian Treaty." But he believed the Soviet Union would not precipitate a global war over Taiwan and would "not openly" come to China's assistance. Like Dulles, he did not see any immediate danger to Taiwan or Quemoy and Matsu. The purpose of the Formosa Resolution was simply to "stabilize" the situation and balance the bad psychological effects of a "very disastrous defeat" in the Dachens. Called to argue for the defense treaty on February 1, Dulles claimed it would "provide firm reassurance to the Republic of China [on Taiwan]."[116] Both the resolution and the treaty were approved expeditiously: Congress passed the resolution on January 29; the Senate approved the mutual security treaty a week later.

What were the Chinese to make of these American reactions? Washington had capitulated in the face of China's only direct military action and had mobilized the Congress to authorize the use of armed combat against a nonexistent threat. Thereafter, Washington announced its readiness to fight a major war over Taiwan, but took no serious steps to bolster its conventional military forces in the region. According to the political scientists Morton Halperin and Tang Tsou, the U.S. military "appear to have counted on the right to use nuclear weapons if necessary to defend Quemoy and in this sense saw the [Offshore] Islands as defensible,"[117] and the Chinese appear to have concluded that the Americans were preparing to fight a nuclear war against them. The question is: did the Chinese have real cause to come to such a conclusion? Hindsight suggests that the answer is no, but the Chinese could not have known this and would have felt justified in reaching the opposite conclusion in light of what was to come. Dulles was right: what leaders believed outweighed any other reality.

We believe that Beijing's leaders reacted to, rather than initiated, the Taiwan crisis of 1954-55. In the main, Chinese leaders confronted the United States in order to communicate the fierceness of their opposition to the U.S.-Taiwan treaty negotiations. They acted from a position of military weakness. They had no assured capability of invading the principal offshore islands opposite Taiwan, let alone Taiwan itself. Beijing did not expect to halt the treaty negotiations; the Chinese knew they had no power to do so.

Faced with both the treaty and increased American threats to use nuclear weapons against them, however, the Chinese did change their policy. They resolved to acquire nuclear weapons of their own.

THREE

The Strategic Decision and Its Consequences

The events in Korea, Indochina, and the Taiwan Strait constituted the proximate cause of the Chinese decision to build a national strategic force. These events galvanized the leadership to act in the winter of 1954-55 and gave special urgency to the strategic weapons program in the decade thereafter.

Nevertheless, the revolutionary elite under Mao Zedong came to power in 1949 with beliefs that may well have led to the nuclear weapons decision even without the unbroken chain of crises. The leadership's nationalistic ideology and concepts of force and diplomacy shaped its perceptions of the enduring dangers to China and to the restoration of China's international position. Memories of the civil war and fear of aggression by hostile outside powers imbued the top command with a strong military bias and an assurance that its appraisal of China's situation was wholly realistic.[1] In pondering the response to the recurrent American strategic threats after 1949, China's leaders would have weighed fully the vulnerabilities exposed in the Korean battles and their own ultimate reliance on themselves. The decision to acquire a nuclear arsenal rested on fundamental national interests as much as on the immediate security threat.

Mao had long regarded a country's independent capacity to display, deploy, and commit its armies as a vital component of its sovereign independence. He dreamed that China would acquire the unshackled ability to mobilize and use effective military power, for only that power would distinguish the new state from its humiliated predecessors.

Consistent with this dream, Mao's rationale for the acquisition of nuclear weapons centered on destroying the "nuclear monopoly" of China's adversaries. In 1963, the Chinese summarized the perspectives that had guided the nuclear program and caused them to recoil from

Soviet-American efforts to inhibit nuclear testing. Two statements are of particular importance. One sounds the blackmail note: "Nuclear weapons in the possession of a socialist country are always a means of defense against nuclear blackmail and nuclear war." The other adds to this the notion of "never again": "It is absolutely impermissible for two or three countries to brandish their nuclear weapons at will, issue orders and commands, and lord it over in the world as self-ordained nuclear overlords, while the overwhelming majority of countries are expected to kneel and obey orders meekly, as if they were nuclear slaves. The time of power politics has gone forever, and major questions of the world can no longer be decided by a few big powers."[2]

Such Chinese views long before had coincided with the politics of some Western ban-the-bomb advocates, such as the Nobel laureates Frédéric and Irène Joliot-Curie, who also felt motivated to help China break the American nuclear monopoly. In the spring of 1949, shortly after the Communist takeover of Beijing, the Chinese government gave the physicist Qian Sanqiang foreign currency to purchase the country's first nuclear instruments when he attended a peace conference in Europe. The Joliot-Curies, with whom Qian had studied in wartime Paris, helped arrange the purchase in England and France. In October 1951, Frédéric Joliot-Curie in Paris urged the Chinese radiochemist Yang Chengzong to seek out Mao upon Yang's return to China. "Please tell Chairman Mao Zedong," he said, "you should oppose the atomic bomb, you should own the atomic bomb. The atomic bomb is not so terrifying." He noted that the "fundamental principles of the bomb had not been discovered by the Americans." Irène Joliot-Curie then gave Yang ten grams of radium salt standardized for radioactive emissions, because she wanted "to support the Chinese people in their nuclear research." By 1955, this French advice and assistance had helped raise the level of consciousness in Beijing about the bomb and its potential significance for China. Mao characterized that significance for his senior colleagues in 1958, when he told them that without atomic and hydrogen bombs, "others don't think what we say carries weight."[3]

Mao thus understood the importance of nuclear weapons and the power they bestowed. Particularly revealing of his great concern with what he called "U.S. atomic blackmail" are his remarks at the end of January 1955 to the Finnish envoy to China. In this interview, in which the Chairman assailed the United States for "contemplating an atomic war," he began by merely echoing his 1946 pronouncement that atomic weapons are "paper tigers," observing: "The United States cannot annihilate the Chinese nation with its small stack of atom bombs." But then, in a somewhat atypical excursion into hyperbole about nuclear weapons, he added: "Even if the U.S. atom bombs were so powerful that, when dropped on China, they would make a hole right through the earth, or even blow it up, that would hardly mean anything to the universe as a whole, though it might be a major event for the solar system." He concluded the interview with the typical boast that a third world war would mean the end of the American system of government and the global transformation to communism.[4] Yet the tone of Mao's remarks displayed respect, even awe, for the power of the bomb he had already decided to get.

Even as Mao was speaking, the Taiwan Strait crisis was becoming more intense. Artillery fire by the People's Liberation Army (PLA) against the offshore islands flared intermittently and was punctuated by air strikes. In mid-January, air, sea, and ground units launched the PLA's first combined-forces operation, the invasion of Yijiang. In response, as we have seen, President Eisenhower invited Congress "to participate now, by specific resolution, in measures designed to improve the prospects of peace," including "the use of the Armed Forces of the United States if necessary to assure the security of Formosa and the Pescadores."[5] The request was made on January 24. With little debate, the Formosa Resolution became law on January 29.[6]

When the crisis with the United States began deepening in January, the Chinese Politburo expressed ever greater concern about the possible American use of nuclear weapons against China. On January 16, for example, *Renmin Ribao* accused the United States of treating atomic weapons as conventional arms, and the following week, Zhou Enlai claimed that the United States was "brandishing atomic weapons" in an attempt to maintain its position on Taiwan.[7] No sooner was the Formosa Resolution passed than the Chinese press alleged that it included the threat to use atomic weapons against the Chinese people.[8] By this time, the Chinese Politburo had launched the nation's nuclear weapons program.

In the second week of January, as the decision-making process began, Premier Zhou Enlai invited the nuclear scientist Qian Sanqiang to a meeting in his office with Bo Yibo, a minister in charge of economic affairs, and Li Siguang and Liu Jie from the Ministry of Geology.[9] In the discussions that followed, Qian, as head of the Institute of Physics (and later director of the Institute of Atomic Energy), lectured Zhou on the atomic bomb and the status of China's nuclear research and evaluated the country's manpower and facilities in the nuclear field. Zhou questioned Liu Jie about the geology of uranium and, with Qian, reviewed the fundamentals of atomic reactors and nuclear weapons. At the conclusion of this unusual seminar, Zhou instructed those present to prepare for a full-dress meeting with Chairman Mao.

On January 15, Mao presided over an enlarged meeting of the Central Secretariat called to discuss the reasons for and the possibilities of starting a nuclear weapons program.[10] He came to the meeting having been urged on by no less than the Joliot-Curies and with that special air of confidence that flows from superficial knowledge. After all, to the Chairman nuclear fission proved the validity of the basic law of materialist dialectics. As the Chinese leader had once written, Lenin understood that dialectics "is the study of contradiction in the very essence of objects." Lenin also had conceived of development as "a unity of opposites (the division of a unity into mutually exclusive opposites)." These contradictions were illustrated in physics "by positive and negative electricity," and the physics of the atomic nucleus only confirmed what Mao felt he already knew.[11] Indeed, Mao later praised a Japanese particle physicist, Sakata Shiyouchi, for catching up to Lenin, and had the Central Committee publish an article by the Japanese scientist in *Hongqi*.[12] Mao had long since decided that space was both infinite and infinitely divisible, and that nuclear physics was therefore relatively easy to understand. "When one lectures on nuclear physics," he said much later, "it will suffice to talk about the Sakata model; one needn't start from the theories of Bohr of the Danish school; otherwise you won't graduate even after ten years of study. Even Sakata uses dialectics—why don't you use it?"[13] Mao, who appears to have first met the Japanese physicist in 1956, was not intimidated by nuclear physics. Nor, it seems, did the Chinese leader understand what was in fact recondite.

At the Central Secretariat meeting, attended by all the senior members of the Politburo, Qian Sanqiang was joined by Minister of Geology Li Siguang, an eminent scientist, and Liu Jie. Mao quickly turned the meeting over to the scientists, and the conference room in Zhongnanhai, the political and state center in the Forbidden City, became a classroom in introductory nuclear physics and uranium geology. On a table was a sample of uranium, and the members of the Politburo took turns playing with a Geiger counter to hear it click. One source notes, "They all smiled jubilantly."[14]

Having previously attended his own fission seminar with Qian, Zhou Enlai acted as scientific interpreter for his Politburo colleagues.

As Mao posed questions, Zhou sat at the Chairman's side and explained to him what was going on while "frequently reminding the scientists to report in more detail." Mao was ecstatic. After he had heard the scientists out, he made his reply and began by highlighting the promising evidence concerning China's uranium potential and the building of its scientific base. Mao said:

During the past years we have been busy doing other things, and there was not enough time for us to pay attention to this matter [of nuclear weapons]. Sooner or later, we would have had to pay attention to it. Now, it is time for us to pay attention to it. We can achieve success provided we put it on the order of the day. Now, [because] the Soviet Union is giving us assistance, we must achieve success! We can also achieve success even if we do this ourselves.

The Chairman "cheerfully announced that China would immediately devote major efforts to developing atomic energy research" for military purposes. He concluded the session on the same upbeat note: "We possess the human and natural resources, and therefore every kind of miracle can be performed."

Mao then invited the assembled leaders and scientists to dinner, where he raised his glass of *maotai,* saying: "Let's drink a toast to the development of our country's atomic energy cause!" Zhou Enlai then stood and in his toast called on the scientists "to exert themselves to develop China's nuclear program." The Politburo had resolved to proceed with the nuclear weapons project. The program's code name: 02.

Soviet Aid and the Mobilization of Chinese Scientists

In taking this decision, the Politburo did not intend that China should go it alone. Rather, Mao's first impulse, revealed in his statement on January 15, was to enlist Moscow to back a crash program. He had no idea how long the program would take or how much it would cost, but he judged the imminency of war too great to risk a solo race for the bomb. As Mao was to say a couple of months later, "we call for preparedness against a sudden turn of events [and] must envisage the worst possibilities." He defined the worst as a new world war.[15]

This emphasis on the possible outbreak of a global conflict marked a departure from long-standing Maoist considerations of strategic policies. We have long known that in 1955 the Chinese military was preoccupied for a time with questions of nuclear strategy.[16] The previous chapter quoted Chinese writings of that year directed against the possible American use of nuclear weapons. These writings increasingly reflected doctrinal shifts paralleling those in the Soviet Union calling for the preemptive use of nuclear weapons should a nuclear war appear inevitable, and warning about a surprise nuclear attack. The new emphasis in China's doctrines coincided with the Politburo's nuclear decision.

That emphasis also coincided with Beijing's heightened nervousness about American nuclear weapons in the first half of 1955, a period in which American officials deliberately aggravated Chinese fears of nuclear attack. Secretary Dulles, for example, on his return from a meeting of the Southeast Asia Treaty Organization in early March told the president that the situation in the Taiwan Strait was "far more serious than I thought before my trip." [17] He now believed it might be necessary to use "new and powerful weapons" to thwart any Chinese "armed aggression," a message that senior U.S. military and intelligence officers also communicated to Eisenhower.[18] In a news conference on March 16, the president asserted that atomic weapons could be used tactically in Asia without massacring civilians, an assertion that *Renmin Ribao* soundly condemned the next day. Eisenhower's memoirs record his hope that his public commitment to the possible use of tactical nuclear weapons "would have some effect in persuading the Chinese Communists of the strength of our determination." [19]

The president need not have worried, for most Chinese required no convincing of America's willingness to resort to nuclear war. As Soviet and Chinese military doctrines temporarily came into alignment after the death of Josef Stalin, it was clear that both nations had little doubt about that determination. Both expressed great horror and played on the global fears of nuclear war, particularly during a Soviet-initiated, international ban-the-bomb campaign, which began in February 1955. As Guo Moruo, in his role as chairman of the Chinese Peace Committee, said, "The way the whole country sprang to action shows how strongly the Chinese people feel about atomic war." [20] Nevertheless, some senior Chinese seem to have recognized the precariousness of the nuclear course on which they had embarked—to have understood that the very process of developing nuclear weapons, not just their eventual deployment, could jeopardize their country's national survival.[21] Later, as we shall see, Mao Zedong dismissed the dangers of nuclear war and reaffirmed the principles of people's war, but for several months in 1955 the possibility of a preemptive nuclear strike against China, perhaps in the near future, received high-level attention in Beijing.

The parallel developments in Sino-Soviet doctrines reflected another reality, the growing cooperation between the Soviet and Chinese military establishments, and the broader Chinese campaign to learn from the Soviet Union. The toll of the Korean War provided the fundamental motivation for that cooperation. The convergence and articulation of security interests between the two Communist powers in turn profoundly influenced the Kremlin's decisions to support the Chinese nuclear program.

On January 17, 1955, the Soviet government had announced that it would give aid to China and several East European countries to "help them promote research into the peaceful uses of atomic energy." [22] This commitment, confirmed in April, would provide China with a cyclotron and a nuclear reactor as well as fissionable material for research. [23] In exchange, China agreed to provide the Soviet Union with "necessary raw materials," presumably meaning materials for Soviet strategic programs. Moreover, Beijing backed Soviet peace campaigns then in progress. On February 28, the State Council called for the eradication of nuclear weapons in its resolution acknowledging Soviet assistance, and contrasted the Soviet Union's "peaceful use of atomic energy" with America's use of nuclear weapons for blackmail and intimidation, especially "in the East." [24] On the basis of secret deliberations, the two socialist countries also reached an accord to undertake joint exploration for uranium in China.* In time the Soviet government would promise to sell China some of the industrial equipment for uranium hydrometallurgy and for the processing of uranium concentrates and the production of enriched uranium. [25]

Parallel to China's increased dependence on Moscow was Mao's resolution to build up China's own military nuclear capability for the long term, and in the next three years, the Chinese leadership pursued dual paths toward the bomb. Beijing assumed that actions taken along the two paths would be mutually supportive and complementary, and that its dependence on Moscow would be short-lived. Whatever the dimensions of Soviet assistance, any Chinese nuclear program in the

---

* Between 1955 and 1958, the Soviet Union and China signed these six accords related to the development of China's nuclear science, industry, and weapons program: (1) an agreement to undertake jointly run surveys in China for uranium, including China's promise to sell any surplus uranium to the Soviet Union, Jan. 20, 1955; (2) an agreement on Soviet assistance to China for research on nuclear physics and the peaceful utilization of atomic energy, including the supply of a nuclear reactor and a cyclotron, April 27, 1955; (3) an agreement on Soviet aid in building China's nuclear industries and research facilities, Aug. 17, 1956; (4) an agreement to change the form of Sino-Soviet uranium surveys from joint operation to China's independent management with Soviet assistance, Dec. 19, 1956; (5) the New Defense Technical Accord in which the Soviet Union agreed to supply China a prototype atomic bomb and missiles as well as related technical data, Oct. 15, 1957; and (6) an agreement supplementing the August 1956 accord by determining the schedule and scale of Soviet aid, Sept. 29, 1958. Li Jue et al., *Dangdai Zhongguo*, pp. 19-22.

long term would have to create an indigenous capability to manage and use the training and materiel provided by the Soviet Union. The very reasons that drove the Politburo to pursue the nuclear weapons program in the first place made it loath to follow the single path of dependency.

Above all, Mao acted as a realist. China's status as a neophyte in the strategic weapons field forced him to modify his preference for self-reliance. He considered Soviet help a necessary but temporary expedient, and he actively sought it. Soviet knowledge and aid, if adroitly circumscribed, could speed the day toward independence. At the same time, because Soviet assistance would buttress the five-year-old alliance, Moscow could only gain from a stronger bulwark against the United States to the Kremlin's east. In Mao's view, Soviet assistance would reinforce both sides of the defense partnership even though China over time would become less dependent on the Soviet Union. Thus, to him, the Kremlin should willingly work to make China an independent nuclear power.

Within China, the scientific community almost immediately felt the impact of the Politburo's January 1955 decision, and its members, in a country that Mao called "poor and blank," understood the need for foreign support. Nevertheless, this small community would have to shoulder and set the pace for the entire program, whether domestic or imported from the Soviet Union.

The state's new priority on science and technology was immediately reflected in the national budget. Funds for science rose from about U.S.$15 million in 1955 to about U.S.$100 million in 1956; the Chinese Academy of Sciences received three times as much money in 1957 as it had received in 1953, with a large fraction going to purchase scientific literature from the West.[26]

On February 12, the academy's president, Guo Moruo, made oblique reference to the revamped priorities in a discussion of the country's accomplishments in research on atomic energy over the preceding five years. He noted that almost three-fourths of the academy's fellows could read Soviet documents, and that more than a quarter of them could translate these documents into Chinese. With this base, he implied, China's scientists could take full advantage of the Soviet Union's "disinterested gifts and aid" in the nuclear field. Continuing, he said, "our people's democratic system is suitable for the large-scale peaceful utilization of atomic energy," and he called on the country's scientists to go all out "to master the overall technology in order to speed the transition of research from the experimental to the practical and applied phase." If this were done, there could be "decisive accomplishments."[27]

As Guo Moruo's words suggest, the Chinese leaders had already mobilized a credible scientific cadre by 1955.* They had even begun to organize some of the scientific infrastructure that would become critical to the bomb program. The Chinese Academy of Sciences, absorbing the Academia Sinica (Zhongyang yanjiuyuan) in Nanjing and the Beiping Academy (Beiping yanjiuyuan), was founded on November 1, 1949, and the government invited Chinese both at home and abroad to help create a modern scientific research establishment. At that time, the Chinese Physics Association (Zhongguo wuli xuehui), an academic group outside the academy, had a membership of about 570, including ten senior nuclear scientists actively engaged in research.[28] In the next years, major Chinese scientists working or studying abroad began to trickle back to the People's Republic.[29] In these early years, however, the physicists "didn't even have a small-sized accelerator," though within a year or so the government had begun to back the academy's nuclear research efforts.

During the first half of 1950, the Chinese Academy of Sciences gathered various institutes under its wing and reorganized them.[30] One of the new research bodies was the Institute of Modern Physics, with Wu Youxun and Qian Sanqiang as director and deputy director, respectively. In June, the academy leaders decided to give priority to research on the atomic nucleus, and over the year the academy began to receive assistance from Soviet experts in this and other scientific fields. The government also ordered the Ministry of Foreign Affairs to invite specially chosen foreign experts to visit the People's Republic to assist in China's development, including the modernization of its science and technology.[31] In the years that followed, research on nuclear physics continued as a priority and support for that research began to flow to the Institute of Modern Physics.[32]

The Soviet Union played a critical role in this early development of China's nuclear science. For example, from February to May 1953, it played host to a delegation of 26 experts led by Qian Sanqiang, sent to "study Soviet experiences on the development of scientific research and to exchange views on extending scientific cooperation between

---

* As early as 1944 in Yan'an, the Politburo issued a directive to send technical cadres abroad, including the United States, for advanced training in the sciences and engineering. Those sent to the United States later helped form an organization to persuade all Chinese technical personnel in America to return to China after 1949. Huang Fengchu and Zhu Youdi, "Hou Xianglin," pp. 18-19.

the two countries."[33] The Chinese lectured their Soviet colleagues on the state of modern sciences in their country, and in his remarks Qian paid special attention to his academy's preparations for research in nuclear science, dealing with both basic and applied physics.[34]

By the year of the nuclear decision, China had available some key scientists who had been trained abroad in nuclear physics, notably Qian Sanqiang, who with his wife, He Zehui, had studied nuclear physics at the Curie Institute in Paris, and Peng Huanwu, who had studied with the Nobel laureate Max Born in Edinburgh. As early as 1932, the Chinese had established the Chinese Physics Association with some 70 members, and the following year *Zhongguo Wuli Xuebao* (Chinese Journal of Physics) had been launched. Over the next 20 years or so, the journal published hundreds of articles reporting the results of the research conducted at six main centers: the Institute of Physics of the Academia Sinica (nuclear physics), the Institute of Physics and the Radium Science Institute of the Beiping Academy (radium science), Beijing University (molecular spectroscopy and quantum field theory), Zhejiang University (nuclear physics), and Qinghua University (nuclear physics). In the same period, a number of outstanding Chinese scientists made contact with their counterparts in Europe and America. Among them were Qian Sanqiang and He Zehui, who had investigated fission in uranium; Peng Huanwu, noted for his research on cosmic rays; Wang Ganchang, a specialist on radioactivity and bubble chambers, who had studied with Lise Meitner in Berlin; and several missile experts, including Qian Weichang, who had spent the war years at the California Institute of Technology's Jet Propulsion Laboratory.[35]

Because we will encounter many of these Chinese scientists throughout this study, we may profitably pause here long enough to look at the backgrounds of two of the most famous ones the better to understand the small but impressive scientific group that manned the upper echelons of China's nuclear program. These are Peng Huanwu and Qian Sanqiang, who were graduated from the Physics Department of Qinghua University in 1935 and 1936, respectively. In 1938, Peng went to Scotland to become Max Born's first Chinese student. In the next ten years, he earned two doctorates and acquired a reputation for leading work on quantum field theories.[36] Meanwhile, Qian, as noted, spent the war years in German-occupied France working on theoretical physics with Irène Joliot-Curie. His interests focused on gamma and alpha rays and the fissioning of uranium. He received his doctorate from the Curie Institute in 1943.[37] Before returning to China in

1948, he and Peng met in Paris and discussed how they would serve China after returning home.

Once there, Peng began teaching university students in Beijing and was invited by the Ministry of Education to design summer courses to train physics teachers in quantum mechanics. Two of his students from this period, Zhou Guangzhao and Huang Zuqia, later participated in the weapons program.[38] Qian, a leader of that program, was to recommend Peng Huanwu, his longtime friend and schoolmate, to head the theoretical unit that designed China's first fission and fusion bombs.

From 1953 to 1955, it should be noted, the Chinese Communist Party had enjoined all cadres to study key Soviet reports, ideological studies and histories, and works on science and technology.[39] On April 23, 1953, while the Qian delegation was in the Soviet Union, the Central Committee called on all cadres to pursue an eighteen-month study program on Soviet experiences in socialist construction.[40] The leaders of the Chinese Academy of Sciences focused this study campaign on Soviet literature (instead of American and British).[41] By 1955, after the Politburo decision, one purpose of this study was to ready the scientific community for the anticipated flood of materials on atomic energy following an agreement with Moscow on nuclear cooperation (negotiated and signed in April).[42]

Despite these auspicious beginnings, the post-decision decade in China proved particularly difficult for all intellectuals. In these years, the Party's Central Military Commission called for the recruiting of scientists and engineers who would devote themselves to the strategic weapons program and be innovative, while the Party's Politburo denounced scientists for being apolitical and too independent. Some weapons specialists—such as Qian Weichang—were caught up in one or another of the campaigns criticizing intellectuals as "Rightists" or other "deviationists" from the officially endorsed line.[43]

In January 1956, Premier Zhou Enlai addressed a meeting to end the contradictory signals to intellectuals and to improve the Party's treatment of them. In his report, Zhou concluded that the "peak of the latest developments in science and technology is the application of atomic energy" and called for "mastering the most advanced sciences [so that] we can ensure ourselves of an impregnable national defence."[44]

Unfortunately for the intellectuals, this landmark meeting only temporarily stemmed the tide of criticism against them. The attacks, if they ever stopped, quickly resumed, and scientists a year later complained that the Party still greatly hampered scientific work.[45] As the result of this and similar unguarded complaints, the scientific community in 1957 became engulfed in a so-called anti-Rightist campaign. Within a few short months, many scientists and engineers again fell victim to recurrent denunciations and another round of ideological rectification. Nie Rongzhen, who was then becoming the overall head of the strategic weapons program, has criticized the Party's organizations at that time for "putting forward a slogan of 'removing white flags,'" the Party's jargon for the condemnation of scientists who ignored or disapproved of Communist policies. In his view, "the aftermath of this struggle dampened the enthusiasm of many intellectuals and brought a passive influence to China's scientific programs."[46]

What appears to have protected many scientists in the nuclear program is not their association with the program as such, but the physical isolation that came with it. In the early 1960s, for example, Peng Huanwu, Guo Yonghuai (an expert in mechanics), the chemist Jiang Shengjie, the physicist Wang Chengshu (wife of the noted physicist Zhang Wenyu), and the physicist Zhou Guangzhao "were successively transferred to engage in the development of the strategic weapons program." From then on, "they left the metropolitan areas for the Gobi desert, the snowy mountains and the grasslands, and, keeping their identities hidden, quietly immersed themselves in hard work."[47] In their new roles, the specialists who were assigned to the nuclear weapons program retained only nominal ties with the academy. The moment they passed into that program's secret world, they fell under the military's control. There they remained relatively safe until the Cultural Revolution in 1966.

As good soldiers, these scientists gave their youth and their knowledge to the development of the nuclear weapons program. One source declares that their "names should be written with golden characters" and lists several martyrs: "Famous mechanics expert Guo Yonghuai, famous chemist Cao Benxi, radioactive chemist Liu Yunbin, uranium metallurgy engineer Zhang Tongxing, uranium hydrogeology engineer Chen Jianhua, metallurgy chief engineer Zhang Tianbao, and leading cadre Wu Jilin." (Liu Yunbin, we may note, was the son of then President Liu Shaoqi.) The program commanded the careers and, in some cases, the lives of many of China's best minds.

National Organization of the Nuclear Weapons Program

Within a few months of the January 1955 decision, the nuclear program began to acquire content and direction. The two most important organizational decisions came in July. On the fourth, the Politburo appointed a Three-Member Group under its direct authority to preside over policy making for the program.[48] The three were Chen Yun, Nie Rongzhen, and Bo Yibo. Chen, then just turned fifty, had come to dominate the field of economics and, as a leading Politburo member and vice-premier, held a pivotal position in respect to industrial development.[49] Nie Rongzhen, by contrast, had made his reputation as a senior battlefield commander in the revolution and, from 1950 until 1953, had served as acting chief of the General Staff.[50] In late 1958, he assumed overall supervision of the entire strategic weapons program. The Politburo seems to have selected Bo Yibo, later to become an alternate member of the Politburo and vice-premier, for his managerial talent.[51] Though he had made his mark as a political commissar in the wartime army, his principal post-1949 jobs had dealt with general economic policy and planning. The trio of Chen, Nie, and Bo had the initial authority to approve all work in the nuclear field.

On the day the Three-Member Group was formed, the central leadership empowered the State Council's Third Office (which had been created in November 1954) to serve as the organ responsible for the "concrete affairs" of the nuclear industry.[52] This office, which was led by Bo Yibo, Sun Zhiyuan, and Liu Jie and set policies for heavy industry and construction, established a special administrative body to oversee two organs charged with launching the nuclear weapons program: the Third Bureau and the Bureau of Architectural Technology (Jianzhu jishu ju). The first, a Third Bureau under Li Siguang's Ministry of Geology, had the task of exploring for uranium, and it almost immediately dispatched prospecting teams to Xinjiang and Hunan provinces. Lei Rongtian headed this new bureau, which reported both to the ministry and to the State Council's Third Office.

The second and more important new bureau, created by the State Council on July 1, was the Bureau of Architectural Technology, a name deliberately chosen to obscure the group's true mission.[53] The State Council assigned Liu Wei, then a senior political cadre studying at an army academy, to head this new organ under the State Construction Commission, and picked the physicist Qian Sanqiang to be his deputy. This bureau, as one of its first tasks, would supervise construction of the experimental nuclear reactor and cyclotron being supplied by the Soviet Union. Jointly with the Chinese Academy of Sciences, the bureau worked closely with Qian's Institute of Physics (formerly the Institute of Modern Physics and renamed the Institute of Atomic Energy in 1958), which would be assigned to run the two nuclear facilities.

By 1956, however, the bureau's highest priority mission was to dispatch groups to select locations for nuclear plants. Liu Wei established a Plant Location Selection Office that by mid-1957 had chosen over 60 possible locations. Liu's assignment, moreover, marked the first major step in putting the entire atomic program under the military as well as the State Council.

By the late fall of 1955, the start-up organizations of the nuclear program were all in place, and the Chinese now celebrate November 4 of each year as the anniversary of the establishment of their nuclear industry. Two months before, Bo Yibo had supervised the drafting of a document entitled "Comments on Formulating Our Country's Plan for the Atomic Energy Cause." Revised and renamed the "Outline Plan for Developing the Atomic Energy Cause, 1956-1967 (Draft)" in December, the document stipulated the need to pursue the program "with the vigorous assistance of the Soviet Union so as to enable our country to develop its national economy and consolidate its national defense." It proposed the creation of a nuclear industry with a complex of production facilities, including military reactors for plutonium.[54]

About a year later, on November 16, the State Council formed the Third Ministry of Machine Building to assume directorship of China's nuclear industry.[55] Until 1958, the nuclear program had its own specially integrated system (*xitong*), but in that year it was restructured as the Second Ministry of Machine Building, becoming part of the general industrial system organized into several numbered machine-building ministries.

We still know very little about the early history of this ministry, even less than we know about the strategic missile ministry. For years it was enveloped in extreme secrecy. The Chinese specialists we interviewed explained these tight security precautions in terms of the ministry's dual functions. The fact that it oversaw civilian programs as well as military ones caused the Central Military Commission to build an especially strong wall of security around the work involving weapons research and development. Furthermore, Soviet advisers imposed extremely rigid security regulations wherever they touched the nuclear program. Even missile specialists received only the most basic information—such as weight, center of gravity, external dimensions, and safety controls—about the nuclear warheads to be attached to the missiles. Because the missile ministry managed no civilian programs, its military personnel escaped the Kafkaesque world of their colleagues at the nuclear ministry.

With the creation of the Third Ministry of Machine Building in November 1956, the Politburo's Three-Member Group became inactive. The new ministry now assumed some of the responsibilities previously assigned to the then Second Ministry of Machine Building, which to that point had exercised overall responsibility for China's war industry.[56] In 1952, when Zhao Erlu became head of the newly formed Second Ministry, the engineering and technical personnel in the nation's defense industry numbered fewer than 1,000.[57] Zhao devoted special attention to training a much larger cadre of defense specialists and selected the best of them to study in the Soviet Union. He set his sights for the army to catch up with the advanced industrial countries, and as a later participant in the strategic weapons program, he is credited with leadership in the development of the nuclear program's industrial base.

Once nuclear weapons entered the picture, the Second Ministry's concern was limited to conventional weapons, and direct responsibility for all development in the nuclear field passed to Song Renqiong, the head of the Third Ministry. A senior commander and political commissar in the revolutionary armies, Song brought important national-level administrative experience to his office.[58] Before coming to the ministry, he had served as deputy director of the Central Military Commission's General Cadres Department and as a deputy secretary-general of the Party Central Committee. In the latter post, he continued to build on a long-term relationship with then Party Secretary-General Deng Xiaoping. With his ministry appointment, Song seemed destined for increasing national stature. But for unexplained reasons, he is reported in the official sources to have left the ministry in 1960, and then Beijing itself in 1961, when he became first secretary of the Party's Northeast Bureau. Nevertheless, his name remains firmly attached to the early successes of the program, and *Mimi Licheng* and *Dangdai Zhongguo de He Gongye*, the histories of the nuclear weapons effort, carry introductions by him.

In March 1956, the State Council had created a temporary body called the Scientific Planning Commission, headed by Marshal Chen Yi. Its charge was to complete the Twelve-Year Plan for the Development of Science and Technology, which, Zhou Enlai had announced in January 1956, would make "our scientific and technological level" in defense, production, and education approach that of the Soviet Union and "other great powers."[59] Preliminary discussions on the plan had already begun in December 1955, when some 200 scientists and technicians assembled in Beijing.[60] By the next June, segments of the program had been drafted, and, as part of the overall evaluation, the Chinese "made a fairly systematic analysis and study of the levels reached in the progress of world science and technology and the prevailing tendencies."[61] Earlier, the Central Military Commission had taken steps to accelerate scientific research and development on conventional weapons and thereby laid some of the foundation for work on strategic weapons.[62]

China's strategic missile program also began to take shape during early 1956.[63] Zhang Aiping later remembered 1956 as the year the Party's "Central Committee decided that developing missiles and developing atomic energy were the two key projects in China's defense modernization," and gave priority to strategic missiles, the construction of nuclear fuel production plants, and the development of the atomic bomb.[64] This came at a time China was struggling to limit its defense expenditures for conventional weapons and to upgrade its basic science and technology. To help guide the missile-development effort, Qian Xuesen, shortly after his return to China from the Jet Propulsion Laboratory in California, became head of a new Fifth Academy (Diwu yanjiuyuan) under the Ministry of National Defense.[65] In February, Qian published an article on defense aviation, and within a few weeks, the Party's Central Military Commission created under its aegis the Aviation Industrial Commission. Nie Rongzhen rose to head the new body, with Huang Kecheng and Zhao Erlu as his deputies.[66] The commission appointed An Dong, a key figure who was to become a top administrator in several military industrial programs under Nie, office director.[67] All of these men were coming into Nie's orbit and would later play significant roles in the strategic weapons program.

In October 1956, Party General-Secretary Deng Xiaoping approached Nie Rongzhen about Nie's future career.[68] By this time, the Party had already charged Nie with the development of China's military industry and equipment, and he had taken responsibility for the formulation of that part of the twelve-year plan concerning the development of military equipment.[69] Deng offered Nie three job possibilities: to lead the Scientific Planning Commission succeeding Chen Yi, to replace Peng Zhen as mayor of Beijing, or to continue running the defense industry and equipment program. Nie, never hesitating, chose both options one and three. He replied: "I prefer to take charge of the scientific program rather than be appointed mayor. There are close connections between the defense industry and the scientific programs. There is no problem if I concurrently take charge of the defense industry in the future." Deng agreed. Succeeding Chen Yi (even prior to formal State Council action in May 1957), Nie assumed the leadership of the thousand scientists and seventeen Soviet advisers working on the twelve-year plan and, at the same time, as a new vice-premier, he became Zhou Enlai's principal agent in all scientific and technical policy making.[70]

In these final months of 1956, Nie appears to have stressed the military aspects of the twelve-year plan and to have begun forging the scientific base for the strategic weapons program. By December, the Chinese scientists and their Soviet advisers completed the plan, which gave "the peaceful utilization of atomic energy" the highest priority.[71] In Nie's discussion of the contents of the plan, it is clear that for him the concept of "peaceful utilization" of atomic energy by no means excluded military uses; by definition everything undertaken by the socialist states, including China, had a peaceful intent, whether civilian or military. Nie notes that the Aviation Industrial Commission, the Equipment Planning Department of the General Staff, and the defense-related ministries all had helped draft parts of the plan.[72] These parts, he writes, defined the preliminary targets, with the following items listed as having the highest priority: the development of ground-to-air and air-to-air guided missiles and the development of atomic reactors for military use.[73] Nie further records that Zhang Aiping, in charge of planning scientific research for conventional weapons, had supervised this phase of the drafting process, a job that was to lead in due course to Zhang's command of the First Atomic Bomb Test Commission.

While high-level planning moved ahead on the strategic weapons program, the lingering effects of the anti-Rightist campaign in 1957 and the broader policies of the Great Leap Forward launched in 1958 began to warp the organization of science and technology from below. As Nie Rongzhen has reported, the Second Session of the Eighth National Congress of the Party in May 1958 "put forward the great tasks of technical revolution" and outlined a whole set of principles of "walking on two legs."[74] As a result, a broad national campaign of technical revolution engulfed the technical-scientific community and downgraded "the special nature of science and technology." Central Committee directives inspired ordinary citizens to act like scientists and experts, and ill-equipped Party cadres became embroiled in the heady world of advanced science and technology. The walking-on-two-legs political line transformed the authority of Nie's organizations by calling for "the simultaneous development of science and technology by the central and local authorities, the simultaneous development of modern and indigenous methods, the combining of special scientific and technological research institutes with popular activities, . . . and the combining of production with teaching and scientific research."[75] The new line not only complicated centralized leadership within the emerging nuclear establishment, but also had disruptive and sometimes bizarre effects on the program.

Top-down control and local populism thus placed competing pressures on a program of a magnitude equal to America's Manhattan Project and postwar missile program combined. We shall examine the costs of the Chinese nuclear program in Chapter 5 and of the missile effort in a separate volume. As we shall see, these costs ran high, and everything was being done with what one Chinese leader has called "great and deliberate redundancy."[76] Prior to late 1958, coordination occurred under the presumed direction of the Central Military Commission and the State Council. These senior bodies had to manage a variety of commissions and ministries, though each had substantial autonomy in practice. From 1956 to 1958, Nie Rongzhen's name appeared high on the rosters of many of these organizations, but he by no means exercised overall command in these years. His authority in many situations remained ambiguous and limited.

In the nuclear program, to be specific, the Third Ministry seems to have played the dominant leadership role immediately after its establishment in 1956.[77] Its mission, as we will often note in this study, was to set in motion a string of simultaneous projects covering all the elements for a comprehensive nuclear program, from uranium prospecting to weapons design and testing. To this end, in the summer of 1957 Song Renqiong asked Li Jue, one of his bureau chiefs, to assume overall initial responsibility for the yet-to-be-built nuclear weapons research academy, the uranium enrichment facility, the weapons assembly plant, and the nuclear test base.[78] At about the time Li accepted this assignment, Wu Jilin and Guo Yinghui joined him in working with Soviet experts. Li promptly dispatched survey teams containing these experts into the western interior to identify appropriate locations for nuclear weapons design-and-assembly facilities and the main plant for uranium isotope separation. On October 15 and 17, 1957, Song placed his seal on the initial orders for the construction of the first nuclear energy plants.[79] These stories of individual-by-individual selection of senior cadres and detailed supervision of construction plans were repeated in the nuclear program throughout most of the next decade. In the formative years, 1956-58, the ministry dominated the day-to-day decision-making process.

Parallel to these domestic developments, the government's negotiations with the Soviet Union for defense assistance progressed throughout 1956 and 1957, and by 1958 Soviet equipment, technical data, and experts had begun to arrive in numbers, and Song's ministry had inaugurated many parts of the strategic program.[80] But just as the ministry's efforts were gathering momentum, the Party's Great Leap Forward began to undermine both the domestic strategic program and Beijing's negotiations with Moscow.

To Nie Rongzhen's dismay, the Great Leap Forward policy line from the fall of 1958 on created what he calls "several problems" in China's defense industries. The mass movement immediately led to a downgrading of the importance of Soviet advice and to a false or exaggerated sense that unqualified workers could improve on Soviet-supplied prototype weapons and blueprints. Swayed by the cacophony of slogans and capricious ideas, cadres would not readily obey higher orders as written and felt obliged to design even the most advanced weapons from scratch. Several units attempted to initiate independently "some sophisticated military-technical programs, such as the development of a guided missile, an artificial satellite, atomic energy, and a new type fighter plane." All defense programs teetered on the brink of disaster, and in the advanced aircraft program, at least, fatal design errors were introduced.[81]

By the end of 1958, key members of the Politburo became convinced that the time had come to impose a central, military-type organization on the entire strategic program and thereby eliminate, or at least minimize, the overlapping leadership systems and populist interference. The chapters that follow will illustrate how critical many phases in the nuclear program had become at this time and how inadequate the existing mechanisms for coordination and control were proving. The travail of the Great Leap Forward had exposed the frailty of those mechanisms. In October, Nie Rongzhen, to reverse the chaotic tide, "suggested" to the Central Committee that the Scientific Planning Commission and the State Technological Commission be merged into a State Science and Technology Commission for overseeing the civilian side of science and technology.[82] The Central Committee approved, and Nie became director of the new commission.*

---

* The disbanded State Technological Commission had been set up under the State Council on May 12, 1956, to develop national policies on technology and to manage

More critical to our discussion, Nie at the same time proposed to the Central Committee that the Aviation Industrial Commission and the Fifth Department under the Ministry of National Defense be merged into a new Defense Science and Technology Commission.[83] Nie also became director of this new defense commission, with Chen Geng, Liu Yalou, Zhang Aiping, and Wan Yi as his deputies. Taking charge of the staff office of 200 was administrator An Dong.[84] Where it touched strategic weapons decisions, Nie's new team could control the scientific and technical resources of the PLA, the State Council's military industrial system, and the defense-related sciences of the Chinese Academy of Sciences. Jointly with the Ministry of National Defense, the defense commission oversaw the ministry's Fifth Academy, a body that ran the strategic missile program in its early years.[85] Nie's chain of command passed through the Central Military Commission to the Politburo itself (Fig. 1).

By this time, the Second Ministry had been reconstituted from the former Third Ministry of Machine Building, and Figure 2 outlines what we know of its structure within the nuclear weapons program.[86] Between late 1959 and mid-1964, the following system had been created:

1. Ninth Bureau, Li Jue, director. This bureau, reporting both to Nie's Defense Science and Technology Commission and the Second Ministry, was the most secret organization in the entire nuclear program. It was set up in 1958, and had no other name at first, but later became the Nuclear Weapons Bureau. The Beijing Nuclear Weapons Research Institute, the Qinghai-based Northwest Nuclear Weapons Research and Design Academy (Ninth Academy), and the Nuclear Component Manufacturing Plant in Jiuquan Prefecture came under this bureau, and Li Jue also led the academy. Teams from the academy had a direct leadership role in the test program located at the Lop Nur Nuclear Weapons Test Base and maintained liaison with the First Atomic Bomb Test On-Site Headquarters led, in the first half of the 1960s, by Zhang Aiping and Liu Xiyao.

2. Geological Bureau, Lei Rongtian, director. This bureau, formerly the Third Bureau of the Ministry of Geology, had primary responsibility for uranium prospecting.

---

technical personnel. The new State Science and Technology Commission was established on November 23, 1958. All scientists who had been leaders in the former commissions were dropped, and all five of the new vice-chairmen were senior Party leaders. *Renmin Ribao*, May 13, 1956; *Liaowang*, July 23, 1984, p. 21; *Nie Rongzhen Huiyilu*, p. 782.

3. Mining and Metallurgy Bureau, Sun Yanqing, deputy director; from 1959 Su Hua, director. This bureau, formerly the Third Department of the Ministry of Metallurgy and often referred to as the Twelfth Bureau, controlled the major mines and facilities for processing uranium concentrates and uranium oxides.

4. Fuels Production Bureau, Bai Wenzhi, director. This bureau had great power. It was in charge of the uranium tetrafluoride, uranium hexafluoride, and plutonium facilities and the nuclear fuels bases in Lanzhou, Jiuquan Prefecture, and Baotou. It was also in charge of the Lanzhou Gaseous Diffusion Plant, code named the Fifteenth Bureau and later Plant 504.

5. Construction Bureau, Liu Hua, director. This bureau was in charge of both building factories and installing equipment. It had close associations with the Design Academy (later the Design Bureau), which designed all nuclear plants and other facilities.

6. Equipment Manufacturing Bureau. Created in 1961, this bureau took charge of the manufacture of materials, instruments, and equipment.

7. Sixth Bureau, Jiang Tao, director. Created in 1957, this bureau appears to have controlled equipment and materials supply. It ran the entire supply and transportation system in the nuclear program and oversaw a network of material transfer stations.

8. Science and Technology Bureau, Deng Zhaoming, director.

9. Information Bureau, Wang Yalin, deputy director. The name of this bureau was changed to Information Institute in 1964. Also known as the Eleventh Bureau, this unit had the task of collecting all available technical books, journals, reports, and articles on nuclear energy. It obtained essential information for the nuclear weapons program from the *Proceedings of the Second United Nations International Conference on the Peaceful Uses of Atomic Energy* (published in 1958)[87] and from declassified technical reports of the U.S. Atomic Energy Commission as well as Soviet sources. All such materials were indexed, abstracted, and selectively translated for use in the program.

10. Security and Protection Bureau, Deng Qixiu, director. This bureau's main concerns were safety, environmental protection, and health. For each weapons test, it assigned special officers to monitor radiation protection and general safety. It also oversaw nuclear waste disposal. The bureau cooperated with the Ministry of Health in running special hospitals and maintaining personnel health records. When, in 1960, the State Council passed codes (*fagui*) for the protection of personnel in the nuclear industry, they fell under its purview.

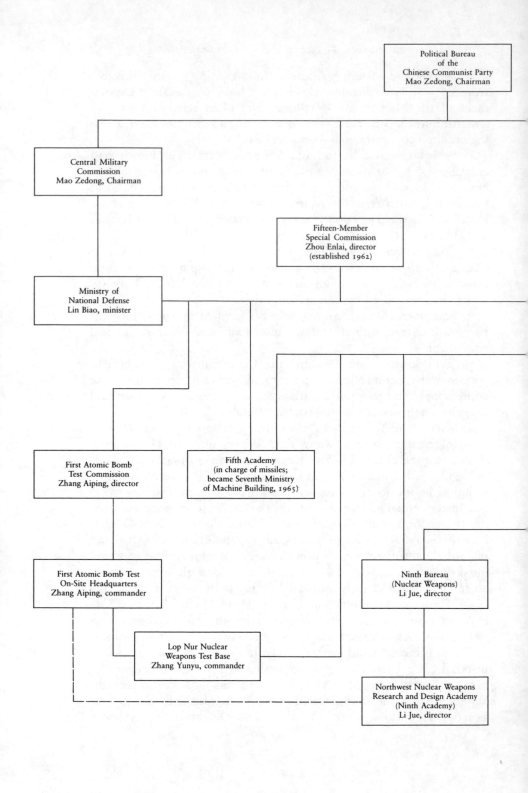

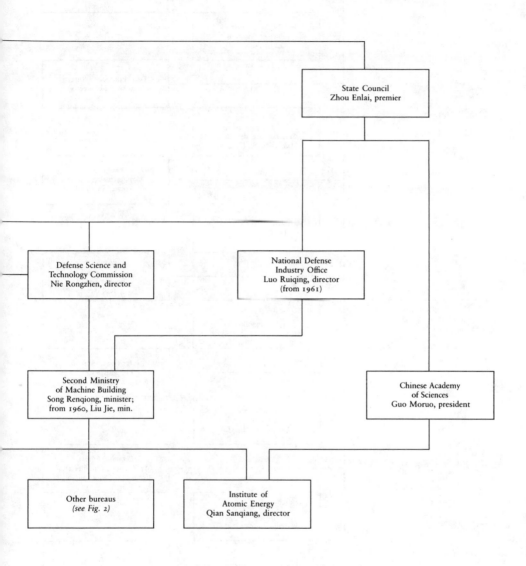

Fig. 1. Organization of the Chinese nuclear weapons program, 1959-1964. The dotted line indicates liaison between the units.

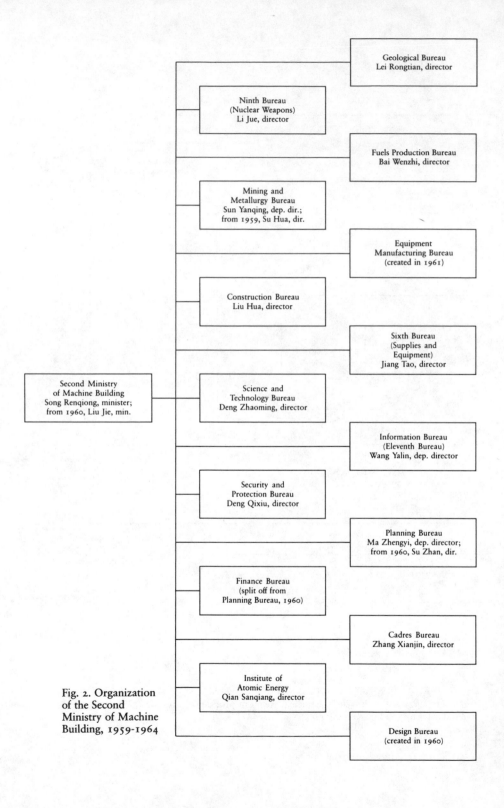

Geological Bureau
Lei Rongtian, director

Ninth Bureau
(Nuclear Weapons)
Li Jue, director

Fuels Production Bureau
Bai Wenzhi, director

Mining and
Metallurgy Bureau
Sun Yanqing, dep. dir.;
from 1959, Su Hua, dir.

Equipment
Manufacturing Bureau
(created in 1961)

Construction Bureau
Liu Hua, director

Sixth Bureau
(Supplies and
Equipment)
Jiang Tao, director

Second Ministry
of Machine Building
Song Renqiong, minister;
from 1960, Liu Jie, min.

Science and
Technology Bureau
Deng Zhaoming, director

Information Bureau
(Eleventh Bureau)
Wang Yalin, dep. director

Security and
Protection Bureau
Deng Qixiu, director

Planning Bureau
Ma Zhengyi, dep. director;
from 1960, Su Zhan, dir.

Finance Bureau
(split off from
Planning Bureau, 1960)

Cadres Bureau
Zhang Xianjin, director

Institute of
Atomic Energy
Qian Sanqiang, director

Design Bureau
(created in 1960)

Fig. 2. Organization
of the Second
Ministry of Machine
Building, 1959-1964

11. Planning Bureau, Ma Zhengyi, deputy director; from 1960, Su Zhan, director.

12. Finance Bureau. This bureau was originally part of the Planning Bureau but was split off as a separate bureau in 1960 or thereabouts.

13. Cadres Bureau, Zhang Xianjin, director. This bureau was in charge of personnel, labor conditions, salaries, and education and training.

14. Institute of Atomic Energy, Qian Sanqiang, director. This institute was under the dual leadership of the Second Ministry and the Chinese Academy of Sciences.

15. Design Bureau. In 1960, the ministry merged its Design Academy and other design organs into this bureau, which lasted until 1965. During the years 1960-65, the design system changed frequently and became quite complex.

All of these bureaus (except the Equipment Manufacturing, Finance, and Design) and the Institute of Atomic Energy were set up by the end of 1959 and lasted through at least 1964.

The Second Ministry was under the dual leadership of the Central Military Commission (via the Ministry of National Defense and Nie Rongzhen's Defense Science and Technology Commission) and the State Council (via the National Defense Industry Office, NDIO, after its creation in 1961). Nie Rongzhen's commission commanded the nuclear weapons and missile program, which functioned as a civilian industrial system under strict military control. After the National Defense Industrial Commission was merged into the NDIO in 1963, the NDIO was placed in charge of the production of nuclear fuels, plant construction, and industrial production.

People who worked in the nuclear weapons program did not wear uniforms and were not considered part of the military, though they were under military control. By contrast, until 1965 personnel in the missile program were assigned to the People's Liberation Army and wore uniforms. Specialists in aircraft and warship research and development (not production) reported to Nie's defense commission and wore uniforms. Those in production were under the NDIO and did not wear uniforms.

The symbolism of dress codes effectively dramatized the growing complexities of the military procurement system. Those who participated still recall what they wore and the importance of the symbolic refinements. Yet, years later, they find it difficult to explain why this mattered.

The Rise and Fall of Sino-Soviet Strategic
Cooperation, 1955-1959

The Chinese pursued their quest for organizational coherence in the strategic weapons program as they labored to bind their security interests closer to the Soviet Union. The Politburo, as its overriding security task, had to oversee and keep consistent the foreign and domestic avenues to the bomb. Nevertheless, Mao had placed his bets for quick results on Soviet assistance, and the story of Sino-Soviet military cooperation following the Korean War illustrates the complicated relations that at once promoted and inhibited the creation of an effective strategic partnership.

In his somewhat self-serving and one-sided memoirs, Nikita Khrushchev says that he had come to distrust Mao Zedong early in his rule of the Soviet Union.[88] Much of this distrust stemmed from a deep personal dislike for China's revolutionary commander. Khrushchev records: "Mao Tse-tung [Mao Zedong] has played politics with Asiatic cunning, following his own rules of cajolery, treachery, savage vengeance, and deceit. . . . I was always on my guard with him."[89] While Khrushchev insists that he "always bent over backwards to be fair and friendly" in his dealings with Mao, by 1954 he had concluded that conflict with China was inevitable.[90]

What particularly rankled Khrushchev were the Chinese leader's views on nuclear war and American power. He found Mao's slogan "Imperialism is a paper tiger" to be "incredible." In the Soviet view, the United States hardly constituted a "paper tiger," but was "in fact a dangerous predator." The two leaders argued about the strategic implications of nuclear weapons, with Mao trying to reassure the bewildered Khrushchev that "the atomic bomb itself was a paper tiger."[91] Their differences over the implications of nuclear weapons persisted throughout the years and, as later revealed in the bitter polemics of the 1960s, became a major cause of the Sino-Soviet split.[92]

Yet even while Khrushchev registers his horror at Mao's views on fighting a people's war in the nuclear era, he states: "Before the rupture in our relations, we'd given the Chinese almost everything they asked for. We kept no secrets from them. Our nuclear experts cooperated with their engineers and designers who were busy building an atomic bomb. We trained their scientists in our own laboratories." He also says that "[Soviet] specialists suggested we give the Chinese a prototype of the atomic bomb," and that only a last-minute reversal halted the bomb's delivery: "They put the thing together and packed it up, so it was ready to send to China. At that point our minister in charge of nuclear weapons [Minister of Medium Machinery Yefim Pavlovich Slavskii] reported to me. He knew our relations with China had deteriorated hopelessly. . . . In the end we decided to postpone sending them the prototype."[93] The "we" in this case was the Soviet Politburo, where Anastas Mikoyan reportedly argued against supplying the educational bomb and Nikolai Bulganin (whom Khrushchev forced from the Politburo in March 1958) argued in favor.[94] Khrushchev wavered for a time before siding with Mikoyan, and the sequence of events, including Bulganin's ouster, suggests that the initial (and secret) Soviet decision to renege came in early 1958.

Kept in the dark about the actual Soviet intentions, the Chinese continued to work on a room suitable for exhibiting the prototype for instructional purposes, but for months Soviet security men judged the room to be inadequate. The room's "lack of security" became the excuse for repeated delays, and Moscow's on-site advisers forced the Chinese to make one modification after another. When these advisers were finally satisfied, the Soviet government, in October 1958, told Beijing that "the prototype and technical data will be delivered . . . in November." Yet, the deadline came and went with no bomb in sight. Right up to mid-1959, according to one source, the designers of China's bomb counted on the Soviet delivery and repeatedly sent workers to the railroad station to pick the prototype up, but each time they "came back empty handed."[95] When the Chinese responded to Moscow's reversal by "exploiting our decision [to delay delivery of the prototype bomb to China] for all it was worth," Khrushchev says he was outraged: "All the modern weaponry in China's arsenal at the time was Soviet-made or copied [from the Soviet Union]. We'd given them tanks, artillery, rockets, aircraft, naval and infantry weapons. Virtually our entire defense industry had been at their disposal."[96]

Khrushchev's memory of Soviet generosity ignores an important point, namely, how directly that "generosity" was tied to the Hungarian uprising of October 1956. To be sure, the Soviet government had authorized substantial assistance to upgrade Chinese armaments and to provide training in missile science and engineering before the uprising, but from the Chinese perspective, it was slow in coming and had strings attached. Only Soviet self-interest, the Chinese believe, had kept even that aid flowing. For example, further advances in the Soviet nuclear program depended on obtaining uranium ores from China, and to get those ores the Soviets, in the winter of 1955-56, had pledged unofficially to provide China "full-scale assistance." In August

1956, after months of haggling, the Soviet government finally re-deemed that pledge and signed an agreement to help build China's nuclear industries and research facilities.[97]

Nevertheless, the Chinese felt dissatisfied: Moscow's promised aid, whether generous or not, had excluded assistance on the advanced weapons themselves. As late as September 1956, following the visit of a delegation led by Vice-Premier Li Fuchun seeking technical aid on missilery, the Soviet leaders would only go so far as to train "the back-bone scientific and technical personnel, with only 50 Chinese college students permitted to major in this specialty in the Soviet Union." This decision convinced China's policy makers that "the Khrushchev leading group was unwilling to help China develop sophisticated weapons."

Moreover, the Soviet Politburo had imposed even greater restraints on any aid to China's nuclear program. In contrast to its at least half-hearted backing of the Chinese missile efforts, the Soviet Union had up till now confined its support of the nuclear program to industrial applications. The events in Eastern Europe in October proved a turn-ing point in Beijing's favor.

Just as the Chinese were putting in place the start-up organizational system to oversee the strategic weapons program, the anti-Communist (and anti-Soviet) eruptions in Poland and Hungary increased China's bargaining power. As Nie Rongzhen recalls the sequence of events, Khrushchev, in trouble, became "more flexible on the matter of giving sophisticated technical aid to China."[98] The Soviet Union, in an at-tempt to consolidate its deteriorating position in the socialist com-munity, purportedly needed the support of the Chinese and was forced to make concessions to them in their demands for strategic weapons aid. Additionally, the Chinese believe that Mao's actions in mid-1957, during the anti-Rightist campaign, lessened Khrushchev's worries about the fate of communism in China. These actions forcibly dem-onstrated Mao's opposition to the kind of anti-Communist and anti-Soviet tendencies that had swept Poland and Hungary.

Nie adds that the Chinese "seized the opportunity," and with their new leverage, were able to negotiate the Sino-Soviet New Defense Technical Accord, in which the Soviet Union agreed to supply China the prototype atomic bomb and missiles as well as related technical data. Moscow also undertook to sell industrial equipment for the pro-cessing and enrichment of uranium and to provide enough uranium hexafluoride for the initial operation of the Lanzhou Gaseous Diffu-sion Plant. One Beijing demand went unfulfilled: Moscow refused to supply data on its nuclear-powered submarines.[99]

Nie recounts his role in these events. He says he guessed that Khrushchev might prove more flexible in light of the Hungarian uprising and linked that possibility to China's obtaining strategic weapons from Moscow. He relayed his conclusion to Premier Zhou Enlai, who gave him permission to make the overture to the Soviet adviser in charge of aid to China, I. V. Arkhipov. After checking with Moscow, Arkhipov informed him on July 20, 1957, that the Soviet government had agreed to negotiate. Nie, Chen Geng, and Song Renqiong then went to the Soviet Union in September and, on October 15, the two sides signed the New Defense Technical Accord.[100]

Beijing's victory, if one can call it that, proved to be both Pyrrhic and short-lived. Khrushchev, in his revealing memoirs, confirms the long-held view of American specialists that differences over the implications for grand strategy and the actual sharing of advanced weaponry fueled the growing Sino-Soviet estrangement.[101] Recent Soviet writings also confirm this view.[102] However, Khrushchev goes further than just accusing the Chinese of absurd ideas on nuclear warfare or ingratitude. He denounces them for their failure to accede to Moscow's request, made in 1958, to build a radio station in China "so that we could maintain communications with our submarine fleet operating in the Pacific."[103] He holds that Mao tricked the Soviet government into providing aircraft, long-range artillery, and air force advisers for a projected invasion of offshore islands held by Chiang Kai-shek in the Taiwan Strait, which the Chinese had no intention of carrying out.[104] Moreover, when the Chinese somehow obtained a copy of an advanced American missile and gave it to the Soviet military, it was missing some essential parts. By early 1959 at the latest, Khrushchev had concluded that China was an unreliable partner, and that the Sino-Soviet alliance was one-sided and dangerous.

Not surprisingly, the official Chinese version of the period from 1957 to 1959 is markedly different from Khrushchev's. However, Chinese specialists have told us that some senior Chinese still alive in 1985 confirmed that account and blamed Mao for the collapse of the Sino-Soviet alliance. One military leader reportedly told his successors, "We should apologize to Moscow."[105] Whatever the truth of the matter, Nie, among others, acknowledges the strategic weapons program's immense debt to the Soviet Union and provides some detail on the aid, especially in the missile field, that the Chinese received. "After the conclusion of the . . . accord," he writes, "both countries carried it out relatively smoothly at the initial stage. During the period from 1957 to 1958, the Soviet Union provided us with prototypes of several kinds of outmoded guided missiles, aircraft, and other military equipment and relevant technical data." Starting in the first half of 1958, Soviet nuclear specialists, two of whom were nuclear weapons designers, began to arrive. On September 29, Moscow and Beijing signed a supplementary accord setting forth the general plant specifications and firm deadlines for the equipment to be delivered to the nuclear industry.[106]

Nie himself confirms Khrushchev's tale of China's refusing to accept the radio station but labels the Soviet demand to install the station an infringement on Chinese sovereignty. He adds that in July 1958 Khrushchev called for the establishment of a joint Sino-Soviet Pacific Fleet, an assertion Soviet authors flatly deny.[107] Contradicting some Western accounts of the quarrel at this time, Nie has told us: "It is not true that in 1958 the Soviet Union demanded the deployment of nuclear weapons in China under its own control."[108]

The strains in the Moscow-Beijing alliance began to take their toll in 1958. The Chinese account of the Taiwan Strait crisis that year ignores Khrushchev's accusation of trickery but stresses his worry about a possible Soviet-American nuclear confrontation. Mao's intra-Party speeches suggest his early hope that the bombardment of the offshore islands of Quemoy and Matsu would force a Nationalist withdrawal without risk of direct American involvement. When, as the political scientist Allen Whiting has noted, Washington responded "with an unprecedented concentration of nuclear force," Mao adopted a less ambitious stance and tried to explain his caution by saying that those critics did not "understand the paper tiger problem" who wondered why, since the United States was a paper tiger, China did not attack Taiwan. He had already implied an answer in September: "I simply did not calculate that the world would become so disturbed and turbulent."[109] Moreover, when the Soviet leaders reacted to the possibility of their involvement in a nuclear confrontation, the Chinese assured them that China would "make national sacrifices in case of war . . . and the Soviet Union would not have to participate."[110]

The Chinese also explain why Beijing later expressed such abhorrence of U.S.-Soviet arms control deliberations, especially those relating to a nuclear weapons test ban.[111] In mid-1959, the Chinese Party's Central Committee received a formal letter of June 20 from its Soviet counterpart saying that because of negotiations on a test ban then under way in Geneva, Moscow would not supply the prototype bomb or blueprints and technical data on the bomb. Nie has elaborated on Khrushchev's version of these events and quotes the letter as saying, "If the Western countries learn that the Soviet Union is supplying China with sophisticated technical aid, it is possible that the socialist community's efforts for peace and relaxation will be seriously sabotaged." [112] Nie denounces this linking of aid to the test ban effort as a pretext and as amounting to a "unilateral tearing up of the Sino-Soviet New Defense Technical Accord" to provide China with a prototype bomb and relevant technical data. His denunciation agrees with the well-known Chinese statement of August 15, 1963, which added that "this was done as a presentation gift at the time the Soviet leader went to the United States for talks with Eisenhower." [113]

When the Chinese Politburo received the "perfidious" Soviet letter, it met within days to express its anger and defiance. It vowed that China would produce its first atomic bomb within eight years and shifted the balance from reliance on the Soviet Union to an ever more independent course. [114]

Mao's Move Toward Self-Reliance

While Mao Zedong appears to have focused most of his attention on the nonsecurity aspects of international and domestic affairs after the nuclear decision in January 1955, many of the current security issues were not lost on him. We referred in the last section to some of his conversations with Nikita Khrushchev on nuclear war, and from time to time the Chinese Chairman expanded on his notions of imperialism and reactionaries as "paper tigers." [115] His main line of thinking on global strategy, however, continued to reflect his belief that nuclear weapons had not changed the basic weakness of the imperialist forces.

Mao had said in 1955 that the atomic bomb was not "an invincible 'magic weapon.'" [116] It had not altered the calculus of strategic conflict. Thus, he could reaffirm a policy line carried over from the revolution that one should despise the enemy strategically but take it seriously tactically. [117] His essential reasoning was that U.S. imperialism only appeared to be strong; it was in fact "very weak politically because it [was] divorced from the masses of the people." This fundamental weakness meant that the United States was "nothing to be afraid of . . . and [was] unable to withstand the wind and the rain." However, the Chinese leader differentiated between this basic political weakness and the fact that militarily the United States had "claws and fangs." The task of the socialist states would be to remove the claws "step by step." [118]

From the structural perspective, Mao's thinking on the political and military sides of strategic doctrine paralleled that of Soviet military strategists writing at the time. In the prevailing Soviet view, as the political scientist David Holloway notes, "military doctrine has two closely connected aspects: the political (which is supposed to be dominant) and the military-technical. The former sets out the political purposes and character of war. . . . The military-technical aspect deals with the methods of waging war."[119] Holloway further states that the political aspect of Soviet doctrine stressed war prevention while the military-technical dealt with the conduct of war should it occur. Similar to the different war-prevention and war-fighting emphases in Western strategic theories, the strategic argument always centered on how to balance the two aspects and on the comparative advantage of one mix over another.

From 1955 on, Khrushchev's combination shifted to the politics of prevention and repudiated the "law of the inevitability of war" enunciated by Lenin and Stalin. As revealed in the authoritative Soviet journal *Military Thought* in April 1955, the "new line characterized the doctrine of permanently operating factors as inadequate."[120] Because of the rising military power of the socialist camp and of the "forces of peace," Khrushchev proclaimed that the forces of aggression could be thwarted: "Today there are mighty social and political forces possessing formidable means to prevent the imperialists from unleashing war, and if they actually try to start it, to give a smashing rebuff to the aggressors and frustrate their adventurist plans."[121]

In 1955-56, Chinese strategists, including Mao, would have accepted these basic Soviet formulations, not excepting Khrushchev's emphasis on the political aspect of military doctrine. Chinese and Soviet specialists would have agreed that a nuclear war would cause immense casualties, would lead to the destruction of capitalism, and was not inevitable. But beyond this level of consensus, Soviet and Chinese doctrines parted company. Where they differed, as the Khrushchev-Mao encounters make clear, was on the ultimate consequences of a nuclear world war. According to Khrushchev, he himself seriously questioned the fate of civilization following such devastation, whereas Mao argued that "war is war. The years will pass, and we'll get to work producing more babies than ever before."[122] In later polemics, the Chinese lashed out at what they considered to be statements taken out of context to slander their Chairman. They quoted Mao as having said, "If the worst came to the worst and half of mankind died, the other half would remain while imperialism would be razed to the ground; . . . in a number of years there would be 2,700 million people again."[123] The Soviet leaders would hardly have seen the difference between the two renditions.

By 1957, Soviet-Chinese disagreements had become somewhat ritualized and repetitive, focusing on what would happen if a U.S.-Soviet war did break out and how to avoid it in the first place. However, during Mao's visit to Moscow in November 1957, he extended the dispute to the future conduct of a war. Khrushchev has recorded his conversation with Mao.[124] The issue they clashed on was whether giving an aggressor against the Soviet Union a "smashing rebuff" constituted a sound strategy. Specifically, Mao found fault with a statement made by Defense Minister G. K. Zhukov that Soviet forces would "strike a counterblow" against any aggressor that attacked a Soviet ally. Astonished, Khrushchev replied, "If we don't take the position he [Zhukov] stated, the aggressive forces will destroy us bit by bit. . . . They want to divide and conquer." Mao rejoined, "Not so. . . . If the Soviet Union is attacked from the West, you shouldn't engage the enemy in battle; you shouldn't counterattack—you should fall back." When queried on the meaning of "fall back," Mao said, "I mean retreat and hold out for a year, two years, even three years." Khrushchev asked Mao where these retreating forces would go and dismissed Mao's comparison to the Second World War. The Red Army's retreat in 1941, he reminded the Chinese leader, was not "motivated by either tactical or strategic considerations. . . . The enemy *forced* us to retreat."

What unnerved Khrushchev was that Mao imagined a future global war as a mirror image of the Communist Chinese forces' struggle against Japan and Chiang Kai-shek. Khrushchev himself was moving toward a position that the prevention of nuclear war required the capability of inflicting "the same damage on our enemy as he can inflict on us"—a mutual-hostage relationship dominated by offensive technologies. Where Mao saw the prevention of war as largely a matter of mobilizing global political forces—the political side of doctrine—Khrushchev now gave greater weight to the military-technical side. Politics had other uses, of course, but could not alone carry the burden of keeping the peace. Step by step and reluctantly, the Soviets were coming to accept a doctrine of mutual retaliation, though the political side of doctrine still played an important role in war prevention.

The Soviet shift in emphasis implied, however, that global war could erupt in many different quarters, and that only the Soviet Union's strategic forces could prevent the escalation of lesser conflicts into a direct attack on its homeland. Khrushchev noted that Mao denigrated the change in the Soviet policy line. To Mao, it would render the Soviet Union more cautious in its support of national liberation struggles in the Third World. More on the political defensive as a consequence of that line, Moscow would shy from mobilizing and backing the "forces of peace" against U.S. imperialism. Khrushchev states that after reaching this conclusion about the Soviet line, Mao began "to torpedo our policy of peaceful coexistence, claiming outright that it was un-Leninist and bound to give way to pacifism."

For their part, the Chinese in 1957 had concluded that the military achievements of the Soviet Union had drastically lessened the threat of overt American aggression and the use of nuclear blackmail. Relatively speaking, the Soviet Union's rapidly growing strategic might had freed the forces of national liberation to act with impunity. The rise of Soviet power had liberated the forces of revolution from any credible threat of American intervention, and hitherto weak revolutionary forces could take to the offensive. Mao said: "To seize state power by armed force—this is a strategic slogan. If you insist on peaceful transition, there won't be any difference between you and the socialist parties." [125] Inevitable revolutionary warfare would crush the elements in the United States or elsewhere that would launch a nuclear war. Mao proclaimed his own strategic line to be the correct path to preventing nuclear conflict under the new conditions.

Mao's greater emphasis on the role of national liberation struggles stemmed principally from the key event of 1957, the launching of the Soviet satellite, Sputnik I, in October. As he saw it, this event heralded a shift in the "correlation of forces" and placed the United States in a paralyzed position of strategic inferiority. Its far-reaching implications greatly intensified Mao's interest in strategic problems and caused a change in his own doctrines. In November, when he ventured to Moscow for the first time since 1950, he addressed an audience of Chinese students: "The direction of the wind in the world has changed. . . . At present, it is not the West wind that prevails over the East wind but the East wind that prevails over the West wind." In another speech during the same visit, Mao talked about a "new turning point," by which he meant that the "socialist forces are overwhelmingly superior to the imperialist forces." [126]

In the months that followed the Moscow visit, Mao from time to time returned to the themes that so riled Soviet officials—the likely consequences of nuclear war and the need for national liberation struggles in the Third World—though he gave principal priority to launching the Great Leap Forward. Yet even here he made the link to nuclear war. In the spring of 1958, at the landmark Second Session of the Eighth Party Congress that decided the "general line" and sparked the Great Leap, Mao clearly had in mind that the rural communes then forming would be a helpful part of the "preparation for the final disaster," if war should break out.[127] The communes, according to the Central Committee organ *Hongqi*, would be "organized along military lines" so that "the whole population would be citizen soldiers ready to cope with the imperialist aggressors."[128] Characteristically, Mao equated the need for such preparations with the need to prepare for splits in the Party, but argued that both war and Party splits could be avoided if matters were handled correctly. Still, preparations were needed because "the maniacs" might launch a war. If the bomb was then resorted to, he thought the duration of atomic war would be short, "three instead of four years.... We have no experience in atomic war. So, how many will be killed cannot be known. The best outcome may be that only half of the population is left, and the second best may be only one-third." In any case, atomic war would mean the "total elimination of capitalism" and usher in an era of "permanent peace."[129]

Furthermore, in late 1958, Mao told a group of local officials that it was better not to be afraid of war, not to fear ghosts.[130] On balance, Mao concluded in that speech, the purpose of imperialist policy was exploitation, and "the object of exploitation is man." Mao believed in boldness, often bordering on bravado, when facing the nuclear ghosts: "If man is killed, what's the use of occupying soil? I don't see the reason for the atom bomb. Conventional weapons are still the thing." For the moment, conventional weapons were all Mao had, and he was making the most of them.

Even as Mao spoke, however, he had set the nation on a firm course to acquire nuclear weapons. After the decision in January 1955, he expected Moscow to support China's strategic ambitions, though his policies, as always guided by a philosophy of contradictions, made comparable demands on the nation itself. For Mao, the struggle of "opposites" would settle the ultimate path, though he could not admit that the dualism of his ideology struck others as duplicity in practice. From the Kremlin's viewpoint, his actions did not just mirror the struggle, they caused it.

In his memoirs, Nie Rongzhen records his conclusions concerning the causes and course of the deteriorating Sino-Soviet ties after the signing of the New Defense Technical Accord. At first, he says, the two countries carried out the accord "relatively smoothly," though, as we noted earlier, the Soviets only supplied "outmoded" equipment.[131] In reflecting on the changes during this initial period of cooperation, he places the principal blame on Khrushchev's demand to set up a Sino-Soviet joint fleet and radio station for the Soviet submarines in the Pacific. Later Chinese denunciations would refer to these as "unreasonable demands designed to bring China under Soviet military control."[132] Nie laments, "the good times didn't last long."[133]

A turning point seems to have been reached with the decisions taken at a conference of the Central Military Commission held from May 27 to July 22, 1958, just after the session of the Party congress.[134] The commission, chaired by Mao, dealt directly with the nuclear weapons program, and its protracted deliberations, we believe, produced "The Guidelines for Developing Nuclear Weapons."[135] The eight guidelines read in part as follows:

1. Our country is developing nuclear weapons in order to warn our enemies against making war on us, not in order to use nuclear weapons to attack them. . . .

2. The main reason for us to develop nuclear weapons is to defend peace, save mankind from a nuclear holocaust, and reach agreement on nuclear disarmament and the complete abolition of nuclear weapons.

3. To this end, we have to concentrate our energies on developing nuclear and thermonuclear warheads with high yields and long-range delivery vehicles. For the time being we have no intention of developing tactical nuclear weapons.

4. In the process of developing nuclear weapons, we should not imitate other countries. Instead, our objective should be to take steps to "catch up with advanced world levels" and to "proceed on all phases [of the nuclear program] simultaneously."

5. In order to achieve success rapidly in developing nuclear weapons, we must concentrate human, material, and financial resources. . . . Any other projects for our country's reconstruction will have to take second place to the development of nuclear weapons. . . .

6. It is time for science and scientists to serve the Party's policies, not for the Party's policy to serve science and scientists. Therefore, we must guarantee the Party's absolute leadership of this [nuclear weapons] project. . . .

7. The task of training successors [for the nuclear weapons program] is as important as the manufacture of nuclear weapons.

8. We must set up a separate security system so as to guarantee absolute secrecy.

In a speech to his military colleagues, Mao Zedong amplified on the point of not imitating other countries. The matter of the "mechanical application of foreign experience" had already been raised in a January newspaper article, though the Chinese were still talking at the time about selectively "incorporating Soviet advanced experience" into their military science.[136] Now Mao adopted a stricter line. "It is not practicable to execute orders in accordance with the Soviet army ordinances in wartime," he said. "It is necessary to strive for Soviet aid, but what is of primary importance is self-help." He criticized "blind faith in foreign dogmas and in the Soviet Union," and said that China must apply a "developmental viewpoint" to the study of the Soviet army. Consistent with what he had told Khrushchev concerning a response to an American attack, he said: "At present, the things worked out by the Soviet military advisers (such as operational plans and thinking) are all of an offensive nature, based on victory; no provision is made for the defensive and for defeat. This is not in conformity with practical conditions." Mao concluded that China had its own system of military science and tactics, whereas Stalin had had none at all, and so, "we don't have to learn from the Soviet Union."[137]

While moving toward self-reliance, Mao did not explicitly reject the continuation of Soviet aid and, when the cutoff began, the Chinese voiced their outrage at Moscow's betrayal. Hence, when Mao said in June 1958 that China would have nuclear weapons in ten years—"Let us work on atom bombs and nuclear bombs. Ten years, I think, should be quite enough"[138]—he still expected the Soviet Union to abide by its October 1957 commitment. All the same, Mao and his colleagues were getting ready for a reversal. In August 1958, Marshal He Long expressed regret that China had mistakenly "tried to solve our problem purely from the military point of view and had hoped for outside aid instead of relying on the mobilization of the masses."[139] Nie Rongzhen at this same time argued that China "should and absolutely can master, in not too long a time, the newest technology on atomic energy in all fields."[140] He added that "there is in fact no possibility of our making wholesale use of the existing experiences of other countries." The duality of dependency and self-reliance in China's quest for nuclear weapons endured right to the moment Soviet aid terminated.

The denouement stretched out for more than a year. Following the letter of June 20, 1959, in which the Soviet Central Committee informed China of its two-year suspension of assistance on strategic weapons, the Soviet Union "deliberately took sides with India" in the

Sino-Indian border dispute.[141] Furthermore, Nie Rongzhen writes that on at least two occasions, once at the United Nations and once in Beijing itself, Khrushchev attacked the Chinese by innuendo in his speeches.[142] On front after front, the partnership was breached.

Khrushchev visited China in the autumn of 1959 for the celebration of the tenth anniversary of the People's Republic. In his major speech, on September 30, he reportedly warned China's leaders not to "test the stability of the capitalist system by employing armed force." Two days later, Foreign Minister Chen Yi told Khrushchev to his face what he thought of charges that China was being adventurist and belligerent. The exchange was personal and bitter. Angry, Khrushchev left Beijing with the belief, now openly expressed, that China was "keen on war like a bellicose cock. . . . It is not reasonable."[143]

On the basis of these events, Nie Rongzhen says, China had the "ominous premonition" that Moscow was preparing to "tear up unilaterally all Sino-Soviet accords." During the latter months of 1959, a number of critically placed Soviet specialists, on the pretext of "going home on furlough," had "gone never to return." Nie adds that by early 1960 Moscow had begun "to refuse China's demands for the supplying of technical data on key projects, the production of raw materials, and design and theoretical calculations. They made excuses to turn down China's demands or to delay the delivery of precision instruments and special raw materials." Meanwhile, Moscow increased its demands on Beijing's leaders for much more technological data from China.

In response, Nie alerted the Central Committee to the downward spiral of Sino-Soviet ties and recommended that it be "fully prepared" for possibly drastic changes in the relationship between the two countries. In January, the Politburo convened an enlarged meeting in Shanghai to review its collapsing alliance with the Soviet Union. In his report to the meeting, Nie itemized the problems in Sino-Soviet technical cooperation and charged the Soviet Union with failing to provide the promised data on the MiG-21 fighter plane and on guided missiles. "Soviet technical aid," he said, "has become untrustworthy," and the Soviet aim is "to maintain a considerable gap between China and the Soviet Union in scientific research on the development of new types of weapons and military equipment."[144] In the end, 40 percent of the nuclear-industrial equipment and raw materials pledged to China by the Soviet Union never arrived, and step by step, the 233 Soviet advisers in the nuclear-industrial system were pulled out. By August 23, 1960, with about a month's warning, Moscow had fully withdrawn the last of its technical experts from the Chinese strategic program.[145]

FOUR

The Uranium Challenge

Only a tiny number of Chinese scientists, working mostly in Europe and the United States, participated directly in the prewar and early postwar scientific races to study uranium or to find uranium deposits.[1] There was little question of them working in China itself. Reeling from the blows of Japanese aggression, the Chinese had transported what scientific equipment could be salvaged and moved to makeshift campuses in the southwest by the end of the 1930s. In the chaos of civil war after the defeat of Japan, few Chinese could even dream of China's joining the nuclear club. The events in Korea and the Taiwan Strait crisis of 1954-55 were to transform that dream into a national crusade.

Finding uranium and then mining and processing it were among the first great tasks that confronted China's leaders in the quest for the atomic bomb. By 1955, a year of international meetings on the peaceful uses of atomic energy, the West had begun to publish basic works on uranium exploration, and the Chinese had worked closely with Soviet and East European geologists in the search for non-ferrous and rare metals in Xinjiang Province. What the Chinese scientists and experts knew from their reading and from consultation with Soviet colleagues was that finding substantial deposits of uranium posed a formidable challenge. Even in the West, the science of uranium geology and exploration was still young and unsystematic, though the forced-march effort under national-security sponsorship had made it clear that certain belts in the earth's crust could be regarded as uranium "provinces."[2] Between 1945 and 1956, hundreds of such provinces had been discovered. None of any consequence, according to the Chinese, had yet been found in China.

Uranium in the earth's crust occurs in far greater abundance than mercury, silver, or gold and in about the same amounts as lead. How-

ever, its chemical and physical properties—"its polyvalence, its large
atomic radius, its chemical reactivity, [and] the relative solubility of
many of its hexavalent compounds in aqueous solutions"—permit the
metal to form compounds with numerous elements, so that concentrations of high-grade, minable metal in the uranium "provinces" are
relatively rare.[3]

Western scientists had long recognized that three general geological
factors were critical to the identification and evaluation of any potential uranium provinces: the nature of mineralizing solutions, the geological structures through which they moved, and the character of the
rocks in which the ore had been deposited.[4] Largely by trial and error,
American geologists, for example, had determined that the "most important uranium-bearing vein deposits are found in either granitic
rock, continental sedimentary rocks, or volcanic flows."[5] As we shall
see, the Chinese allege that the Soviet geologists they worked with
dismissed the possibility of finding major uranium deposits in granite.
But by the 1950s, Australian, French, and Spanish prospectors had
found deposits of considerable capacity in ore bodies close to the mar-
gins of granite, in areas underlain by granitic intrusive rocks, and in
fractures in granite. So if Soviet specialists did downplay the likelihood
of finding uranium in granite, their advice would have constituted an
important misdirection away from China's first substantial uranium
finds.

For their explorations, Chinese scientists did have access to the required radiation detection equipment and did understand the recommended steps in a well-planned uranium exploration program. They
could buy or manufacture Geiger and scintillation counters and had
assembled what crude geological maps they could find for various
parts of the country.[6]

Their prospecting task was complicated by the fact that both types
of radiation counters react to beta and gamma rays from all sources,
including cosmic rays and the radioactive material present in all rocks.
The uranium prospector must thus make periodic determinations of
background count away from any more-intense radioactive source.
The prospector working in granite experiences the highest background
radiation. The rule of thumb for the uranium explorer in the 1950s
was that a field would be worth painstaking exploration only if the
count recorded exceeded the normal background by ten times. Finding
uranium even within a region of known promise thus presented problems for experienced prospectors.

For Chinese, the more daunting challenge was to locate possible uranium provinces in a poorly mapped land area roughly the size of the United States, including Alaska. The three normal phases of an exploration program in the West were "(1) preliminary reconnaissance, (2) detailed geologic studies, and (3) physical exploration."[7] Preliminary reconnaissance in theory should begin with a review of existing information on areas geologically suited for exhaustive exploration. Out of the nation's vastness, such an evaluation would narrow the search to places containing promising formations and geological structures. Part of this analysis would pair aerial photographs, if available, with maps, and from this pairing cartographers would draw new detailed prospecting maps. The reconnaissance to this point would pave the way for direct-ground observation at high-saliency sites. Western sources in the 1950s recommended the supplementary use of aerial reconnaissance at this point to facilitate broader regional coverage.

Assuming phase one revealed a promising area, phase two then required the preparation of even more-detailed, large-scale geological maps, and phase three, if it was called for, physical exploration, including trenching, stripping, mine opening, or drilling.

The Search for Uranium

Chinese trace the origins of their hunt for uranium to the year 1934, when the geologist Zhang Dingzhao spectroscopically analyzed tungsten, tin, bismuth, and molybdenum from southern Jiangxi Province and found traces of uranium.[8] Four years later, another geologist, Zhang Gengsheng, found uranium daughters in the alluvial sands of Guangxi Province, and in 1943 two other scientists discovered the first real uranium ores in Zhongshan County, Guangxi. With the tides of war swirling around them, such events made barely a ripple among Kuomintang leaders, and more than ten years were to pass before the government, now Communist, put uranium prospecting on the official priority list.

The change, almost serendipitous, began in 1954, when a geological prospecting team found traces of uranium in Guangxi during a comprehensive mineral survey. The vice-minister of geology, Liu Jie, reported the find to Mao Zedong and Zhou Enlai, and Mao asked Liu to send him ore samples and a Geiger counter. After testing the samples, Mao reportedly said, "We still have much undiscovered ore! We have great hope; continue to search! You definitely will discover many uranium mines. . . . Our country also wants to develop atomic energy." In the winter of 1954, Beijing ordered the Ministry of Geology to dispatch uranium prospecting teams to Guangxi and Liaoning, and arranged for a Soviet adviser to accompany them.

However, the Chinese date the beginning of the official campaign to locate uranium provinces in the following year. On January 20, 1955, just as the Politburo was deciding on Project 02 (and three days after it announced the Soviet decision to provide China with a cyclotron and a nuclear reactor), Beijing and Moscow signed a secret protocol to create a joint uranium prospecting organization.

In March 1955, the State Council established the Third Bureau under the Ministry of Geology and, as noted, assigned bureau chief Lei Rongtian to organize a nationwide uranium prospecting program. Lei immediately created two prospecting teams, for China's Central South Region and for Xinjiang Province, to which he assigned the code numbers 309 and 519, respectively. These teams, composed of ten survey units with over 1,000 members, faced the formidable problem that China's geological records contained information only on "abnormal geological phenomena concerning uranium." Moreover, there were no systematic geological studies for the entire nation.[9] To the extent possible, however, the geologists of teams 309 and 519 followed the textbook: preliminary reconnaissance, detailed geological studies, and physical exploration. The exploration process took almost two years, and China could not begin construction on its first uranium mine until May 1958.[10]

While we know little about the quality and scientific basis of prospecting in China in the mid-1950s, most sources stress that by then Soviet, Polish, and Hungarian experts had taught a cadre of Chinese the "majority of well-tested geophysical and geochemical methods."[11] In January 1956, Premier Zhou Enlai reported that, whereas "in the early days of the liberation" there had been fewer than 200 geologists in China, the number of "geological engineers alone had [now] increased to 497, while there were 3,440 technicians who had graduated from institutes of higher learning."[12] By the end of the year, the number engaged in uranium prospecting exceeded 20,000. Furthermore, in the early 1950s the Chinese had progressed from importing prospecting instruments to copying and manufacturing them on a trial basis. By 1958, they were producing "highly sensitive" scintillation and Geiger counters, and had accelerated development of radioactive methods for prospecting.

Our understanding of the timetable and organization for uranium exploration contradicts reports that China's prospecting for uranium had begun much earlier and as a joint venture under the Sino-Soviet

Non-Ferrous and Rare Metals Corporation.[13] As one U.S. government publication asserts, "In the communist idiom 'non-ferrous' metal is frequently used as a euphemism for uranium or thorium."[14] Such assumptions about the hidden meaning of "non-ferrous," when juxtaposed with acknowledged Soviet assistance in many parts of China's atomic programs, convinced the U.S. intelligence community that a joint Soviet-Chinese quest for uranium had begun in the early 1950s.[15] Nevertheless, it is clear from subsequent authoritative Chinese accounts that the uranium for the first Chinese bombs came from sources in the south and northeast some years later.[16]

The Chinese, in working alongside Soviet specialists in the Non-Ferrous and Rare Metals Corporation, undoubtedly did gain experience that was useful to their overall nuclear weapons program.[17] According to one official Chinese account, in 1954 the corporation had put into production the largest ore-dressing plant for non-ferrous metals (but not uranium) in Northwest China, and thousands of technicians had been trained with young Chinese in charge of most key operations.[18] The Soviet Union gave these technicians scientific information and equipment for excavation. Many Chinese experts had moved into management positions and would have acquired the competence to run similar plants in the uranium field. On January 1, 1955, China gained complete control of the corporation.

The country's geologists used the three years from 1955 to 1958 to develop the minimal scientific and technical foundation for their all-out hunt for uranium and came relatively close to the international standard for exploration.[19] This incipient program began to pick up tempo in late 1955; with the establishment of the Third Ministry of Machine Building a year later, in November 1956, the Chinese were ready to organize a comprehensive nuclear industry and to launch the nationwide search for uranium.[20] The following month, on December 19, the Chinese took full control of the joint Sino-Soviet uranium prospecting organization.

Even before the ministry was created and acquired Lei Rongtian's bureau from the Ministry of Geology, teams 309 and 519 had conducted preliminary air and land surveys in the mountain ranges of South and West China. Working according to the book, the bureau had simultaneously mounted an exhaustive search of the literature on the country's most accessible, geologically mapped counties. The spotty history of uranium exploration in China to date gave them little to go on. Later explorations would fan out throughout the country, but the urgency of the uranium quest made it necessary to narrow the immediate search to areas where some relatively large-scale geological maps already existed. This search included selective aerial surveys, and from the knowledge assembled, the bureau prepared detailed maps that divided the areas chosen into grids and rated each grid zone according to its potential for uranium. The maps thus identified places in several southern and northwestern provinces deemed ripe for ground exploration. The Chinese geologists discovered in this mapping, for example, that most of the uraniferous granite massifs found throughout the region were in an older uplifted zone and closely related to Mesozoic-Cenozoic block basins. In these granites, future prospectors would find uranium in crystals and leached in microfissures, where the ore had been absorbed by a variety of altered minerals.[21]

Some of this hard scientific work paid off. By late 1955, teams 309 and 519 had identified large numbers of potential uranium provinces. Team 519, about which we know relatively little, had discovered three rich deposits in western Xinjiang (in Daladi, Mengqiku'er, and Kashi) early in the year. Team 309 had equally good luck, and its first pinpointed aerial reconnaissance took place over southern Hunan Province in 1956.[22] So remote was this area, apparently, that the local people were reportedly thoroughly amazed at the sight of a plane diving toward a local mountain peak, Jinyinzhai, near the town of Xujiadong, Chenxian County. As far as the villagers could recall, the only time such an event had occurred before was when Japanese bombers had attacked Kuomintang fortifications on the mountain more than a decade ealier. For the next year, ground parties continued the survey near Xujiadong; and in April 1958, a mining cadre named Liu Kuan led a preparatory team of about 1,000 miners to settle at the bottom of Mount Jinyinzhai. Their task, which began in May, was to build, near the county town, the Chenxian Uranium Mine, the first such mine in China (Map 2).[23]

Lei Rongtian's bureau picked the location for this mine and seven others from a report submitted in early 1958 on eleven promising uranium sites. All eight mines were put into operation between 1962 and 1965. According to the Sino-Soviet accord of August 1956, the Soviet Union agreed to design the Chenxian mine, the Dapu Uranium Mine in Hengshan County, Hunan, and the Shangrao Uranium Mine in Shangrao County, Jiangxi. The Zhuguang mountains on the Jiangxi-Hunan border (at Dongkeng and Lanhe) and the Xiang mountains on the Jiangxi-Zhejiang border (at Quzhou and Fuzhou) eventually produced the richest veins. As the number of major uranium sites grew to

Map 2. Southeast China

26, their distribution proved to be 41 percent in granite, 20 percent in volcanic rock, 21 percent in sandstone, and 18 percent in other rocks and sediments. By 1967, the ministry had built two major mining and metallurgical complexes in Guangdong and Jiangxi (at Fuzhou), and the "backbone" of the country's uranium mining and metallurgy was located in these two provinces plus Hunan.

The quality of the uranium ores was always a problem. The average grade of uranium often fell below the Western norm for commercialization; even in the 1980s, China operated uranium mines in which more than 70 percent of the ore was deposited in small ore bodies of relatively low grade.[24] Such ore bodies occur irregularly and are widely dispersed, and miners found conditions mostly unfavorable to the use of large-scale mining equipment. In general, the so-called tunneling ratio in South China's uranium mines was 40-60 meters or more per 1,000 tons, which is two to three times the ratio for ordinary mines. Virtually all of the mines were underground, and expensive horizontal cut-and-fill techniques had to be used in almost 60 percent of them. Other techniques evolved after 1966, and mining experts differ on what criteria to use in evaluating them. (See Tables 1-3.)

The Nanling mountains in the northern part of Guangdong Province also became a center for uranium prospecting in these pioneer years. Here dogged legwork combined with serendipity led to the initial discoveries. In the autumn of 1957, team 309's unit 11, headed by Luo Pengfei, ran into strong gamma ray emissions in the vicinity of Lianxian.[25] Up to this point, the unit, equipped with Geiger counters, had slogged for days through thickly forested mountains. Tired and discouraged, a young unit member implored Luo to abandon the area. He told Luo: "Lao Luo, we have spent several days prospecting for uranium in this area and so far have found nothing useful to our work. It is better for us to leave as soon as possible."[26] Luo rejected the plea and ordered a three-hour trek to another mountain, but upon arrival a team member realized that the unit had wandered into a restricted area demarcated in red on their map. Concluding that they had gone several kilometers off course, the unit members discussed their predicament and finally agreed that they might as well conduct a survey. A unit member, the story continues, recorded a reading with a gamma coefficient of over 1,000 from a huge rock, the first of nearly 100 high-grade veins found in an area of about 100 square kilometers.[27]

The Chinese make much of what happened next. When Luo Pengfei reported these findings to a higher body, he received a brusque reprimand—not, however, for the unit's entry into the restricted zone.

Rather, a Soviet expert attached to the higher body censured Luo for violating standing orders not to search for uranium in granite. The expert allegedly stated: "Why didn't you prospect for uranium ore in the widely dispersed sandstone areas? Why did you just want to rush into a granite area? Can you bear responsibility for all the serious consequences if your reckless behavior interferes with your work?"[28] Luo's Chinese superior thereupon charged him with insubordination and ordered him to withdraw immediately from the granite areas.

But the matter did not rest there, for in China's bureaucracy official reports keep moving up the chain of command. When word of Luo's discovery reached the Geological Bureau of the Third Ministry in Beijing, the officials decided to investigate Luo's find more fully. Following procedure, they dispatched a team of over 20 Chinese and Soviet experts to Guangdong to assess the uranium potential in the Nanling granite structures. Based on their conclusions, the central authorities then approved a plan to begin core drilling in the granite vein, numbered 86, that had been discovered by Luo's unit. The Lianxian mine, opened in 1963, was China's first uranium mine in a granite region.[29]

Another success story is told of a prospecting team working in an isolated ravine in a sparsely populated region of the same area of Guangdong's Nanling mountains. After spending the night in a small village, the team members were unable to adjust their scintillation counters. They quickly realized that the stone walls that had housed them during the night were highly radioactive. When they located the piece of granite emitting the most intense signal, they persuaded the owner of the house to show them where he had dug the rock.[30] The spot where he took them on a nearby mountain became the site of another large uranium mine. Explorers continued to unearth major uranium deposits in Guangdong Province for the next quarter century.[31]

The Chenxian Uranium Mine

We know less about the Guangdong and Jiangxi mines than about the Chenxian Uranium Mine in Hunan, though it is clear that they produced substantial quantities of ore for China's early nuclear explosives.[32] The Chenxian miners found uranium deposits in quartzite, "both as simple ore and in the form of scattered, impure deposits," including iron pyrites.[33] Natural leaching of the uranium ore had occurred.

Shrinkage stoping is the mining technique used at the Chenxian Uranium Mine. A stope is an excavation underground for the removal

TABLE I

Characteristics of the Important Uranium-Mining Techniques Used in China

Technique	Principal equipment	Means of support	Monthly stope output capacity (tons)	Output per shift (tons)
Cut-and-fill methods				
Horizontal cut-and-fill	Roof drills, electric hoes[a]	–	630–1,200	2.4–5.2
Inclined slice cut-and-fill	Electric coal drills	Wooden supports	490–670	1.3–1.6
Open-stope methods				
Shrinkage stoping; open-stope with shrinkage	Jackhammer drills, roof drills, electric hoes	–	720–1,000	2.0–6.3
Completely open-stope and room-and-pillar	Jackhammer drills, electric hoes	Rock or ore pillars	250–800	2.9–4.2
Caving methods				
Wall caving	Jackhammer drills, electric hoes	Wooden supports	940–1,080	1.8–2.0
Slice caving	Jackhammer drills, hand-guided pneumatic hoes	Wooden supports, metal nets, metal pillars	530–750	1.7–2.0

SOURCE: Wang Jian, Li Yunfeng, and Liu Enchong, "Outline of Development of China's Uranium Mining Technology and Suggestions for Its Improvement," *He Kexue yu Gongcheng* [Nuclear Science and Engineering], 3 (Sept. 1982), in *China Report: Science and Technology*, 24 (JPRS-CST-84-024, Aug. 23, 1984), p. 9.

[a] In a few cases, rock-drilling platforms, pneumatic loader-conveyors, and other kinds of equipment are used.

TABLE 2

Principal Defects of the Uranium-Mining Techniques Used in China

Technique	Ore loss rate (percent)	Ore dilution rate (percent)	Other
Cut-and-fill methods			
Horizontal cut-and-fill	2-5%	10-18%	Much time and labor required to find side slopes and clear bottoms
Inclined slice cut-and-fill	6-10	24-30	Low production capacity and efficiency; high wood consumption
Open-stope methods			
Shrinkage stoping; open-stope with shrinkage	3-8	30-36	High degree of ventilation needed
Completely open-stope and room-and-pillar	5-20	20-25	High loss rate
Caving methods			
Wall caving	1.4-2	30-36	Inefficient; high wood consumption; high ore dilution
Slice caving	7-8	10-16.5	High wood consumption; low efficiency; ventilation difficult

SOURCE: Same as Table 1.

TABLE 3
*Comparative Yields of the Uranium-Mining Techniques
Used in China, 1966-1980*
(percent)

Technique	1966	1976	1978	1979	1980
Cut-and-fill methods					
Horizontal cut-and-fill	58.0%	58.5%	54.1%	57.3%	50.9%
Inclined slice cut-and-fill	–	–	–	–	3.3
Open-stope methods					
Shrinkage stoping	12.0	10.2	8.2	9.2	7.1
Completely open-stope; open-stope with shrinkage	–	7.1	11.8	9.3	4.4
Room-and-pillar	–	–	–	–	2.8
Caving methods					
Wall caving	29.3	16.4	15.4	15.6	17.7
Slice caving	–	7.6	9.7	8.4	13.2
Square-set stoping	0.7	–	0.8	0.2	–
Other	–	0.2	–	–	0.6
TOTAL	100.0%	100.0%	100.0%	100.0%	100.0%

SOURCE: Same as Table 1, p. 7.

of ore. Formed by mining the ore from a block of ground, the stope is usually a step-like excavation for removing the ore in successive layers. In shrinkage stoping, the roof of the stope remains unsupported but broken ore shores up the walls.[34] Mining engineers consider shrinkage stoping more dangerous than other stoping methods, where "the men are under solid ground that is tested every day."[35] At Chenxian, the angle of the ore veins was steep, and unlike cut-and-fill mines, the mine had no sublevels.[36] In general in stope mines, "horizontal sections of the ore are mined from the bottom upward.... Because the ore expands when it is broken, about 35% of it must be drawn off at the bottom.... After the stope is completed, the ore is drawn out and the space left open."[37] Chinese sources hold that the stoping process is simple and efficient, even though, on average, China's miners at Chenxian and elsewhere recovered only about 70 percent of the total ore body.[38] The average area of one stope, which may have been atypical, was 350 square meters; 17,000 tons of ore were extracted annually from the top 13 slices of a stope 21 meters high.[39]

At many of the shrinkage-stoping uranium mines in China, the miners used the worked-out space for what is called a leaching heap.[40] They stored low-grade ore there, drenched the ore heap with leaching water, collected the leaching liquid in a tank at the bottom of the shaft, and then pumped it out to the surface, where the metal was separated. At Chenxian, the miners had to rely on a magnetic separator, but in both the mechanical and the hydrometallurgical system China's miners ran into problems. Chinese technicians recognized the challenges involved in alternative separation systems, and the nuclear ministry founded the Hengyang Institute of Uranium Mining and Metallurgy to devise more effective methods for mine-level ore processing. The Bureau of Mining and Metallurgy also built the Hengyang Uranium Hydrometallurgy Plant (Hengyang you shuiye chang) to process all the ores from Chenxian and other mines using magnetic separators.

The story of the Hunan and Guangdong mines is a tale of intense personal commitment and struggle against hardships in the pursuit of national objectives. J. Robert Oppenheimer once remarked that making the atomic bomb took 1 percent genius and 99 percent hard work. In China, at least, much of that 99 percent began in the mines. The site selected for the Chenxian Uranium Mine, for example, lay in the heart of southern Hunan's malaria belt.[41] Once, when the project leader, Liu Kuan, was making a trek across the mountain, a companion chanted the local folk rhyme, "In Chenzhou [Chenxian], boats stopped, horses died, and people suffered malaria." Endemic malaria, however, was only one of the problems facing Liu. Housing his 1,000 miners also presented a great challenge, for the local area contained few building materials and no paths to any construction sites. The newcomers had to pitch tents and spend most of their first weeks there building suitable dwellings; in a half year they had 20,000 square meters in floor space at their disposal.

Many of Liu's miners, mostly from the north, found the intense heat and humidity almost beyond endurance and compared the place to a dumpling steamer. In the summer rainy season, disease spread, and the displaced northerners grumbled, "We would rather die in North China than in South China, because the coffins made in the north are bigger than those in the south."

After capital construction began at the mine in the spring of 1958, Liu soon discovered he would have to make do with little in the way of tools and heavy construction equipment. In the second work area, which had the task of industrial-type capital construction, for example, the original benchwork crew had only two ordinary hammers, and the forge team had to pump air by crude two-man bellows. The construction crew did all the ground preparation by hand.

China remembers the years that followed, 1960-62, as the "three hard years." All units, including some in the military, suffered from malnutrition.[42] Because many units attached to the strategic weapons program were stationed in remote areas, their suffering was unusually high, and few escaped the hardship. At Chenxian in these years, Liu put all miners on short rations. Drillers received 16.5 kilograms (33 catties) of rice per month, and the others 12.5 kilograms (25 catties). All received about one-tenth of a kilogram (2 taels) of edible oil and one-quarter of a kilogram (half catty) of pork. Most went hungry, and to ward off the spreading dropsy, the miners sent their families in search of edible wild herbs and bark. The Party enjoined its members, always the exemplars, not to eat any meat at all.

Yet despite these handicaps, the work proceeded, and the Chenxian Uranium Mine became partially operational on September 1, 1960. The Chinese followed a Soviet blueprint in designing the mine but had to solve endless problems on their own. One of the most serious arose when the drillers constructing shaft number 1 hit a stream of hot water that flowed at the rate of 350 cubic meters per hour. Liu Kuan feared for the mine itself and, in the emergency, accepted the offer of a young woman, Zhang Guizhi, to halt the water flow. Standing regulations prohibited young women from entering the mine because the radon gas there emitted toxic levels of gamma rays of particular risk to women in their childbearing years. Zhang, however, had already worked underground, and with mining engineers flown in from the Ministry of Coal Industry, she helped organize an emergency crew that devised a way to halt the water.

The example of Zhang's heroism illustrates the dangerous conditions at Chenxian and other uranium mines. Some health hazards stemmed from the lack of water for the drilling operations, a condition revealed during a high-level inspection trip in February 1963.[43] The first miner to die, Hou Qi, succumbed to silicosis. He was far from the last; others fell victim to cave-ins, radiation poisoning, and disease. Hou lies buried in a cemetery of martyrs located in the northwestern corner of a storage yard next to the mine rail tracks. As noted above, large releases of highly radioactive radon gas were a serious problem in the Chenxian mine from the start. Two decades were to pass before sustained and effective national-level efforts were made to reduce the radiation hazards to uranium miners.[44] In the late 1950s and early 1960s, the government program emphasized finding and extracting the uranium ore, not safety.

Uranium Mining and Mass Movements

Despite the shortcuts, the race to find and mine uranium took years, and by a twist of fate the material for the first bomb did not come from the regular mines. The eight major mines put into construction in the late 1950s did not reach full operational status until 1962-63, too late in the program for their product to be used in the first atomic bomb. Beijing desperately looked to other sources for the required uranium, and it found them by turning to the Chinese people.

The years from 1958 to 1959 were a time of surging "ultra-Leftist" radicalism: the establishment of the people's communes, the Great Leap Forward, and the heavy emphasis on popular participation.[45] This was a time of mass movements, a time when all China succumbed to the popular hysteria of the Great Leap Forward. Backyard steel furnaces and large water conservation projects provided the high drama for that period. Local cadres drew ordinary peasants and workers into advanced industrial production, often with disastrous results. These mass movements also influenced the uranium mining project, but in this case they had some positive results as well.

By 1958, when the Great Leap Forward commenced, the first two geological prospecting teams, 309 and 519, had gone to Hunan and Xinjiang; 309 had already made the finds at Chenxian. Other military-led geological teams had fanned out to Guangdong, where the Lianxian discoveries had occurred, and to Liaoning. These teams located a number of small, shallow uranium deposits of good quality ore.

In the spirit of the Great Leap Forward, the Second Ministry in mid-1958 issued the slogan "the whole people should engage in uranium mining" (*quanmin ban youkuang*). The challenge was quickly taken up in Hunan, where the provincial Bureau of Metallurgy in July called for a Great Leap in the production of all types of non-ferrous metals. By September 1959, the head of Hunan's Bureau of Geology could note that for the past year the Party had been stressing the "principle of walking on two legs . . . in prospecting for big, medium-sized, and small mines. Both professional geology teams and the broad masses have been mobilized." The Party had created geology teams in many counties and rural communes, the bureau chief said, and had promoted the development of "a broad network" for prospecting. Tens of thousands of peasants had joined the hunt, which within a few months had covered 95 percent of the province.[46]

When the mass-prospecting movements began in the summer and

autumn of 1958, the expert teams provided the peasants with scintillation and Geiger counters, and instructed them in very simple yet effective methods for mining ores in the small deposits found close to the surface. The teams showed the peasants how to process the ore into uranium yellow cake (heavy ammonium uranate with a uranium content of between 40 and 70 percent). Many received short courses on both mining and ore-processing technology. By the thousands these peasant miners exhausted the uranium located in the top couple of meters of the Hunan, Guangdong, and Liaoning fields.

The use of local methods cost a great deal, wasted uranium, caused a major depletion of raw materials used in processing uranium, such as soda and acids, and produced serious pollution because of the near-absence of environmental protection equipment. But the Chinese recall the pluses as well as the drawbacks of this episode. The mass-prospecting movement, they say, spread the knowledge of atomic energy, dispelled the notion that atomic energy is magic, and made full use of ores from otherwise uneconomic deposits. The major advantage was that, in the quest for nuclear weapons, mass-based methods produced the first 150 tons of uranium concentrates. The curious mass-based link to the weapons program was to last into 1961, until the overall system from mines to bomb design had begun to take shape. The timely acquisition of this uranium is credited with shortening the race for the bomb by one year. In this limited sense, the first Chinese bomb was a "people's bomb."

Mine-Level Processing and First-Stage Procurement

By the late 1950s, the basic knowledge about concentrating uranium ore by a series of mechanical and chemical steps was widely understood and openly published both in the West and in the Soviet Union. The Chinese had ready access to works that described authoritatively and in detail the technology of the treatment of uranium concentrates and of uranium in general.[47] In uranium mining, concentration is an essential part of the recovery process because of the low percentage of uranium ore per ton of rock mined. A number of methods are used, such as mechanical ore treatment and upgrading, and alkaline or acid leaching. The Chinese turned to Soviet advisers in managing the ore recovery and concentration and did not find this phase of the nuclear project especially taxing. The end result of this stage in the refining process was uranium yellow cake. This crude uranium compound on average contained perhaps 60 percent uranium by weight and the rest various impurities.

As the uranium ores from the mass-mining operations became available and the mines in South China began operating, the Chinese constructed a series of plants to process the ore. The miners at the Chenxian Uranium Mine, for example, built a magnetic separator facility (the first of its kind in China) adjacent to the mine. They used Soviet industrial equipment. By the time they finished the main building, however, the Soviet team had pulled out, and the Chenxian workers found it hard to install the separator's machines. The head engineer for the project, Wang Wenyuan, and his crew worked from September 1960 to April 1962 installing and making operational this equipment "of originally authoritarian design." [48]

In sophistication, the Chenxian facility far outshone the one set up at Lianxian, Guangdong, where several straw sheds were converted for use in first-stage uranium processing. "Five huge pots, which were full of crushed uranium, nitric acid, and sulphuric acid, stood side by side in one straw shed. The solution was poured from the first pot to the second and then the third." Each time, physical settling would further concentrate the uranium compounds. "The final solution flowed into an underground container from which it was poured into the first pot" so as to be concentrated to the maximum degree possible. The miners then hauled this solution to another shed for drying "by the same indigenous method." [49] The dried uranium compound was then trucked to a more modern mill for advanced processing.

Long before either the Chenxian or the Lianxian ore separator had gone into full operation, other plants farther along in the production chain leading to uranium enrichment were completed. In the program's first half dozen years or so, any unit, whatever its position in the production chain, could press ahead somewhat independently to meet its own deadlines and to make demands for materials yet to be produced. This situation directly resulted from the central-level decision in the winter of 1957-58 to build all facilities for the production of nuclear weapons simultaneously. Consequently, facilities in the sequence from uranium mining to the ever more advanced processing of uranium might come on line out of order, though on the basis of the New Defense Technical Accord of October 15, 1957, the Chinese expected the Soviet Union to deliver uranium hexafluoride, allowing them to proceed directly to their first bomb without using their own raw materials.

Until the Soviet withdrawal in August 1960, then, the chronological order in which each facility became operational was deemed to be inconsequential, so long as that facility's product preceded uranium hexafluoride in the mines-to-enrichment production chain. As the Soviet withdrawal became a reality, however, the Chinese suddenly had to complete all steps in the production sequence on their own and in the right order.

The history of yellow-cake production illustrates the chaotic situation that followed the Soviet departure. When the Chinese completed the uranium oxide plant that needed uranium yellow cake, workers were still building the facilities where it was to be produced. As we have learned, the system for the production of uranium yellow cake not only had gotten out of phase with the program to build more advanced processing plants, but had fallen temporarily into the hands of peasants. In the throes of mass campaigns and commune organization, the peasants in southern Hunan, for example, had acquired uranium from nearby uranium fields, refined it into crude yellow cake, and taken it to the market town of Zhuzhou. The nuclear ministry had designated a number of such towns as collection points for the yellow cake, and assigned, first, the Geological Bureau and, later, the Mining and Metallurgy Bureau to oversee the purchasing process. When the authorities from the uranium oxide plant sent a cadre to look for yellow cake, he found it at a shop near the Zhuzhou railway station. Here the peasants could sell the uranium to the ministry. The cadre reported: "What a place it was! Yellow cake . . . was piled at random and could be seen everywhere." The cadre sent to buy yellow cake was amazed and immediately "explained and publicized among the masses technical knowledge relating to safety in the production and storage of yellow cake."[50]

The Production of Uranium Concentrates at the Sixth Institute

Whatever the methods used to obtain uranium yellow cake, the next stage required a much more sophisticated treatment of the uranium concentrates in specially designed plants. This stage aimed to produce one of two products: Metallic uranium or uranium hexafluoride. For both products, the processing plants had to convert the uranium concentrates first to an oxide and then to tetrafluoride.[51]

As the Chinese labored to mine uranium and to put plants into operation for the mass production of uranium oxides, the central authorities under Nie Rongzhen recognized the need to set up a central-level institute to conduct research on uranium ores and their processing. To this end, in 1958 the Second Ministry created the Sixth Institute[52]—the Uranium Mining and Metallurgical Processing Institute (Youkuang xuanye yanjiusuo)—in Tongxian, a few miles east of Beijing. (It was subsequently renamed the Fifth Institute.)* Two years later, in 1960, under orders to produce two tons of uranium oxide "within the shortest time," the institute selected eight key scientific and technical cadres as a preparatory team to build a production plant at the Tongxian site.

The Uranium Oxide Production Plant (Plant Two) was formally established on August 12, 1960. The ministry promptly dispatched 148 skilled technical cadres and university graduates to staff it, and the plant leader assigned them to experimental, design, machine repair, and construction teams. In this small one-story building and in the later Plant Four (for manufacturing uranium tetrafluoride), Chinese technicians processed the uranium used in China's first atomic bomb.

The institute's immediate task was to locate sufficient quantities of uranium yellow cake on which to experiment. To this end, the institute leaders ordered a cadre, Tian Zhaozhong, to Hunan to obtain enough of the concentrate to make the two tons of uranium oxide, and it was he who found the piles of yellow cake in the Zhuzhou market. As Tian surveyed the town's radioactive trove, he realized that he would have to test the highly impure uranium concentrate for quality, and indeed for downright fakes. "Yellow cake brought to him was of all colors of the rainbow: yellow, black, gray, and dark red. Tian knew how to distinguish between real and fake yellow cake. He put drops of reagent on the samples; those that produced an orange precipitate were genuine." Yet how could Tian test enough uranium for two tons of uranium oxide? He had neither the instruments nor the time to do that on his own. Enlisting the help of specialists from Beijing, he managed to get the needed raw material to the institute in Tongxian. He did so within the time limit set by the ministry.

In 1958, the nuclear ministry assigned a student who had returned from the Soviet Union, Lu Fuyan, to head the institute's Fourth Research Section. Subsequently, he was made head of the institute's Production Section and given the task of directing the trial production of uranium oxide and tetrafluoride. In one of his inaugural meetings with his research team, Lu discussed the various processes known to yield the oxide compounds, and the team members set out to develop the simplest and quickest possible methods. After comparing the processes used in the United States and Great Britain with those used in the Soviet Union, they decided, because of the low grade of the Chi-

---

*This institute should not be confused with the Fifth Academy (sometimes referred to as the Fifth Institute) under the Ministry of National Defense, which was set up in October 1956 and dealt with missiles.

nese uranium concentrates, to adopt the U.S.-British method for the first phase of the refinement process and the Soviet one for the second phase. However, they suggested ways to modify the Soviet process by adding further steps to the treatment stage.

Preliminary to the conversion of yellow cake to an oxide, the Chinese technicians removed all impurities in the yellow cake that absorb neutrons, such as cadmium, boron, and certain rare-earth metals. This had required several steps: dissolving the yellow cake in nitric acid to yield a solution of uranyl nitrate and silic acid; separating out the silic acid; purifying the uranium nitrate solution by solvent extraction; and then further refining this solvent to recover the uranium as an aqueous solution. This done, technicians could extract uranium oxide by heating the solution or using a precipitate. In discussing their modification of this process, the Chinese state: "In comparison to the process used at the Soviet Lermontov Plant, which began operation in 1958, [we] could process five to eight times more uranium concentrate and, per unit of equipment, could produce four to eight times more." The implication is that the process was designed according to revised drawings from the Lermontov Plant.[53]

With the design finished, the Chinese raced to build Plant Two. As the structure began to go up in August 1960, they placed priority orders for its equipment, and in case of shortages, the ministry empowered its Sixth Bureau to commandeer equipment from other units for the plant. The plant cadres gave the Sixth Bureau a detailed list of the equipment required. More than ten factories, for example, were instructed to help in the manufacture of a special furnace made of high-grade steel, and the Beijing Iron and Steel Academy took on the job of producing the steel. The Shanghai Electric Furnace Factory was charged with making the furnace itself, and the Shanghai Glass Machinery Factory with the design and manufacture of an automatic control system for it.

Despite the great authority given them, the cadres at Plant Two encountered serious problems in quality control. One order for a small stainless-steel pump, for example, went to a machine repair factory attached to the institute. The diameter of the pump's piston had to be strictly limited, but the pump had to have a high "lift capacity." The factory technicians experienced sixteen failures in design during the first three weeks but finally fabricated a pump that would meet Lu Fuyan's specifications.

In September 1960, the Hunan yellow cake arrived, but a full year was to pass before Plant Two became fully operational, even though the recent Soviet defections had accentuated the spirit of urgency. Again, there were unforeseen problems. Equipment deficiencies caused some of the difficulties. Once these were overcome, the plant appeared to function well, but the final uranium product fell short of the required standard. The amount of uranium still left in the tailings proved too high, and the intermediate products contained unacceptable levels of impurities. The plant assigned one of its managers to tackle the problems, and he and his team members lived beside the equipment for the next month. The Chinese recall how, as in so many other parts of the nuclear program, technicians had to run endless experiments to pinpoint the flaws in the system. To help them solve the problems and to persevere, the ministry sent senior officials to give the team personal encouragement, and in time the repeated experimentation paid dividends.

On November 18, 1961, a little over two months after commencing operations, Plant Two had produced 2,377 kilograms of uranium oxide. The institute had met the ministry's production goal for delivery of the compound. Equally important, it had fashioned a national model organization to help develop, validate, and improve the processes needed in mass-production plants.

Meanwhile, sometime in the autumn of 1960, the Party committee of the Tongxian institute had instructed the Fourth Research Section under Lu Fuyan to build a plant (Plant Four) for the production of uranium tetrafluoride. This toxic, radioactive, corrosive compound forms as green crystals and is sometimes known as green salt. The Fourth Research Section assigned one of its cadres to settle on a site for Plant Four. This cadre chose a work shed in the institute yard and picked his key personnel from the scientists and technicians with whom he had worked before. The shed storeroom became the plant workshop; the initial equipment consisted of odds and ends taken from other parts of the institute. The workers did most of their work by hand and with minimal safety gear, but the official history alleges that all personnel appreciated the extreme dangers they faced: "Although the production of uranium tetrafluoride was radioactive, corrosive, explosive, and inflammable, the scientific and technical personnel of Plant Four did not care a bit. Heedless of their personal safety, they . . . persisted in their course, engaging in design work, the installation of equipment, and production as they proceeded so as to race against time."[54] After two months of "desperate efforts," the workers finished the construction, and "soon afterwards" Plant Four produced 80 kilograms of tetrafluoride.

Technical difficulties had already caused long delays in the production of uranium oxide at Plant Two, but what set back Plant Four after its initial spurt of progress were the crises plaguing the broader society in 1961. Every staff member felt the impact of the "three hard years" following the Great Leap Forward. Many suffered from dropsy and could not work. Even more serious, key cadres and workers were caught up in the *xia fang* ("to the lower levels") movements of the time and were transferred to the countryside to perform manual labor. As far as we know, this plant is the only facility in which key personnel in the nuclear weapons program had to leave their units for the countryside, and at one point only fourteen people remained at the plant. Morale plummeted. It is no surprise, then, that six months elapsed between the initial success of creating the 80 kilograms and the production of "a considerable quantity of uranium tetrafluoride." The plant thus achieved full operational status only in the fall of 1961, about the same time Plant Two began to manufacture the quantities of uranium oxide required to yield the tetrafluoride.

To attain the institute's production goals, the personnel of Plants Two and Four paid a high price. The quality of the tetrafluoride product from Plant Four remained unstable, and safety fell below standards. A "sour mist" and dust filled the plant: "The glass in the windows was so corroded as to be indistinct. Objects were hazy in the dim light of the electric light bulbs. Off duty, the workers would take off their gauze masks and find that their noses and nostrils had become green."[55] Despite these human health hazards, the plant cadres in 1962 gave first priority to ensuring the quality of the product by eliminating dust pollution and installing automatic devices for fluorinating the uranium oxide. Only later were plastic air ventilation pipes installed, and, gradually, the plant was able to improve both the workers' safety and the quality of the uranium tetrafluoride.

The Mass Production of Uranium Oxides and Tetrafluoride

In their pursuit of nuclear weapons, as we have often remarked, Chinese planners simultaneously initiated work on all phases from mining to bomb design, but created national lead centers to anticipate and solve critical problems before mass-production units became operational. The centers would give guidance and training to the emerging mass-production facilities and serve as their backup research units. It was in line with this strategy that Beijing had created the Tongxian institute and linked its work to ore-processing factories then being built in Hunan, Jiangxi, and Guangdong.

The story of mass production of the oxides can be traced to March 1956, when Bo Yibo, one of the Politburo's Three-Member Group, recommended the establishment of scientific and industrial facilities for this production. Two years later, the Second Ministry formed the Uranium Mining and Metallurgy Design and Research Academy at Hengyang in addition to the Tongxian institute. It charged the academy with the design of hydrometallurgical plants and research on the mass production of the oxides, while the institute increasingly turned its attention to research on technological processes.

In 1958 the ministry, adopting a plan developed in 1957 for a uranium complex in Hunan Province, authorized construction of the first such plant, the Hengyang Uranium Hydrometallurgy Plant (No. 414, later 272); a second plant would later be approved for Shangrao because of differences in ore composition. In time, to save on transportation and reduce contamination of the ores, plants would be built at most of the mines.

Construction at Hengyang began in August 1958. The next year as the work was just beginning, the Second Ministry picked a cadre from the Central South Institute of Mining and Metallurgy to tackle the job of organizing it. With its usual secrecy, the government assigned this task to Sun Zongjie, a technical specialist. The Party secretary of a department of the institute summoned Sun and told him: "Comrade Sun Zongjie, the Party is assigning you a very important task, which has been classified as top secret. From now on, don't write any letters to your family, and don't bid farewell to your relatives and friends. We want [you] to build a highly classified and important plant in southern Hunan Province for the purpose of producing nuclear industrial products." Sun's job was to construct Plant 414, to be located near the city of Hengyang on the banks of the Xiang River, at the site of a defunct reform-through-labor camp.[56] Along with the uranium mining and metallurgy institute and the research and design academy, Plant 414 helped make Hengyang one of the nation's major uranium centers.

In 1960, the uranium phases of the program encountered innumerable obstacles as a result of nondelivery of Soviet aid and just at the time the central leadership was giving all aspects of the production of fissile material priority. In January, Sun received orders to proceed to Shanghai to hire qualified scientists for the new plant. Chinese students who had studied in Moscow or in leading Chinese universities had been summoned to Shanghai to await assignment. Some of them had studied the chemistry and chemical engineering related to uranium purification. As Sun conducted his interviews on the fifth floor of the Shanghai Hotel, Chairman Mao Zedong was presiding over a Politburo meeting on the thirteenth floor. Security officers evicted everyone from the hotel who did not hold a Project 02 (nuclear program) clearance.

Sun understandably expected his clearance would be an open sesame to his nation's best minds, for a project cadre had told him of the instructions Mao issued in 1956, when the program was being launched: "Priority should be given to the development of Project 02. Provision for all the leading scientists, technicians, and equipment necessary to the project should be guaranteed and take precedence over all other claims."[57] But as it turned out, Sun did not have his pick of the interviewees. When he submitted a wish list of his selections to the leading cadres of the Party organization department, they rejected it. He was informed that "none of these experts could be transferred to Plant 414 because they would be sent to work in the plants and institutions that had been assigned more important tasks." As alternatives, they gave Sun the personnel files of 29 young technicians and assured him that these were "qualified personnel with an enterprising spirit even though they have not studied in foreign countries." Disappointed at first, Sun finally got more than 30 technicians and skilled workers assigned to the plant. On reflection, he considered the personnel allocation to Plant 414 no small feat at a time when engineers and skilled workers were in short supply.

One of the technicians assigned to Plant 414 was twenty-eight-year-old Liu Jingqiu, a graduate of the Higher Industrial Vocational School in Suzhou. Since Liu had taken the Nanjing Institute of Technology's correspondence course in metallurgy, Sun placed him in the plant's central laboratory. Liu's immediate problem was to master the technical data on the production of nuclear fuel. Following the accord of August 1956, the Soviets had designed Plant 414, and virtually all such data were written in Russian, a language he did not know. His task was made somewhat easier because the ministry had already modified the Soviet-designed processing plan in order to shift from imported chemicals to ones manufactured in Shanghai and Tianjin.

Liu's story from then on stresses his zeal and the price he paid in the loss of his health. To help him, the ministry endorsed Plant 414's request to send Liu to the Tongxian institute. As Plants Two and Four were going up, Liu completed his training in uranium purification. By this time, moreover, the frustrations of ill-equipped technicians were rising elsewhere, and the Sixth Institute at Tongxian was increasingly functioning as a training base to assist others like Liu throughout the nuclear weapons program.

While Liu was mastering the details of the purification process, another cadre, Li Hao, struggled with the problem of installing 414's industrial equipment. From the summer of 1960 into the winter of 1961-62, Li camped out at the work site, but progress, especially in installing the purification tower, proceeded slowly. Again part of the problem was spreading food shortages, this time for the equipment installers, and Li used the lure of increased rations to spur his workers forward: "Everyone will be given a catty of tinned pork as a material reward if the tower is installed on schedule." Because "in the early sixties, even several pieces of biscuit or a piece of sweet potato meant a gleam of hope for life," the bribery worked. The tower was installed on time.

But then one setback after another bedeviled the plant, and it was not until the spring of 1962 that Li's crew could get it to operate properly. The first uranium oxide produced contained impurities and did not measure up to the prescribed standard. Back in the laboratory, Liu Jingqiu conducted eleven experiments and still met with failure. He then looked beyond Plant 414 to the waters of the Xiang River and thought that adding a precipitate to the river water might remove the impurities. He asked for approval to use the precipitate, and, on the next runs, in November 1963, Plant 414 succeeded in producing uranium oxide at the acceptable standard of purity and went into mass production. From then on, the Chinese continued to use this same method "for over 20 years to eliminate impurities in the process."

Next came the industrial manufacture of uranium tetrafluoride. The Second Ministry selected Baotou, in Nei Mongol (Inner Mongolia), as the location for a large plant to produce the tetrafluoride, nuclear fuel rods, and materials for the hydrogen bomb. In 1956, China elected to build this Nuclear Fuel Component Plant (He ranliao yuanjian chang), known as 202, with the help of Soviet specialists, who soon arrived to advise on the plant site and to plan the construction. The ministry approved a site in the Baotou suburbs and the preliminary design plan the following year.[58]

Administrative and technical cadres under the leadership of Director Zhang Cheng and Deputy Director Wang Huanxin began to arrive in Baotou in early 1958, and that October they commenced construction on Plant 202's Uranium Tetrafluoride Workshop. When the Soviet advisers departed in mid-1960, the tempo at the workshop accelerated, and Zhang Cheng, following Soviet advice, adopted a liquid process (with hydrogen fluoride) to produce the tetrafluoride. The Chinese used the standard filtration, washing, drying, and calcination techniques as they pressed the workshop into operation.

The abrupt Soviet withdrawal, however, halted the delivery of some key equipment made with nickel alloys, and delays ensued. To meet the emergency, Premier Zhou Enlai authorized the Baotou plant to draw 40 tons of these alloys from the central state reserves. The ministry asked a factory, a welding research institute, and an iron and steel academy to cooperate in fabricating the missing apparatus, and the result was the manufacture of China's first heavy equipment made with nickel alloys.

Several additional workshops and plants helped build other machinery, and on December 12, 1962, over a year after Tongxian's Plant Four had begun full operation, the Uranium Tetrafluoride Workshop commenced mass production. Its product, however, did not contribute to the fuel for China's first nuclear weapons. That fuel came as tetrafluoride from the institute at Tongxian and as hexafluoride from the Institute of Atomic Energy.

The Production of Uranium Hexafluoride

With the fabrication of uranium tetrafluoride, the Chinese could proceed on to the next step in the refining sequence—producing uranium hexafluoride. The process in most common use at the time involved compacting a compound of uranium tetrafluoride and magnesium chips, then heating it in an electric furnace with an argon atmosphere. In the reaction, the compound separates into a slag of magnesium fluoride and uranium. Technicians could fluorinate the metal to produce uranium hexafluoride or chemically produce the hexafluoride from the uranium tetrafluoride; the evidence suggests that the Chinese used the direct method and worked with the tetrafluoride in crystalline form. The hexafluoride could be handled as a gas, as solid colorless crystals, or as a liquid.

Like all fluoride compounds, uranium hexafluoride is highly corrosive and toxic, and its manufacture demands the use of advanced chemical-engineering techniques, standards, and equipment. Production of the hexafluoride was the penultimate step in the production of the enriched uranium fuel for a fission bomb. At each stage, the reaction processes engineered by the Chinese required close control and a sophisticated understanding of the limits of structural materials and equipment. Achieving these levels of control and understanding proved immensely challenging.

Nie Rongzhen's Defense Science and Technology Commission deemed the attainment of the necessary standards in this domain a "weak link" in the development of advanced weapons and, by 1959, had begun to give urgent consideration to this and other weak links in industrial quality control.[59] The commission had identified the three most problematical areas of quality control as the "new types of raw materials, precision instruments, and heavy duty equipment," and conscious of the gravity of the situation, made the introduction of improvements in all three areas a top priority. Finding a solution to the problems of quality control became even more urgent late in the year, when the Soviets began to refuse China's demands "for the supply of technical data on key projects, the production of raw materials, and design and theoretical calculations."[60]

The nondelivery of precision instruments and "special raw materials" hit China's plans for the production of nuclear fuels at a crucial moment. In June 1960, Moscow, having already reneged on its promise of October 15, 1957, to ship the Chinese the hexafluoride to be enriched for the first bomb, halted further technical assistance on China's hexafluoride production. The production of uranium hexafluoride thus became the "weakest link in the chain of China's nuclear industrial production."[61] As the uranium program rushed forward on other fronts, the hexafluoride plants could not be readied in time to meet the ministry's timetable.

Persuaded that the situation was urgent enough to warrant going outside the military-led system, Nie Rongzhen again looked for a stopgap facility. At some point before July 1960, the Defense Science and Technology Commission turned to the Institute of Atomic Energy to initiate research and development on the needed uranium hexafluoride. Located in the town of Tuoli some 35 kilometers south of Beijing, that institute (first code named 601 and later 401) had grown out of the Institute of Physics and formally came under the dual leadership of the Second Ministry of Machine Building and the Chinese Academy of Sciences.[62] Although its director, Qian Sanqiang, had participated in the Politburo's decision in January 1955 to launch the Chinese nuclear weapons program, the institute's scientific and technical personnel had concentrated largely on basic research and the development of peaceful uses of atomic energy. When, in June 1958, the Soviet-supplied atomic reactor and cyclotron had begun to operate,[63] some of the staff were assigned to reactor research projects. Because many of the major scientists, such as the deputy director Wang Ganchang and the physicists Peng Huanwu and Zhu Guangya, worked outside the institute on other projects, Qian assigned principal responsibility for the manufacture of the uranium hexafluoride to a less-well-known deputy director, Li Yi.[64]

In July 1960, the Second Ministry assigned the code name Project 161 to this research-and-development effort. By this time, the institute's roster had grown more than 25-fold since 1954 and now numbered 4,300. It took the project workers three years to produce the target amount of over ten tons of the hexafluoride compound, which became the raw material for enrichment in the Lanzhou Gaseous Diffusion Plant.[65]

The Second Ministry chose Professor Wu Zhengkai, the British-trained chairman of the Fudan University Chemistry Department in Shanghai, to supervise the production of the hexafluoride. As director of the Institute of Atomic Energy's new Research Department 615, Wu's job included research and the trial production of uranium hexafluoride and research on the engineering requirements for enriched uranium. Up till now, Wu had had no experience in radioactive chemistry.[66]

Even before Wu was selected, engineers at Section 615A of the research department had begun research on some key theoretical problems in gaseous diffusion, but now the priority shifted to Section 615B to make the hexafluoride. The day the Soviet advisers pulled out of the institute in August, one engineer, Huang Changqing, stayed awake all night, struck by the awesome task ahead of him. His colleagues included technicians sent from the planned Lanzhou facility to receive training in gaseous diffusion, and, after burying themselves in books following the Soviet withdrawal, they embroiled the entire section in an argument on flow charts for the trial production of the hexafluoride. In his sleepless night, Huang had devised his own chart, but departmental representatives wanted to adhere to a previously adopted Soviet design. Huang's argument was that the institute should switch to his simplified system in order "to shorten the time even though there would be risks."[67] Using the Soviet system would mean a loss of several years because it was designed for a large plant and was "overly elaborate." The institute took the competing designs to ministry officials, and they decided in favor of Huang's system.

The Second Ministry then assigned the Beijing Nuclear Engineering Research and Design Academy to work with Section 615B on a simplified installation. It also told the academy and 615B to concentrate simultaneously on the techniques of mass production, and for both these jobs the academy's deputy director, Cao Benxi, moved to Tuoli. Cao, a nuclear chemist and concurrently chief engineer of the ministry's Fuels Production Bureau, joined forces with Wu Zhengkai for the next fifteen months.

After entering the research department, Huang had started theoretical work on chemical reactors in Laboratory 131. With the higher-level approval of his flow-system design in hand, he received orders to build a reaction chamber for the trial production of the uranium hexafluoride. Now his problems were practical ones: heat and corrosion. Huang knew from Soviet and Western literature that he should make the reaction chamber from Monel alloy, which could withstand several hundred degrees centigrade and the corrosive fluorine gases.[68] At that time, however, China had no Monel alloy, and even with the alloy, construction of the chamber would take several months and cost over 100,000 yuan (at least U.S. $50,000).[69]

True to the prevailing spirit of innovation, Huang improvised. He found some red copper pipe lying around the institute and shipped it to a factory in Shandong Province for nickel plating.[70] At the same time, he asked the head of a design group of a factory attached to the Institute of Atomic Energy to design a reaction chamber made of the pipe, and, more, he wanted the designers to come up with it in three days' time. Meanwhile, an unexpected snag arose when the factory in Shandong, after nickel plating the pipe, could not find a freight train to transport it back to the institute. Leading cadres from Beijing got involved and finally had to write a letter to the Ministry of Railways for assistance. Within days, the ministry authorized an express passenger train to ship the pipe to Tuoli.

Engineer Huang, using the local factory's design, then fabricated the novel reaction chamber in Laboratory 131, but on the eve of testing it, he began to have qualms about his invention. Whereas the Soviets used a three-staged reaction chamber, his was designed to produce the hexafluoride in a much less complicated, two-staged process at a considerably lower temperature.[71] Huang's boldness now gave way to panic. The radical new reaction chamber might fail. Moreover, as his comrades made painfully obvious to him, the test of the chamber, coming in late 1960, was scheduled at a pivotal point in the bomb-building program.[72]

His mind was awhirl with doubt. Yet at Huang's worst moment, the night before the first experiment with the reactor, Qian Sanqiang, the institute director and vice-minister of the Second Ministry, boosted the nervous Huang's morale: "Xiao Huang, what about the preparations for the reaction experiment? Have you met any trouble? Please don't worry about a possible failure. It will be a victory for you if you can trial produce only one gram of uranium hexafluoride."[73] Huang, it is reported, felt reassured, especially since Zhang Aiping, deputy director of the Defense Science and Technology Commission, had recently given him similar encouragement.

The test of Huang's reactor seemed to go well at first. As the experiment began, a worker lit a flame in the reactor and gradually built the temperature up to 300 degrees centigrade. The test crew then released a valve for fluorine gas to combine in the reaction chamber with measured amounts of uranium tetrafluoride crystals injected by a propeller fan. The chemical reaction had begun, and after six hours Professor Wu, Huang, and their co-workers checked the reactor and found 3.3 kilograms of the precious hexafluoride. One veteran worker could finally manage a smile after the long ordeal, though he reportedly felt ashamed that he had needed the encouragement of his comrades before he had fed the raw material into the chamber. At the onset of the experiment, when he had brought in a tray filled with the tetrafluoride crystals, he had felt as though he had been "holding a bomb."[74]

Minister Liu Jie and Qian Sanqiang also heaved sighs of relief and relaxed into smiles and mutual congratulations, but their jubilation was short-lived. The reactor began functioning intermittently as "hard, highly radioactive lumps formed [in it] resulting in a blockage of the passage of the raw material. It was a thorny problem." Even in the fourth experiment, a young worker had to remove the lumps of hexafluoride with an electric hand drill. To protect himself from the intense heat, he wore a special rubber suit. Unfortunately for him, the suit conducted heat, causing him untold suffering as he worked.[75]

The burden of solving the production problem fell to the illustrious Professor Wu, who gathered a small group to discuss the recurrent lumping. Weighing the advice he received, Wu decided to experiment to find the right ratio between fluorine gas and uranium tetrafluoride. He ordered the reactor crew to start trial-and-error tests, and eventually a mix was found that would not produce lumps.

Yet technical problems continued to plague the experimental program, and in late 1961 there was a major breakdown: "After a year [of] success in solving problems of feeding material, fluorination, hermetical sealing, anticorrosion, . . . the operation had to be suspended because of the blockage of a [reactor] condenser."[76] This breakdown occurred in October, and Wu, now under ever greater pressure to produce the hexafluoride, turned to Soviet technical data for help. Even though the condenser being used was rated for a maximum capacity of 200 kilograms, it could barely reach 30. The blockages recurred, and Wu assigned Huang's team to manufacture a replacement condenser based on the Soviet design.

The breakdown coincided with the national food crisis, and the Second Ministry was forced to evaluate what that might do to the hexafluoride project. During a meeting at the ministry, Huang corroborated the reports of widespread hunger that were coming into Beijing from all parts of the nuclear program. "Many of the staff and workers of Research Department 615 suffer from dropsy and cannot work," he told Minister Liu Jie, cautioning: "When [we] begin continuous operation, we will be short of hands."[77] Liu promptly called the institute's deputy director, Li Yi, and told him that though the ministry had no power to secure the needed food for the hungry and sick workers, he could order Li to transfer technicians from other sections to fill the gaps. Li Yi complied with the ministry's order, and the acute personal suffering persisted.

Nonetheless, the additional manpower eventually solved the reactor problem, and in July 1962 the experimental stage gave way to sustained production of the hexafluoride. A second reactor went on line within a few weeks. Only a few months before, Plant Four at Tongxian had moved to quantity production of tetrafluoride, and the entire program was now able to adhere to the ministry's timetable. On October 6, 1963, the Institute of Atomic Energy attained its own goal of producing over ten tons of uranium hexafluoride. In the process it had trained Ding Shufan, the engineer who had been assigned to mass-produce hexafluoride.* Most importantly, the institute could provide Lanzhou with its precious fuel ten days ahead of schedule.

*In April 1960, Ding Shufan and a picked crew had broken ground for the Uranium Hexafluoride Plant (Liufuhuayou shengchan chang) to mass-produce uranium hexafluoride in the Jiuquan Atomic Energy Complex, Gansu Province (discussed in the next chapter). During the construction of the facility, the ministry sent Ding to the Institute of Atomic Energy's Research Department 615 for training. He subsequently became a central figure in the production of nuclear fuel for China's strategic arsenal. The plant used the testing data from Section 615B to conduct five of its own test runs from November 1962 to April 1963. Mass production began on Nov. 27, 1963. Niu Zhanhua, "Shanliang de Ganghuan," p. 142; Li Jue et al., *Dangdai Zhongguo*, p. 185.

FIVE

The Production of Fissionable Material

The atomic bombs exploded at Alamogordo, New Mexico, and Nagasaki, Japan, in 1945 used plutonium and the so-called implosion triggering method. The Soviet Union also used plutonium for its first nuclear bomb, detonated on August 29, 1949. Both Washington and Moscow developed an alternate weapon that would employ a rare isotope of uranium, 235. The gun-barrel-type bomb dropped on Hiroshima was a U^{235} device; the Soviet Union exploded a U^{235} bomb in October 1951.[1] Technically, it had been easier to manufacture plutonium than highly enriched U^{235}, but it proved to be harder to design the plutonium than the uranium bomb.

One path for China thus required the enrichment of uranium in the ratio of the 235 isotope to the much more abundant 238. In nature, the two isotopes occur at a ratio of about 0.7 (235) to 99.3 (238). The necessity in the military program for the enrichment to more than 90 percent U^{235} posed a formidable challenge because the two isotopes, though chemically identical, differ very slightly in weight, and the membranes used to diffuse the uranium hexafluoride gas must meet severe industrial standards. The second path required uranium dioxide to fuel a plutonium-production nuclear reactor; either natural or slightly enriched (at a level of about 3 percent U^{235}) uranium could be used in the reactor. The final reactor product had to be treated chemically to remove the plutonium. Both gaseous diffusion (for U^{235}) and chemical separation (for plutonium) depend on advanced industrial technologies and demand the highest production standards.

Technological Choices and Costs

Because of the parallel development of the different parts of the nuclear weapons program, we must return once again to the 1950s. As late as 1958, the Chinese lacked the required technological base to manufacture any fissionable material, and they could not have known whether their technicians could design either type of atomic bomb. The equipment sold to China by the Soviet Union, however, presumed that the Chinese would enter the nuclear arms race by accepting the need for high redundancy and by building production lines for both enriched uranium and plutonium.[2]

The small-scale production of uranium oxides and uranium hexafluoride had given China the basic chemical compounds to follow either path. The Chinese had progressed this far in the nuclear weapons program with some Soviet help, but to fabricate the fissionable materials for weapons they had to rely even more on the Soviet Union. It was at this point in the process that the Chinese became most dependent on Soviet assistance to surmount the countless technological obstacles—and that assistance grew steadily over the years.

In October 1954, the two countries had signed a Scientific and Technical Cooperation Accord, by which the Soviet Union, for a price, agreed to send China "technical data, relevant information, and experts"; a companion agreement granted China substantial credits.[3] On January 17, 1955, about the time the Politburo was making its decision to acquire strategic weapons, the Soviet government announced it would help China and other socialist states "promote the peaceful utilization of atomic energy" and would provide them with atomic reactors and fissionable material for research.[4] This announcement led to the Sino-Soviet atomic cooperation accord of April 1955, by which Moscow agreed to supply an experimental nuclear reactor, a particle accelerator, and "the necessary quantity of fissionable material."[5] These and follow-up bilateral agreements paralleled the Chinese government's decision to give added priority to training and research in nuclear physics.[6]

In the start-up years of the strategic weapons program, Chinese planners focused on the country's dearth of qualified nuclear engineers and scientists. They anticipated that the need would be the greatest from the enrichment phase through the bomb-design phase. The planners began by sending some 5,800 students, all but about 1,200 of them undergraduates, abroad to study in the years 1954-56. Then, in the subsequent two years, Beijing gave priority to sending postgraduate students abroad.[7] Moreover, several hundred advanced trainees received general instruction in nuclear physics after the Joint Institute for Nuclear Research was set up in Dubna, near Moscow, in March 1956.[8] Beijing further defined as a strategic task the wooing back to

China of top foreign-trained scientists and engineers still abroad and gave the Ministry of Foreign Affairs the job of "winning them over."[9]

In general, the Chinese confronted the problem of simultaneously cultivating technical personnel, establishing advanced facilities, and coordinating a vast bureaucratic system while the actual work on the nuclear weapons program proceeded. Zhou Enlai, in particular, took special interest in the shortage of technical personnel. Shortly after the nuclear decision in January 1955, he ordered the Ministry of Education to form a Nuclear Education Leading Group. Professors Hu Jimin and Zhu Guangya left their universities for Beijing University where they organized its Physics Research Section (renamed the Department of Atomic Energy in 1958, and now the Department of Technical Physics). At the same time, Professor He Dongchang set up the Department of Engineering Physics in Qinghua University.[10]

From 1955 to 1958, the ministry recruited hundreds of college seniors majoring in science and technology for these two university departments as well as one started in Lanzhou University. To help train them in nuclear science, the ministry hired Soviet lecturers. The first trainees graduated in 1956.

The shifting priorities also affected Chinese studying in the Soviet Union and Eastern Europe. Over a hundred of them became majors in nuclear science and engineering. Moreover, in October and November 1955, the Chinese Academy of Sciences sent two delegations to the Soviet Union to study the theory and operation of nuclear reactors, cyclotrons, and other equipment. None was given access to military-related research.

Thanks to all these educational efforts, by the time the Chinese had to man the facilities built to produce fissile materials, they had enough qualified technicians to meet the nuclear program's most pressing operational needs. By 1983, the training program begun in 1955 had produced 22,124 graduates.

The Politburo also solved the complex managerial puzzles involved by activating the Party, state, and military command systems. The Chinese state that the units participating in the process of making and testing the first atomic bomb represented 20 provinces, municipalities, and autonomous regions; 26 ministry-level organs; and more than 900 factories, research institutes, and schools. "All together, they formed a most complicated organization. Several hundred thousand people were involved."[11] The government set in motion the construction of the largest plants to fabricate the fissile material and commanded the rest of the industrial system to provide support for these plants. Two critical ones, built with some Soviet assistance, have been mentioned in passing: the gaseous diffusion plant at Lanzhou, in Gansu Province, and the plutonium-producing reactor and plutonium-processing plant in Jiuquan Prefecture.[12]

To create such facilities would cost billions of yuan, and Mao Zedong concluded that to pay the bill other national programs would have to be cut. In April 1956, he wrote: "If we are not to be bullied in the present-day world, we cannot do without the bomb. Then what is to be done about it? One reliable way is to cut military and administrative expenditures down to appropriate proportions." Mao advocated spending more on general economic projects, many of which would provide the industrial base for the strategic weapons program: "In the period of the First Five-Year Plan [1953-57], military and administrative expenditures accounted for 30 per cent of the total expenditures in the state budget. . . . In the period of the Second Five-Year Plan, we must reduce it to around 20 per cent."[13]

The People's Republic of China had the advantage of being a follow-on nuclear state. From a close reading of reports, its scientists knew in general terms which avenues to the bomb were most likely to succeed, and they labored over each step along the way in order to avoid costly mistakes.[14] At each stage in the program, scientists and technicians combined theoretical and experimental missions, taking as their guideline Premier Zhou's slogan "one-time test, overall results" (*yici shiyan, quanmian shouxiao*). That slogan, when operationalized, described a method in which the stages of theoretical research, experimentation, engineering, and production were considered together and undertaken as a whole. Technicians devised each test to prove out multiple unknowns. The Chinese preferred to perform small-scale simulation tests and shunned costly full-scale experiments except when absolutely necessary.

Consistent with Zhou's method, the program's officials endorsed a philosophy of parsimony: "spend less, get more" (*shao huaqian, duo banshi*). They approved each move only after careful comparison to the ones already taken by other nations, with special attention paid to the American program.

As a result, the Chinese say that expenditures on the nuclear weapons program approximately equaled the cost of building one large modern steel facility. Assuming that the completed "stage" of the Baoshan Iron and Steel Complex in Shanghai approximates the scale of the nuclear weapons program (as the Chinese hold), the total cost of the program, from uranium prospecting to a finished bomb, would have been about 12.86 billion yuan in 1981 prices or approximately 10.7 billion yuan in 1957 prices. According to one calculation, this would translate into about U.S. $4.1 billion in 1957 prices.[15] In 1957, total state expenditures for the People's Republic reportedly amounted to slightly over 29 billion yuan, of which 5.5 billion went to national defense.[16] The expenses of the nuclear weapons program were of course spread over a ten-year period, 1955-64, but it is still useful to note that the total sum of 10.7 billion yuan was equivalent to about 37 percent of the entire state budget for the year 1957, and to slightly more than 100 percent of the defense budget for the two years 1957-58. Moreover, the burden of those allocations undoubtedly fell disproportionately on budgets in the 1960s, the years of economic distress.

Yet the price tag could have been much worse. By replacing a Soviet-defined program with one of its own, Beijing lowered the program's total cost by 40 percent. The new principle guiding the nuclear effort was "[keep] the scale small but [build] a complete system" (xiao er quan). It stipulated that a large percent of the "auxiliary" industrial equipment would be made in China in order to save foreign exchange. In general, the Chinese believe that, comparatively speaking, they got by with a quite modest investment for four reasons: (1) they did not have to do much creative work but largely copied what others had done; (2) they did not make costly mistakes; (3) the costs of skilled labor were about 1 percent of the costs of comparable labor in the United States; and (4) they sought to build less sophisticated and less accurate weapons than the ones in the U.S. arsenal.

The Construction of Reactors for Plutonium Production

The Chinese began building a nuclear fuel base to fabricate plutonium at the same time they turned to the task of uranium enrichment.[17] In 1940, four physicists at the University of California had artificially created plutonium by deuteron bombardment of uranium-238, yielding neptunium-238, which beta decays to Pu^{238}.[18] The isotope used in weapons, Pu^{239}, was first produced at the same laboratory by bombarding U^{238} with a neutron. The product after two beta decays becomes plutonium-239. Fissionable, this actinide isotope undergoes an explosive chain reaction initiated by slow neutrons when it achieves a critical mass only about one-third that of U^{235}. Because plutonium rarely exists in nature, it must be made in a nuclear reactor. Of the 233 American-built nuclear reactors in service in 1962, thirteen were dedicated to the production of plutonium.[19] In following the plutonium path, the Chinese also chose to build specially dedicated reactors for the manufacture of plutonium for weapons.

Prospecting team checking map for possible uranium sites, South China

Uranium separator facility, Chenxian Uranium Mine, Hunan Province

First-stage uranium processing at the Lianxian Uranium Mine, Guangdong Province

The ore-grinding workshop of the Hengyang Uranium Hydrometallurgy Plant (Plant 414), Hunan Province

Interior of the Lanzhou Gaseous Diffusion Plant (Plant 504), Gansu Province

Equipment at the Nuclear Fuel Component Plant (Plant 202), Baotou, Nei Mongol

Deng Xiaoping (center left), then General-Secretary of the Chinese Communist Party, visiting the Jiuquan Atomic Energy Complex (Plant 404), Gansu Province, March 1966

The plutonium production reactor of the Jiuquan complex

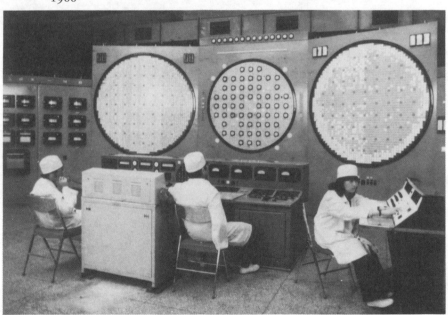

The main control center of the plutonium production reactor, Jiuquan

The Institute of Atomic Energy, Beijing

Zhou Enlai (center) and Chen Yi (right) accompanying Cambodian prince Norodom Sihanouk (left) on a visit to the Institute. Man at far right is Marshal He Long.

Guo Yonghuai (right), with visiting Russian scientist

Chen Nengkuan

The Northwest Nuclear Weapons Research and Design Academy (Ninth Academy)

Li Jue at the Lop Nur Nuclear Weapons Test Base

ang Aiping inspecting prepara-
ns for China's first nuclear test
Lop Nur; the man wearing white
rt is Liu Xiyao

Marshal Nie (right) visiting Lop
Nur, winter 1966; the man at
left is Qian Xuesen, a missile
expert once affiliated with the
Jet Propulsion Laboratory at
Caltech

Officials in charge of China's first nuclear test (from left to right): Zhang Yunyu, Liu Boluo, Liu Xiyao, Zhang Aiping

ABOVE: Marshal Nie (left) and Zhu Guangya at Lop Nur

LEFT: Zhang Yunyu and Marshal Nie at the nuclear missile test of October 27, 1966

Most nuclear reactors hold large amounts of U^{238} plus the fissile U^{235} required to sustain a chain reaction in the fuel elements.[20] During the reactor operation, plutonium-239 forms when a 238-isotope has been hit by the neutrons produced in the fission reaction. Some of these neutrons produce new fissions, thereby liberating other neutrons, and so on and on, in a chain reaction. When the reaction has reached a level approaching one new fission for every neutron liberated, then the number of neutrons remains virtually constant, and the reactor is said to be "critical." Because neutrons continue to bombard the pluto-nium-239, some of the 239-isotopes are converted to plutonium-240 or even higher isotopes. Engineers can design reactors to vary the amounts and ratios of these isotopes. The irradiated fuel elements, which thus contain plutonium, uranium, and fission products, are re-moved from the reactor and allowed to cool under water for several months before they are transported to a dissolver.

The Chinese call the processing in the dissolver "aftertreatment" (*houchuli*). In the dissolver, the jacket surrounding the fuel element is removed, and technicians place the element containing plutonium and other materials in a chemical solution. Chemical processing on an in-dustrial basis, which has been described as a "fantastic scale-up" of an "extraordinary achievement in chemical research,"[21] separates the plutonium and uranium products for additional processing, while the other fission products are sent to storage and later disposal. In the chemical process used in the United States in the 1950s and 1960s, the original solution was precipitated by conversion to a peroxide. From this solution, the plutonium compound was changed chemically to plutonium tetrafluoride. The final steps reduced the tetrafluoride com-pound to plutonium metal, which was then cleaned, melted, cast, and machined. In this complex manufacturing process, the workers were exposed to one of the most dangerous elements on earth.

Soviet plans and experience directed the Chinese toward a pluto-nium bomb as well as a uranium weapon, and the Chinese over time made significant investments in plutonium production and its subse-quent chemical separation. Perhaps because of these investments, an official U.S. national intelligence estimate of December 13, 1960, con-cluded: "Chinese development of uranium resources and their prob-able construction of ore concentration and uranium metal plants cer-tainly would imply an intended use for the uranium in plutonium production. . . . We estimate that a first Chinese production reactor could attain criticality in late 1961, and the first plutonium might be-come available late in 1962." The U.S. government had no conclusive evidence on the construction of reactor and chemical separation fa-

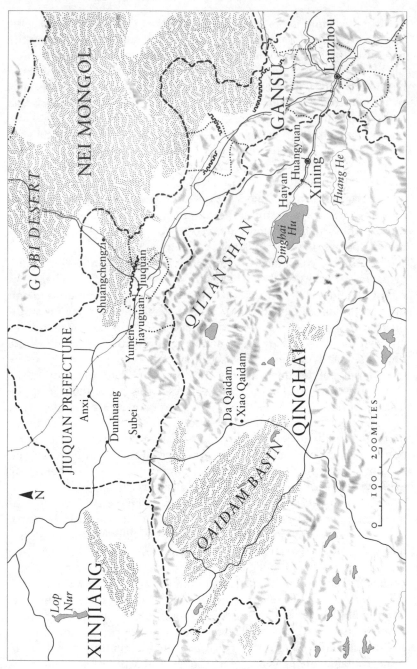

Map 3. Northwest China

cilities, but the assistant chief of the Air Force for intelligence concluded that the "first nuclear device will probably use plutonium."[22]

The Chinese began planning their first plutonium production reactor in 1958 in Jiuquan Prefecture (*diqu*). Nie Rongzhen had personally selected the construction site in an isolated expanse of the Gobi Desert in the western part of the prefecture.[23] The prefecture dated back some 2,000 years, to the Western Han Dynasty, and the prefectural town carried this ancient name until the seventh century, when it was renamed Suzhou. Long a communication hub in the valley corridor between Xinjiang and China proper, Suzhou hosted the Silk Road's dusty travelers until it slid into decline in the eighteenth century. In the late nineteenth century, it had a mere 500 households, compared with more than 18,000 in the Western Han. By the 1950s, however, the town, which once again bore its founding name, began to recover as a transportation and small industrial center. Nie surveyed the desert area over 300 kilometers to the west of this ancient Silk Road town of fewer than 50,000 inhabitants, approved the designs that the Soviet Union had provided for the construction of the plutonium fuel facilities, and arranged to have "thousands of pioneers" brought in to work on the project.[24]

Tucked away at the foot of the tall Qilian peaks, the little-known desert area near the Subei Mongolian Autonomous County (Subei Mongolzu zizhixian) quietly came to house one of China's principal strategic weapons nuclei: the Jiuquan Atomic Energy Complex (Jiuquan yuanzineng lianhe qiye; see Map 3). In the early years, the complex in Jiuquan Prefecture was known only by its code name, Plant 404. The nuclear ministry's Fuels Production Bureau, led by Bai Wenzhi, was put in charge of the large plutonium-production reactor and chemical engineering reprocessing facilities to be built in the Jiuquan complex. In addition to housing the plutonium works, the complex became the home of the processing plant for uranium hexafluoride, the Nuclear Fuel Processing Plant (for converting the enriched uranium hexafluoride to uranium metal), and the Nuclear Component Manufacturing Plant.

Three major facilities compose the plutonium production line in the Jiuquan complex: the reactor, the chemical separation plant, and the Plutonium Processing Plant (for refining plutonium metal for weapons).[25] The nuclear ministry adopted a general Soviet design for the country's first military graphite-moderated, light-water reactor (with natural uranium) to produce plutonium, and broke ground for the facility in February 1960. By August, the excavation and basic foundation work had been finished. As we will see, the ministry then halted the project and did not resume it until June 1962. Formal operation of the reactor began in early 1967.

The Chinese take special pride in Jiuquan's aftertreatment or plutonium chemical separation plant, the largest in China. This facility, designed initially with Soviet help but not completed until 1970, consists of three parts: a working area, a maintenance area, and a so-called "clean area." These parts correspond to the three zones defined by the Security and Protection Bureau's codes passed in 1960.[26] A 1.8-meter-thick reinforced-concrete wall isolates the working area for treating the radioactive materials from the reactor. In the maintenance area, only workers in protective clothing are permitted, and in the clean area, workers use robots and instruments to control the separation operations.

The Chinese created a new satellite city around these facilities and boast of having transformed the "wilderness [where] there were often winds and snowstorms, the sand blocked the sun, and wild rabbits and goats ran rampant." They planted a broad green belt and developed a large residential area with its own shopping and recreational facilities. Writings about the city stress both the safety of its location and the ferocity of a climate that forced construction crews to retreat to Beijing each winter in the early years.[27]

In building their first plutonium-production reactors, the Chinese received Soviet help, and they give due credit to the Soviet Union for its assistance, most of it purchased. Soviet specialists, for example, oversaw the installation of the nuclear research reactor at the Institute of Atomic Energy at Tuoli. On the other hand, for security reasons, those specialists did not allow the Chinese to help insert the nuclear elements in the institute's reactor, and up to the day they left, restricted Chinese access to the reactor.[28]

Moscow gave Beijing "preliminary designs" (*chubu sheji*) for the Jiuquan military plutonium-production reactor and its related chemical separation facility, but little, if any, hands-on aid in building them. "Among the 500 items of equipment required for the reactor, the Soviet Union only delivered about 5%," none of it vital (such as the fuel rods, main pumps, and heat exchangers). They failed to provide 40 percent of the promised design drawings on the undelivered equipment. The Chinese constructed the reactor complex near Jiuquan and several other nuclear reactors on their own—facilities, *Renmin Ribao* asserted in 1985, that "can basically satisfy the needs for the development of scientific research and national defense."[29]

Beyond these few facts, we have almost no reliable evidence on the development of China's plutonium facilities. Several sources assert that the Jiuquan nuclear reactor was one of "at least seven known nuclear reactors in China" in 1976.[30] According to a U.S. government publication, in November 1960, Chinese President Liu Shaoqi "declared that there were at least four atomic reactors in the PRC [People's Republic of China]. At a conference . . . in Moscow in that month, the CCP [Chinese Communist Party] reportedly circulated a document stating that the PRC had four nuclear reactors in operation and that it intended to turn to other than peaceful applications if its security needs were not met."[31] Reactors designed for the weapons program have been reported throughout China,[32] but, in fact, most of these reactors, if they exist at a given location, are dedicated to civilian purposes. Some reports have stated that in the 1960s the Chinese succeeded in extracting sufficient amounts of weapons-grade plutonium to warrant building a number of chemical separation plants throughout the country in addition to the one near Subei in Jiuquan Prefecture.[33] However, we know of no Chinese sources that confirm the location or date of construction of any military plutonium facility other than the one in the Jiuquan complex and one in Sichuan Province.

At the outset, Chinese leaders did not give precedence to the enrichment of uranium and the production of a U^{235} bomb over the plutonium path to the bomb.[34] Until the twin crises of 1960—the economic calamity of the "three hard years" and the anticipated withdrawal of Soviet assistance—the manufacture of enriched uranium and plutonium received the same high priority. At the moment of truth in 1960, however, the reactor program lagged behind the building of the uranium enrichment facilities, and the Chinese chose to delay the plutonium program and give sole priority to uranium enrichment after weighing the economic and technical variables. As a result, plutonium did not show up until an explosion on December 27, 1968, the eighth in the Chinese nuclear test series.[35]

The initial decision in favor of uranium over plutonium came in April 1960. Beijing decided to concentrate on the production of enriched uranium at Lanzhou and, in August, when the Soviet specialists headed home, to halt work on the plutonium reactor and separation plant in Jiuquan Prefecture. The construction of the reactor had barely commenced. Only the excavation and foundation work for the buildings had been completed, and the separation plant had not gotten beyond the blueprint stage. With so little progress on the plutonium front and so many problems confronting the entire nuclear program, none could argue for continuing to give the plutonium bomb high priority. With plutonium production temporarily on hold,* the leadership attached maximum urgency to the enrichment of uranium.

The Lanzhou Gaseous Diffusion Plant

The critical mass of uranium that must be assembled for an explosion varies inversely with the ratio of U^{235} to U^{238}, or the percentage of enrichment.[36] There was only one efficient path known in the 1950s for increasing that percentage: isotope separation through gaseous diffusion.[37] The problem for the Chinese was to assemble a gaseous diffusion plant that would achieve the high levels of enrichment needed for making weapons. The Chinese plant would have to approach the scale of the Oak Ridge gaseous diffusion plant in Tennessee or of a comparable Soviet gaseous diffusion plant in the Urals.[38]

Gaseous diffusion exploits the slight difference in mass between U^{238} and U^{235}. In a thermal equilibrium, the lighter U^{235} moves faster than its sister isotope. The objective of gaseous diffusion separation is to diffuse uranium hexafluoride gas through porous barriers within the thousands of separation elements making up the plant. The lighter molecules containing U^{235} come in contact with the walls of the barriers more frequently than the heavier molecules because of their higher average speed, and are therefore more likely to encounter and pass through a pore. Diffusion systems contain thousands of separation elements because the degree of separation (or separation factor) in each element is minute. In the United States, the production of good nickel barriers had proved a daunting industrial challenge because of the high probability of corrosion and plugging.[39]

With the support of the Soviet Union, Beijing elected gaseous diffusion as one path to the first bomb; in the spring of 1960 it became the sole path. Chinese leaders understood that the cost of an enrichment plant would be staggering. The financial investment for China would not equal the American level of spending, all things considered,

---

* In December 1960, the Second Ministry convened a meeting on the manufacture of the reactor's equipment. The Second Ministry (with the First Ministry), in 1962, formed a joint Reactor Equipment Leading Group, which charged 37 plants with manufacturing the key items, and then, in June, restarted the reactor project. A year later, plants under the Third Ministry (in charge of conventional weapons) joined the effort. After the detonation of the first atomic bomb, the Fifteen-Member Special Commission determined that the follow-on hydrogen bomb project would need plutonium. Construction on the Jiuquan plutonium facilities was accelerated, and an "intermediate pilot plant" was built for the initial chemical separation. The Chinese used the plutonium from this plant in the thermonuclear test of December 1968. Li Jue et al., *Dangdai Zhongguo*, pp. 207-9, 229-31.

but it would be high.[40] Moreover, constructing the plant would place a heavy burden on China's industry. The Oak Ridge plant contained thousands of converters (for the diffusion cascade) and pumps, a half million valves, thousands of coolers, three million feet of corrosion-resistant piping, and tens of thousands of instruments.[41] The Chinese would have to manufacture this equipment or purchase it from the Soviet Union. And since the massive pumps and compressors for such a plant would consume enormous quantities of electrical power, they would also have to build giant power installations to serve it.

This was one of the considerations that led to the siting of the plant at Lanzhou. Located on the ancient Silk Road in Gansu Province, the city had the advantage of interior remoteness (from U.S. reconnaissance planes), the nearby Huang He (Yellow River) for cooling water and electrical power, and an important industrial base. A traditional garrison town and transport center since ancient times, Lanzhou grew into a large industrial city during the 1950s. In 1958, it had a population of about 732,000, almost double the 1953 number.[42] A large thermal power plant, put up with Soviet aid in 1957, used coal from two nearby coalfields, and the city spawned a host of machine-building, metallurgical, and chemical factories at about the same time. As the nuclear ministry's Bureau of Construction started work on the gaseous diffusion plant, plans proceeded to expand Lanzhou's electrical power and industrial base.[43] The location seemed ideal for this Oak Ridge of China, as one Beijing writer has called it.[44]

At the same time the Sino-Soviet New Defense Technical Accord was concluded (October 15, 1957), the Party leadership decided to accelerate the development of China's nuclear industry.[45] Moscow committed stocks of uranium hexafluoride for the Chinese from its strategic reserve, and sustained Soviet backing seemed assured. In November, confident of this commitment, Beijing's Third Ministry appointed the director of a preparatory office to plan the establishment of the Lanzhou enrichment facility.

By then (February 1957), a site-selection unit had settled on a stretch of ground for the facility in a U-shaped valley on the bank of the Huang He about 25 kilometers northeast of Lanzhou. A newly constructed aircraft factory stood on this place, but the Central Military Commission, following Soviet advice, gave the site to the nuclear program. When Wang Jiefu, a future leader of the plant, surveyed the proposed location the next winter, Minister Song Renqiong's words sprang to mind: "All the plants under the administration of the Third Ministry of Machine Building are located in desolate places where even the rabbits won't defecate." More to the point, Wang realized that the river water would be a major problem for the plant's cooling system: "When you ladled out a bowl of water from the Huang He, you would find half was sediment."[46] It was clear that special filtration systems would have to be installed. Obstacle piled on obstacle.

To overcome such multiple problems, experienced officials at the nuclear ministry knew they would have to rely on personal teamwork, and a four-man directorate chosen to build the Lanzhou plant is a textbook example of personal networking in Chinese politics.[47] Shared experience, proven competence, and total loyalty linked individuals in the revolution, and as a cadre's senior colleagues moved up the ladder of success, so did the cadre. The reverse was often true as well. In the Lanzhou case, such links appear to have been forged ten or more years earlier, when some members of the directing group were comrades-in-arms in the Communist Second Field Army.[48] The Second Field Army's military and political commanders, notably Liu Bocheng and Deng Xiaoping, had made war together for at least a decade by the late 1940s. The military historian William Whitson writes: "The stability of leadership achieved by late 1940 is an important aspect of present-day loyalties in Communist China, for all the leaders who later served with distinction and success under Liu Po-cheng [Liu Bocheng] and Teng Hsiao-p'ing [Deng Xiaoping] in the Second Field Army came together . . . to plan, organize, and execute the Hundred Regiments Campaign [in 1940]."[49] Launched by the Eighth Route Army (a precursor of the People's Liberation Army, including the Second Field Army), the campaign committed regiments from the army's three divisions. When the Japanese retaliated with a savage "burn all, kill all, destroy all" campaign of their own, the Communist army suffered heavy losses. Ever after, critics stingingly reminded leaders of the Hundred Regiments Campaign of the campaign's costly errors,[50] but this trial by combat proved to be the source of strong personal bonds among the campaign's veterans.

These men of the Second Field Army had eventually fought to victory through North China and thence in the landmark Huai-Hai campaign that basically ended Nationalist control of China.[51] After the founding of the People's Republic, many of these veterans were assigned to posts in the southwest or Beijing; and in the years to come they established their leadership in the conquered land. The four leaders of the Lanzhou enrichment plant—Wang Jiefu, Zhang Pixu, Liu Zhe, and Wang Zhongfan—emerged from or had close ties to these Second Field Army veterans and consistently returned to their network leaders for guidance and support.

Song Renqiong, a political commissar under Liu Bocheng and Deng Xiaoping at the time of the Hundred Regiments Campaign, became the pivotal figure for these four as Song's own mentors, Liu and Deng, moved on to higher posts in Beijing. Song himself rode his connections to the Central Military Commission's General Cadres Department (as deputy head) and, in 1956, to the Third Ministry. In 1950, the Party's Southwest Bureau appointed Wang Jiefu to a prefectural Party committee in the southwestern province of Sichuan, and about the time Song became first deputy secretary of the bureau, Wang accepted the directorship of its Urban Work Research Section. One year after Song moved on to Beijing, Wang was picked to investigate industrial projects in the Soviet Union; soon thereafter he became political counselor in the Chinese embassy in Hungary.[52] Song Renqiong chose Wang to direct the Lanzhou Gaseous Diffusion Plant.

During 1950-54, the years Song served as the secretary of the Yunnan provincial Party committee, Zhang Pixu, later the Lanzhou plant's Party secretary, was political commissar of an army division there and secretary of one of the province's prefectural Party committees. Liu Zhe, the third member of the Lanzhou directorate, became the deputy secretary of the plant's Party committee. In the revolution, he had fought alongside Wang Jiefu and worked with him in Shandong Province until the Party sent Wang to Sichuan. Finally, Wang Zhongfan, a veteran logistics officer from the Korean War, became a director of a Third Ministry factory (number 120) and had worked in another nuclear plant (number 403). In Song's first year in the ministry, he had put Wang Zhongfan in charge of planning the building of the nuclear plants in the northwest. Song personally selected him to become the Lanzhou plant's first deputy director.[53] When the Lanzhou four acted, they could call on network contacts all the way up to the Politburo.

On May 31, 1958, Deng Xiaoping, a Second Field Army leader and by then the Party's general secretary, approved one of their first recommendations, the siting of the plant. The timing of this approval could not have been worse. In this same month, the Second Session of the Eighth Party Congress endorsed the new "general line for socialist construction," thereby giving Mao Zedong the green light for his Great Leap Forward. The impact on Lanzhou was immediate, like "a gale pounding at the construction of the plant." Perhaps the most damaging blow came during Wang Jiefu's absence on a trip to Beijing. He had barely arrived when he received a call from his Lanzhou colleagues demanding that they be authorized to "launch satellites," which in the idiom of the day meant to let the general citizenry, "the masses," improvise. Wang refused and forbade them to tamper with the plant's original designs. But his refusal ran counter to the political movement being fanned by Mao Zedong, and the Lanzhou staff and workers, snubbing their inexperienced director, began to launch their satellites. These actions, it is alleged, resulted "in over 290 accidents, of which more than 20 were ranked as major." Twenty-six years later, Wang Jiefu bitterly remembered the heavy economic costs and concluded that the mass movement had set the project back six months.[54]

The increasing high-level disagreements with Moscow that we discussed earlier intensified the impact of the domestic upheaval on the plans for Lanzhou. Soviet specialists from May 1958 on had advised on the preliminary design of the gaseous diffusion facility, but then Moscow refused to accept Chinese trainees in comparable Soviet plants. As a result, the nuclear ministry had to establish a gaseous diffusion laboratory in the Institute of Atomic Energy's Research Department 615. In October, the department had outfitted the lab, and the Soviet Union agreed to provide "relatively comprehensive training" in it.[55]

As the turmoil of the Great Leap Forward continued, technicians at Lanzhou and other such plants "wantonly planned to modify the nuclear industrial equipment provided by the Soviet Union." They disassembled and modified Soviet-supplied electrical machinery in Lanzhou's main building "on the excuse of performing technological innovation." The Second Ministry of Machine Building, worrying about the spread of the "innovation" campaign, tried to stop such mischievous independence—and finally Minister Song Renqiong decided to appeal to Mao Zedong.

In the Song-Mao interview, the Chairman said: "Just like children learning writing, first write in regular script and then write in cursive script [*xian xie kaishu hou xie caoshu*]."[56] Contrary to its use in other settings, Mao's "first write" instruction this time was interpreted to mean that Chinese workers must not modify Soviet-supplied equipment until they had mastered the technology and operation of that equipment. Song returned to the ministry and issued an instruction forbidding any further unauthorized revisions of the Soviet plans or installations. In the same spirit, Premier Zhou Enlai endorsed this interpretation and issued the so-called *san-gao* (Three Do's) order to the nuclear industry, instructing it to work in line not only with political principles but also with scientific discipline.

In late 1959, with the mass turmoil still engulfing the plant, Wang Jiefu proceeded to develop more appropriate security and administrative systems for the projected enrichment plant. He began by criticizing the stringent security regulations imposed on the plant, and the anecdotal history of the nuclear program suggests that much of the problem came from his Soviet advisers during the building of the "main technological process workshop," which would perform the actual enrichment. During the summer, Sino-Soviet relations had worsened, and soon after China's National Day, October 1, the ministry directed Wang to concentrate on completing the workshop. The harsh Gansu climate worked against the Lanzhou workers, however, as they installed "the dense spider web of pipes with the required vacuum." The workshop's interior had to meet exacting temperature, cleanliness, and humidity standards, and the region's extreme temperatures, pervasive dust, and low humidity placed maximum stress on the plant's marginally skilled construction force. Finally, on December 18, 1959, the workers finished the external shell of the workshop, and then, just as the rush to install its critical equipment began, the Soviet advisers imposed tighter security controls.[57]

In essence, the Soviet regulations stipulated that design drawings supplied by Moscow could not be brought to the construction site, and that all technical data would have to be stored in secure vaults elsewhere. According to the Chinese account, the designs duplicated a set of plans of a Soviet plant in the Urals, leading the Soviet advisers to worry about the secrecy of their own plants at home. These advisers consequently permitted only a few cleared Chinese cadres to have access to any classified data. By and large, the cadres who were cleared, among them the four members of the directorate, had minimal technical competence. The Soviet specialists denied the on-site construction units access to the blueprints vital for installing the enrichment equipment in the main workshop. Recognizing the stupidity of this situation, Wang Jiefu challenged the Soviet-enforced codes and insisted that all necessary data be delivered to the relevant workshops with access given to all who had a need to know. When the Soviet advisers blanched, Wang promptly appealed to his network connections in Beijing for relief, and Song Renqiong obliged by blessing his protégé's proposals.[58]

With the completion of the external frame for the technological process workshop in December, the Chinese expected the Soviet advisers to approve delivery of the workshop's "main machinery" imported from the Soviet Union. This time the advisers stalled because the plant allegedly did not meet minimal standards for cleanliness. Chinese sources do not specify the meaning of main machinery, though it is likely that the definition included critical instrumentation and barrier components within the diffusion elements. As we saw in the account of the work of Research Department 615 at the Institute of Atomic Energy, the Chinese did not possess the high-quality nickel and had not yet mastered the nickel-alloying techniques for the production of dependable barriers. Moreover, the Chinese sources, while itemizing other major problem areas, do not mention the diffusion elements as such as a source of difficulty, but do state that production of replacement barriers began in 1965.* Direct Soviet aid at this stage would have been decisive, and for this reason the Chinese have expressed unusual irritation with the Soviet procrastination.

Whatever the true explanation for their decision to stall, on the day the advisers came to inspect the main workshop, they pronounced it unfit for their equipment. It was dirty. Then, as if to imply that the Chinese might experience a very long wait, because cleaning the workshop would prove an impossible task, they said, "If the cleanliness meets the accepted standard, you may start the installation this very day."[59] Wang Zhongfan and Liu Zhe, ignoring the slight, immediately turned out the work force. They mobilized 1,400 workers on a non-stop shift to render the plant spotless. The next day, when the advisers returned, they expressed amazement at the transformation: "It seems that you have performed wonders. We are convinced you have made magical changes." The flabbergasted advisers ordered the immediate installation of the equipment.[60] The many "drastic steps" taken by the Lanzhou directorate had worked, and the main workshop was but the first of eight major facilities "that stand tall and upright like mushrooms." A new railroad bridge for the plant soon spanned the Huang He to link the plant to the workers' residential area.[61] The "battle of the main workshop" was the turning point.

Wang Jiefu's next major hurdle was to streamline the management of the Lanzhou project. The Chinese, one official history says, "usually built a plant according to a [tripartite] system that put construction administration under . . . the plant, the construction unit, and the design unit."[62] By 1960, Wang had concluded that this system was unduly bureaucratic and time-consuming, and requested permission from the ministry to unify all design, construction, and installation

---

*The Chinese began research on the diffusion barriers in 1960. The scientist Qian Gaoyun organized 14 specialists at Research Department 615 to conduct this research, and over the next years they designed several types of barrier components. In 1965, the ministry approved the production of two types of components that would be used as replacement parts for Lanzhou and to outfit a new gaseous-diffusion plant in Sichuan. Later, the ministry assigned the Physics and Chemistry Engineering Academy in Tianjin to continue research on the technologies of gaseous diffusion. Li Jue et al., pp. 388-89.

work under the plant directorate. He asked to be made the general director of the project, and the new minister, Liu Jie, agreed. Wang's "three-in-one administrative" system "dramatically stepped up the rate of progress" at Lanzhou and became a model for managers at plants elsewhere. His success coincided with a key ministry directive in April giving the highest priority to the Lanzhou project.

During the 700-some days from the date construction began to the installation of the industrial equipment in the main workshop, the plant directors time and again confronted rigid conventions. Each time they altered, sometimes broke, the existing rules and drove themselves and their subordinates harder. To do so, they needed protectors in Beijing, and the powerful network ties formed long before paid off.

National and Local Reactions to the Crises of 1960

As we have seen, the Chinese anticipated the break with the Soviets, and had begun providing for the loss of the Soviet advisers months before the last of them withdrew. At the enlarged meeting of the Politburo in January 1960, the Party adopted an emergency decision on developing the atomic bomb without foreign aid.[63] Another decision somewhat earlier called for solving the key technical problems in three years, building the bomb in five years, and attaining "appropriate [nuclear weapons] reserves" in eight years.[64] Mao declared that the future of the nuclear program would determine the destiny of the state.[65]

Following the January meeting, the Second Ministry ordered the transfer of 106 experts from research and engineering positions to work in the nuclear weapons design facility (Ninth Academy), as well as in other facilities in the nuclear industry, such as the plants in Lanzhou and Jiuquan Prefecture.[66] In July, shortly after receiving Moscow's letter announcing the pullout, the ministry had hastily called Zhang Pixu, the Party secretary of the Lanzhou plant, and others involved in the nuclear program to a meeting in Beijing to expose the full story of "the Soviet government's perfidious actions" in abrogating or moving to abrogate agreements, withdrawing its specialists from all ministry programs, and halting deliveries of equipment and technical data. The ministry directed all facilities under its command "to adopt emergency measures." Since there were still some Soviet advisers at Lanzhou, the technicians there were ordered "to master new skills before the complete withdrawal of the Soviet experts from the plant." But they had little chance to do so: the last five advisers left only a few days later; Wang Jiefu personally escorted them to the airport.[67]

That same afternoon, Wang called a meeting of all section-level chiefs and above to gain control of the deteriorating situation. He had just worked out and had the plant Party committee approve a "nine-point law," the details of which are not known, but which was intended to impose rigid control. The Party committee made it clear that no one could violate the specific stipulations listed in the nine-point law without punishment. The issuance of Wang's law had the desired effect. It reportedly helped the Lanzhou workers adapt to the changed environment and restored a "calm, unruffled mood." The iron fist ruled at Lanzhou more than in any other part of the nuclear program.[68]

Wang's problems, however, went beyond those caused by the Soviet defection. Stunned by the advisers' departure, personnel at the Lanzhou plant were already enduring personal hardships brought on by the "three hard years" following the Great Leap Forward. According to one report, "In 1960, China's nuclear industry faced an extremely difficult situation. The construction of most of the workshops was not completed and some equipment delivered there ahead of schedule was not installed. . . . China had very few specialists who had mastered the technology of producing nuclear materials, it had great difficulties in both human and material resources, and the atomic bomb was facing the trial of success or failure."[69] Personnel at the Lanzhou plant especially suffered from the effects of the food crisis then striking all China; dropsy afflicted two-thirds of the staff and workers. In despair, some of the Lanzhou personnel recommended that all the nuclear-program workers located in Gansu and Qinghai, "where they were confronted with the most arduous conditions," move to South China. One cadre argued that they could return to the northwest after the famine ended. Wang rejected the very idea of any retreat and induced Liu Jie, who had been assigned to replace Song Renqiong as head of the Second Ministry, to back him up.[70]

The belt-tightening at Lanzhou disproportionately fell on the women and children and on Party members, for the ordinary staff and workers got special supplementary food allowances. Even then, "the wives and children . . . reduced their own grain ration and saved food for their husbands so that they could engage in hard work." And Wang Jiefu himself sought to set an example for the Party members by joining in workers' expeditions to forage for wild herbs.[71]

In Beijing, Politburo leaders had only words of defiance for Moscow and of encouragement for their subordinates. They could do little else in a practical way to rescue the program. For his part, Mao Zedong on July 18 looked for the positive side of Moscow's break with China:

"It is just as well that Khrushchev doesn't give us sophisticated technical assistance. If he were willing to do so, then it would be very difficult for us to repay the debt."[72] His declaration set the tone for Beijing's emergency response to the crisis.

Premier Zhou in turn reissued his instruction "*Mozhe shitou guohe*" (In crossing the river go stone by stone) to implement the emergency policy. By this instruction, he meant that those responsible should proceed in an orderly and incremental way to formulate a concrete nuclear weapons development plan.[73] Deng Xiaoping as Party general secretary had already promulgated his own instruction to the program: "The relevant line, principle, and policy have been decided. Now it all depends on your work. Just go ahead boldly with your work. You can claim all the credit for yourselves if you achieve successes, and you can ascribe your errors to the Party Central Secretariat if you commit mistakes."[74]

To spread the word of the Party's decisions and instructions, the Politburo directed four senior officials—Deng Xiaoping, Peng Zhen, Chen Yi, and Zhang Aiping—to visit Lanzhou and other nuclear energy plants and "to communicate Premier Zhou Enlai's [previous] instruction to the Party rank and file: 'Starting from the very beginning, we will produce our own atomic bombs in eight years with our own hands.'"[75]

Mao's call for self-reliance struck a nationalistic chord among Chinese living abroad. The physicist Zhou Guangzhao, for example, then at the Joint Institute for Nuclear Research in Dubna, wondered whether a self-reliant path for the atomic bomb project was feasible. Zhou, the head of the Chinese experts' group at Dubna, invited his countrymen to a discussion meeting, after which they composed a collective letter to the State Council. They wrote of their unqualified endorsement of Mao's call for self-reliance in the nuclear program and expressed their willingness to "change their professions to meet China's need even at the cost of giving up research on basic theory in which they had been engaged." Beijing enthusiastically responded and promptly ordered the Dubna scientists back to China to join the weapons program.[76]

As all this was proceeding, the Central Committee, the State Council, the Central Military Commission, and the ministries and commissions in the nuclear program jointly drafted eight major directives on the crisis. We know little about the content of these directives, but their thrust was to "give preferential treatment in every possible way to those devoted to China's atomic bomb development program." They also ordered the relevant provincial and municipal Party committees "to support the [Lanzhou] plant with timely provision of additional food even though they were confronted with a most difficult situation."[77]

The Second Ministry published its own orders to stem the crisis. It reissued an earlier sixteen-character directive: "*Zili gengsheng, guo jishuguan, zhiliang diyi, anquan diyi*" (Solve technical problems using self-reliance and attach primary importance to quality and security). The ministry added Zhou Enlai's stone-by-stone edict for good measure and dispatched one of its vice-ministers, Yuan Chenglong, to Lanzhou.[78] Yuan's stay seems to have continued the flow of imperatives, for he helped the Lanzhou officials "to sum up the productive characteristics of the plant as the 'five major guarantees' and the 'five major continuances' and to teach the staff and workers the absolute importance of production security." The plant's Party committee, for its part, focused on technical problems and ordered its personnel to "go about things steadily and surely; go ahead with your work on the basis of experiments."[79]

What all the high-level policies, instructions, and directives really meant to Wang Jiefu and his colleagues remains unclear, though apparently they considered the outcome relatively positive. They agreed the outlook was promising because the ministry "had made certain to list the enrichment plant as the most important unit of all in the key projects" under its administration. With this priority affirmed, they concluded, or perhaps hoped, that "it was possible to complete the construction of the plant." The Lanzhou four based their assessment on Beijing's implied commitment that the plant would secure the people and material it required, that the equipment on order from other places would still be forthcoming, and that they already possessed the essential technical data from the Soviets. The need for a high priority was real: in early 1961, survey teams organized during Yuan Chenglong's stay had uncovered 1,395 technical problem areas, such as missing equipment and defective valves and pipes. The directors traced the most troublesome unsolved problems to the inexperience and low technical level of most of the staff, but they had confidence in their top scientists and engineers—and in themselves.[80]

Despair and stocktaking gave way to a flurry of activity, a rallying to a great national cause endorsed by Chairman Mao himself. To build the bomb, the Lanzhou scientists now had to assemble and absorb the data on technologies only the Soviet experts had previously possessed and fully understood. They faced a jumbled pile of pipes and machinery that the departing Soviet specialists had ordered dumped in their wake. To begin to make sense of the disarray and then to assemble the main workshop and make it operational, the Chinese had to read the new material, grasp its content, and apply their instant knowledge in the workplace.

In the process of doing so in the succeeding weeks and months, they "acquired over 470,000 [items of] technical data and overcame 157 technical problems."[81] The official ministry history reports that in order to solve the myriad problems that faced them, the plant directors ordered their engineers and technicians to travel throughout China in search of technical help and special equipment. "For example, 81 plants, colleges, and research institutes scattered in 22 municipalities manufactured 833 kinds of various special-purpose equipment, instruments, and components for the enriched uranium plant; 237 plants scattered in 44 municipalities gave priority to the manufacture of 432 kinds of standard equipment, machines, and instruments, and to the production of over 11,600 kinds of material for the enriched uranium plant."[82] This basic construction and installation work was essentially completed by the end of 1961.

Building and properly installing all the machinery represented only part of the immense task.[83] The process of gaseous diffusion would subject that machinery, especially the thousands of pumps, to extreme pressures and the highly corrosive effects of uranium hexafluoride. The pumps would require very special lubricants about which the Chinese knew next to nothing. Prior to their departure, the Soviet security officers at Lanzhou had stored these lubricants in locked vaults to which the Chinese had no access. When the departing Soviets took the lubricants with them, the ministry called on Hou Xianglin, one of the few advanced petroleum chemists in the country, to find a substitute. When his efforts eventually paid off, one more critical bottleneck was removed.

The abrupt shift to complete self-reliance reportedly cost the Chinese an additional 700 days, but the building of the Lanzhou Gaseous Diffusion Plant had become a national crusade. By mid-1963, the Lanzhou engineers and scientists had mastered the key technologies to isolate isotope 235 and had begun to piece together the stacks of diffusion equipment into an intricate system. The timing of their success just fit the scheduled shipping of the more than ten tons of uranium hexafluoride from the Institute of Atomic Energy outside Beijing.

National Controversy and the Fifteen-Member Special Commission

High-level arguments set in motion in 1960 continued even as the work proceeded on the nuclear program. The moment had come for a major national reorganization. The issues in the debate were complex and often involved intense differences among China's most powerful political and military personalities.[84] At the institutional level, the debate masked strong differences of viewpoint between Nie Rongzhen's Defense Science and Technology Commission and lesser-known organizations responsible for the defense industry.

In December 1959, roughly a year after the formation of Nie's commission, the Central Military Commission created the National Defense Industrial Commission (NDIC) to manage the conventional weapons industry. The NDIC lasted until September 1963, when, according to Marshal Nie, "the tasks originally assigned to it and the departments originally under its jurisdiction were all merged into the National Defense Industry Office." That office, known mostly by its abbreviation Guofang gongban, or NDIO, had been established in November 1961. From 1961 to 1963, Nie continues, "The functions and powers of the [principal] organizations were as follows: the Defense Science and Technology Commission was in charge of scientific research on weaponry and military equipment; the National Defense Industrial Commission was in charge of the production of weaponry and military equipment; and the National Defense Industry Office was in charge of coordinating and maintaining relations between the scientific research and production of weaponry and military equipment."[85] Marshal He Long headed the NDIC, and the chief of the General Staff, Luo Ruiqing, with Zhao Erlu as his principal deputy, led the NDIO.* The NDIC and NDIO, controlled by two powerful leaders, had built their support network around China's president, Liu Shaoqi, and the personal ties among these three deepened.

Prior to 1963, the NDIC/NDIO oversaw military production in the conventional weapons field, but shared responsibility with Nie's research establishment for nuclear weapons development. The order establishing the NDIO had given it principal authority over the Second Ministry. After the merger of the NDIC into the NDIO in 1963, the NDIO increased its influence, for, as noted in Chapter 3, it oversaw

---

* Zhao Erlu previously had served as the minister of two machine-building ministries. The other deputy directors were Minister Sun Zhiyuan of the Third Ministry; Fang Qiang, the deputy commander of the navy; and the minister and vice-minister of the Second Ministry, Liu Jie and Liu Xiyao. Li Jue et al., p. 567.

the Second Ministry's production of nuclear fuels, plant construction, and industrial production.

The NDIC/NDIO and Nie's commission had competing as well as changing powers. Because the Soviet Union had provided most of China's conventional weapons and related technologies after the Korean War, the NDIC/NDIO had founded few facilities for military research into these weapons technologies. Nie Rongzhen, however, had stressed the need for such facilities in the nuclear and missile programs, and had pushed for them to be created under his commission and the relevant nuclear and missile ministries. As the priority for the nuclear program increased after 1961, the NDIC/NDIO network controlled (or, in the case of the Second Ministry, partially controlled) many of the military industrial ministries, and, with tensions aggravated by the declining economy, its policies came more and more into conflict with Nie's network.

Many Chinese acknowledge the existence of the rivalry between the two powerful networks, though they differ on the precise points at issue between them. The crises with Moscow and the self-inflicted wounds of the Great Leap Forward intensified the rivalry even as all agreed on the goals of national unity and general military preparedness. The two networks divided the top military elite at a time when the nuclear program required the fullest cooperation. Even minor decisions precipitated tense bargaining and, quite often, the intervention of top-level Politburo figures.

The precise reasons for the sharpening of the debate in the early 1960s remain unclear, but the worries of the regular armed forces about the Indian border and about American forces moving into Vietnam played a role. All military men, remembering the Korean War, wanted more and better conventional weapons. Mao's slogans about "man over weapons" did nothing to modify their viewpoint. Most of the conventional arms in China's arsenal were barely upgraded versions of Soviet-supplied weapons from the early and mid-1950s, and because events on the Indian frontier and in Indochina were becoming more warlike in the early 1960s, the army's commanders fretted about the likely ineffectiveness of that arsenal in modern warfare.

Nevertheless, in the midst of the "three hard years" (1960-62), the leadership had allotted most of the research-and-development budget to the strategic program, and the military industrial ministries outside that program preferred to take few risks, whether economic or political, with next-generation designs of tanks, planes, and guns. At first, officials in the NDIC and its subordinate ministries would merely snipe at the Defense Science and Technology Commission for monopolizing the defense research funds and institutes, but when the pressures mounted within the army for advanced conventional arms, these officials turned to Liu Shaoqi and Mao Zedong to redress the imbalance. The contest between the strategic and conventional weapons networks continued unabated for almost two decades, and even with the purge of Liu, Luo Ruiqing, and He Long in the Cultural Revolution, the tug of war between them for the control of research staffs and budgets did not formally end until August 1982.*

Issues of the army's secret *Work Bulletin* from January through August 1961 provide further clues to the controversy. In October 1960, according to the *Bulletin*, the Central Military Commission resolved that there should be "a correct handling of the relationship of man to weapons."[86] The resolution repeated Mao's injunction that "weapons are important elements of war, but they are not decisive"; asserted that his view would remain true in a future war fought with long-range and nuclear weapons; and stated that "the physical atomic bomb is important, but the spiritual atomic bomb is more important."[87]

The following January, a senior military officer argued that the "problem immediately affecting the will of the men to fight is not mostly the problem of war and peace." The real problem was the disastrous internal situation resulting from the "three hard years": "Concretely speaking, it is the problem of economic life."[88] Nevertheless, as Marshal Ye Jianying said at the end of the month, war might break out, and China "must be prepared to meet any possible sudden incident."[89] So far, Ye continued, Minister of Defense Lin Biao had responded to this threat by emphasizing the strategic weapons (the so-called super) program and the militia. But in respect to the "equipment of our own Army units, their training, and war preparation," Ye thought the state of affairs "very dangerous." He at once praised Lin's new emphasis on training and condemned the emphasis on nuclear weapons:

If a war should come upon us in the next few years, what kind of weapons can we primarily rely on? . . . Here arises the problem of the relationship between conventional weapons and super weapons. . . . Though the power of atomic weapons is great, they can only attack the other party's centers in strategic air raids and destroy its economic potentialities. Afterwards they can

---

* At that time, the Central Military Commission amalgamated the NDIO, the Defense Science and Technology Commission, and the Science and Technology Equipment Committee of the Central Military Commission into the Commission of Science, Technology, and Industry for National Defense. *Renmin Ribao*, Aug. 24, 1982.

only be used primarily according to their power, as firing vanguard before an attack. To resolve the battle, to cut down the enemy's living strength, to capture positions, and to achieve victory, it [China] will still rely upon the ground force, the army, and conventional weapons.[90]

In the final analysis, Ye supported the continued development of "super" weapons, but clearly favored giving increased priority to the conventional forces.

In July 1961, He Long and Nie Rongzhen attended a high-level Defense Industry Conference at the resort town of Beidaihe, east of Beijing. The principal agenda item quickly became the future of the nuclear program. Allegedly backed by Liu Shaoqi, the NDIC group sought to discontinue or delay both the nuclear weapons program and the missile project, a move that was warranted by the effects of the "three hard years," the previous "erroneous policies," and the cutoff of Soviet assistance. Nie writes: "Under such a difficult situation, a sharp contradiction emerged concerning whether to give continuing impetus to this sophisticated defense program." Several members of the NDIC group contended that the program should merely be slowed down; others insisted that it "should be suspended because its immense investment would impede the development of other parts of the national economy." Moreover, in respect to the military program, the nation should concentrate its energies on "the development of aircraft and other conventional equipment."[91]

Some Chinese sources credit Lin Biao with making a key phone call to Mao Zedong in which he intervened on Nie's behalf to preserve the strategic program. For whatever reason, Mao decided to halt any precipitate move to scuttle the strategic program even as the Beidaihe conference proceeded. At his villa in Hangzhou, the Chairman told his secretary to phone Nie with a directive: "China's industrial and technical level lags far behind that of Japan. What should be our guiding principle [for developing the nation's strategic program] deserves to be studied carefully. In August, I will discuss this matter with you."[92] The enigmatic leader had spoken, and Nie knew that no final decision could be reached at Beidaihe.

Nie used the breathing spell to mobilize his own camp. He immediately gathered all the leading cadres from the strategic weapons program who had attended the Beidaihe conference and told them to review the history of the program thoroughly, then assess its future. They had to build a case for Mao and the Politburo.

In doing so, the leaders of the program concentrated on the years 1958 to 1961, years of achievement and a period for which Nie Rongzhen could claim special credit. In this period, Nie's associates concluded, the Second Ministry "had already collected thousands of professional engineers and technicians who had graduated from universities." They reviewed the progress in uranium mining and processing, and noted that even the "assemblers of atomic weapons" were in place. The ministry had mobilized a corps of outstanding scientists to design a prototype fission weapon, and they had already made important breakthroughs. The goal of a completed bomb design by the end of 1963 seemed attainable, and they had in hand a Central Committee directive of July 16, 1961, calling for a strengthening of the bomb program. After Nie made his report on the status of the program to He Long (in his capacity as vice-chairman of the Central Military Commission and member of the Politburo), they jointly submitted the report to Mao and Zhou.[93]

Following procedure, the report was put on the agenda of a special meeting of the central leadership in the summer of 1961. One of the most famous quotes to come from China concerning the atomic bomb can be traced to this meeting. Chen Yi, then the minister of foreign affairs, remarked that the development of the strategic weapons program should continue at any cost, "even if the Chinese had to pawn their trousers for this purpose." Chen added: "As China's minister of foreign affairs, at present I still do not have adequate backup. If you succeed in producing the atomic bomb and guided missiles, then I can straighten my back."[94] This was the kind of argument that would win the day within Mao's Politburo. The Chairman announced his decision: "We should make up our minds to develop sophisticated technologies. We can't relax our efforts or discontinue [the sophisticated defense projects]."[95] Nie left the meeting with a resolution supported by Mao to accelerate the strategic weapons effort. He had won.

In the nuclear part of the program, the Party's decision forced Nie to begin considering how to develop an operational strategic weapons system, not just an explosive device. His commission now needed to formulate guidelines that would stress a more systematic and integrated approach for the overall weapons program. The guidelines, as issued, called for combining scientific research with production and for developing strategic and conventional weapons, but gave first priority to research and strategic weapons. They also called for the completion of the Lanzhou uranium enrichment plant, the design of the bomb "within four years," and the simultaneous development of the missile and aircraft nuclear delivery systems.[96]

As a result of the Party's decision in the summer, the Central Military Commission directed the army to prepare its troops to carry out missions using these future strategic systems.[97] Anticipating the impact of the projected deployments of nuclear weapons, the Military Science Academy commented on the committee's trial "Combat Rules and Regulations." The academy advocated the formation of defensive organizations "against atomic, chemical and bacteriological weapons," and urged the army's higher-level units to "master well and use new technical weapons, . . . utilizing skillfully various kinds of firing power and the effect of new weapons for a surprise attack."[98] In mid-1961, Beijing's military planners began to operate on the assumption that long-range nuclear weapons would enter China's arsenal within a few years.

The policy laid down in the Central Committee's resolution reinforced that assumption and at the same time called for greater coordination between the strategic and conventional forces. In November, as we have noted, the Politburo established the National Defense Industry Office (NDIO) and assigned it the mission of promoting coordination and cooperation throughout the military-industrial system. The NDIO, however, could not reconcile the personal and attitudinal animosities, which continued to fester until late 1962. Those charged with upgrading China's conventional forces felt left behind. Thus, at about this time, Luo Ruiqing, NDIO head and chief of the General Staff, asked Mao to create a new coordinating body to erase the persistent inequities in order to ensure that the strategic program would receive the support it needed from the NDIC/NDIO system. He proposed that "a special leading body should be set up to take charge of China's defense scientific and technological program." In addition, he volunteered for the powerful position of office director. Mao approved this report on November 3, writing on it: "Very good. Act accordingly. [All departments concerned] should vigorously carry out coordination and cooperation [*dali xietong*] so as to complete this [strategic] program."[99] This came to be known as one of the Chairman's "great calls" (*weida haozhao*). In response, Liu Shaoqi quickly convened yet another special meeting of the central leaders involved to consider organizational ways to bring about coordination and harmony among the competing defense establishments.[100]

On November 17, 1962, the Party Politburo, on the recommendation of Zhou Enlai, created a new leading body for this purpose:[101] the Central Special Commission or, as it was usually dubbed, the

Fifteen-Member Special Commission.* Zhou became head of the commission, but Luo Ruiqing, as office director, and Zhao Erlu (who had headed the ministry in charge of conventional weapons and was Luo's deputy at the NDIO) were responsible for the commission's day-to-day business. Many Chinese believe that the creation of the commission put Nie at a disadvantage in the competition with He Long and Luo, and that the enduring political struggle explains important aspects of the inner-Party attacks on Luo just prior to the Cultural Revolution and on He Long soon after it began.[102]

Whatever Nie's personal attitude toward the commission may have been, he dealt with it in a straightforward professional way. The Party's leaders obviously wanted the members of the commission to represent all the principal organizations connected to the strategic program so that, as Mao put it, all the relevant departments would "carry out coordination and cooperation." Mao's stand temporarily stopped the feuding.

The commission's original fifteen members represented virtually all elements of the central military-industrial establishment:[103]

Zhou Enlai, member of the standing committee of the Politburo and premier

He Long, member of the Politburo, vice-chairman of the Central Military Commission, vice-premier, and director of the National Defense Industrial Commission

Li Fuchun, member of the Politburo and Central Secretariat, vice-premier, and director of the State Planning Commission

Li Xiannian, member of the Politburo and Central Secretariat, head of the Ministry of Finance, and vice-premier

Nie Rongzhen, director of the Defense Science and Technology Commission and of the State Science and Technology Commission, vice-premier, and vice-chairman of the Central Military Commission

Bo Yibo, alternate member of the Politburo, director of the State Economic Commission, and vice-premier

Lu Dingyi, alternate member of the Politburo and vice-premier

Luo Ruiqing, secretary general of the Central Military Commission,

---

* There are several names for this commission in the literature. The official name appears to have been the National Defense Industry Special Commission (Guofang gongye zhuanmen weiyuanhui), which was often condensed as Central Special Commission (Zhongyang zhuanmen weiyuanhui). The commission's office was located in the NDIO building, with Luo Ruiqing the director, and the following the deputy directors: Zhao Erlu, Zhang Aiping, Liu Jie, and Zheng Hantao. See Gu Yu, "How to Tackle the Organization of Key Problems"; and Li Jue et al., pp. 47, 568.

member of the Central Secretariat, head of the NDIO, chief of the General Staff, and vice-premier

Zhang Aiping, deputy director of the Defense Science and Technology Commission and later director of the First Atomic Bomb Test Commission

Zhao Erlu, former head of the Ministry of Machine Building (Second and First) and deputy head of the NDIO

Wang Heshou, head of the Ministry of Metallurgical Industry

Liu Jie, head of the Second Ministry of Machine Building

Sun Zhiyuan, second secretary of the Party committee of the NDIC and head of the Third Ministry of Machine Building (at that time the ministry in charge of conventional weapons)

Duan Junyi, head of the First Ministry of Machine Building

Gao Yang, head of the Ministry of Chemical Industry

Backed by the Politburo itself, the commission monopolized the power to make final decisions in the nuclear weapons field.

Still considered a traitor and pariah, Minister of Defense Lin Biao is never mentioned in post-1971 published sources in connection with any of the defense programs of the 1950s and 1960s, though his role was clearly a major one. He had begun to supervise the day-to-day operations of the Central Military Commission in the early 1960s and may have served on the Fifteen-Member Special Commission for a brief period. Also close to the commission, but not on it, was Liu Yalou, commander of the air force. Despite some omissions, the commission drew together virtually all the officials responsible for ensuring the successful continuation of the strategic weapons program and its coordination with other weapons organizations.

The new body exerted great power and influence in the next years, according to Nie Rongzhen. "Afterward," he says, "every important test and all existing problems we encountered in the development of this program would be submitted to this commission for discussion and settlement."[104] The commission ordered all ministries, commissions, and provinces involved in the program to appoint a deputy to have a personal link with the Second Ministry and directed the NDIO and Nie's commission to organize joint teams for periodic inspections throughout the program. During its first two years, Zhou Enlai personally presided over nine meetings of the body.[105] But the commission, which reportedly had fewer than ten permanent staff members, probably did not attempt to play a daily role in guiding the program's operations. And some Chinese have called it a mere skeleton. Moreover, it is clear that whatever influence He Long and Luo Ruiqing gained by the reorganization was balanced by the increasing power of Lin Biao, who not only served as minister of defense but was moving to dominate all decision making on the Central Military Commission. By the mid-1960s, Lin more than any other official was exerting his authority throughout the Chinese military.

At the least, the reorganization of November 1962 succeeded in spurring cooperation among lower-level bodies. Nie Rongzhen recalls that the directives from above put pressure on the "five front armies engaged in tackling the key scientific problems": defense scientific institutes, the Chinese Academy of Sciences, industrial departments, higher educational institutions, and various institutes at the provincial level. The secretary of the Leading Party Group of the academy, Zhang Jingfu, for example, followed the national example and created the New Technological Bureau to enforce cooperation throughout the institutes. Mechanisms of cooperation were also put in place between the Second Ministry and the Fifth Academy in charge of missiles. By the end of 1962, the total number of workers in the science system (excluding defense scientists and engineers) exceeded 94,000, of whom 2,800 were considered senior scientists;[106] as we have seen in the discussion of the Institute of Atomic Energy, many of these served as a reserve force for the strategic weapons program. Furthermore, in the early 1960s, more than one-fourth of the scientists in Zhongguan-cun, the science quarter in Beijing, were transferred to the nuclear weapons program,[107] and the development of effective coordinating mechanisms came just in time for the academy and ministry to assign the newcomers appropriately throughout the strategic program.

For the Second Ministry, the reorganization had established a dependable system of operations for the future. In late 1962, ministry officials felt enough confidence in the system to set firm goals for the Lanzhou Gaseous Diffusion Plant and for the drive to attain the first nuclear detonation. They deemed the time ripe to fight the final "engagement" and directed Lanzhou to acquire its "product" six months ahead of schedule: "The decisive question will be determined within fourteen months."[108]

The Final Push for Uranium-235

At the Lanzhou Gaseous Diffusion Plant, only a few officials in the top command knew of the political battles raging in Beijing. One of them was Wang Jiefu. When he received the ministry's directive, he pondered how to produce the U^{235} by the beginning of 1964. Two questions kept running through his mind: "First, was it possible to produce U^{235} at the required standard? Second, was it possible for the plant to acquire the product at the highest speed?"[109] Wang initiated discussions in the Lanzhou plant on these two questions and received an answer that reportedly surprised him. A young scientist, Wang Chengxiao, indicated that there was no technical difficulty in producing high quality U^{235}, but with the original plan it would be almost impossible to shorten the production schedule. Trained in Moscow, Wang had worked in one of the Second Ministry's design academies until 1960, when he arrived at Lanzhou. He took responsibility for drafting a plan to accelerate the enrichment process.

It took several tries before Wang Chengxiao came up with a plan that would yield the required amount of enriched uranium six months ahead of schedule. His final revision, however, provoked yet another round of controversy. Though Wang's draft satisfied his boss, Wang Jiefu, it met with strong objections from his colleagues. One accused Wang Chengxiao of having run wild, and small groups of technicians organized to rerun his calculations. Even then, the bickering continued, and when Wang Jiefu could not stop the controversy at Lanzhou, he sent Wang Chengxiao's calculations to Beijing for confirmation. At the Institute of Atomic Energy, known by the ministry code number 401, two experts verified the figures, and the debate ended. The ministry approved Wang Chengxiao's revised plan, which moved the production timetable ahead to January 1964.

By the time the ministry had affirmed Wang's plan in late 1962, all the Soviet-supplied equipment had been installed, and the following summer, after receiving Beijing's approval, the Lanzhou technicians conducted their first limited diffusion run to test the main processing machinery before putting the entire plant on line. This partial test was needed to determine whether the feed systems for the uranium hexafluoride and the graphite separation elements could withstand the corrosive gases. The more than ten tons of hexafluoride had just begun to arrive from the Institute of Atomic Energy, and up to now the gaseous diffusion elements had not been subjected to any realistic test run.

Because this run ranked as the most difficult feat to date in the nuclear program, Premier Zhou Enlai personally delivered a speech at the ministry to offer his encouragement. He began by reviewing the achievements of the ministry, especially those made after the Soviet withdrawal, and noted that Song Renqiong's recent departure from the ministry had added to the program's difficulties. Zhou acknowledged the country's economic problems and the fact that the burden had fallen disproportionately on places like Lanzhou, where living conditions had been among the worst. He again urged the workers to follow the Party's standing instructions. The ministry immediately transmitted Zhou's message to Lanzhou, in time for the partial test. The workers reportedly felt encouraged by Zhou's exhortation to press on and credited him for the test's subsequent success.

Following the partial test run, the plant prepared for the trial operation of the entire processing system. For this operation, Minister Liu Jie himself flew to Lanzhou to review alternative techniques with the Party committee. Wang Jiefu used the occasion to point to Wang Chengxiao's revised production plan, which, he said, called for a test of each of the plant's nine production units one by one. The minister and the Lanzhou committee agreed.

When the first processing unit began operating in early 1963, the trial run exposed a number of faults in the plan. To ensure production of the enriched uranium, Wang Jiefu called on his scientists and engineers to study these faults and further refine the plan. They spent half a year meticulously revising it, and in late December the Lanzhou directors approved their revisions. The officials went to the main workshop, and Wang Jiefu issued the order to commence. "After they began the operation, [the workers] spent only 72 minutes to complete 18 working procedures. They had succeeded in operating the key units of industrial equipment to produce the U^{235} in quantity."

On the morning of January 14, 1964, Wang Jiefu and his colleagues went to the central control department. Wang issued an order for the enriched uranium to be drawn off automatically into specially constructed containers. The technicians recorded that event at 11:05 A.M.—the target for uranium enriched to 90 percent 235 had been met. The next day, the ministry sent a congratulatory message to the plant and a personal report to Mao. In his customary way, Mao scribbled a "very good" on his copy. Mao's subordinates now turned their attention to the last stages of the nuclear weapons program, designing and assembling the bomb.

SIX

The Design and Manufacture
of the Bomb

The story of the fission revolution that swept the scientific world has many chroniclers, and we shall long debate the revolution's origins and early history. Let us recall that in 1938 two German scientists, Otto Hahn and Fritz Strassmann, discovered nuclear fission. Building on the research of Enrico Fermi, Frédéric and Irène Joliot-Curie, and others, they attempted to create transuranic elements by bombarding uranium with slow neutrons. Lise Meitner and her nephew Otto Frisch authored the report that explained what the Hahn-Strassmann experiments meant and gave the world the term "nuclear fission."

Niels Bohr and Leo Szilard quickly grasped the implications of the fission experiments; Szilard was also among the first to foresee the peril, along with some other scientists, including Fermi and Eugene Wigner. In 1939, the physicists Yakov Frenkel' in Leningrad and Bohr and John A. Wheeler in Princeton, New Jersey, wrote the fundamental papers that remain the basis of theoretical treatments of fission down to the present.[1] Fermi, an Italian emigré physicist, had a flash of brilliance based on an insight of Szilard's. He suggested that neutrons might be emitted in the fission process and, in a chain reaction, induce further fissioning. He foresaw that such a reaction, if uncontrolled, could cause a nuclear explosion. At about the same time, the German emigré physicist Hans Bethe proposed that the energy of stars derives from fusion reactions. In the remarkable decade of the 1930s, scientists had come to understand that the fusion of two light atomic nuclei into a heavier atomic nucleus was theoretically feasible. In both fission and fusion, the overall reduction in atomic mass in the reactions would release energies on an unprecedented scale.

By 1955, of course, the Chinese had ample evidence that the two types of nuclear reactions could yield immense explosions. Soviet and

American tests of both fission and fusion weapons in the early 1950s made Chinese scientists aware of the possible. As one military specialist from Beijing has told us, "we knew the bomb would work; it was the missile program that was an unknown for us." By the early 1950s, weapons makers in the United States and the Soviet Union had aggressively advanced the science and engineering of nuclear explosives, and, to the consternation of official censors, unclassified publications had revealed many principles of nuclear weapons design.[2] Even without Soviet help, Chinese scientists could have learned these principles and, in general, how to apply them.

The Chinese certainly knew, for example, that all nuclear weapons, fission and fusion, require a fission chain reaction. In order to sustain such a reaction, ending in a continuous and violent release of energy, more than one fission (slow) neutron must be present to induce further fission "for each neutron previously absorbed in fission."[3] To achieve a successful chain reaction, bomb designers must produce conditions under which the loss of neutrons is minimized, and the rate of loss is determined by the surface area of the fissile material. "On the other hand, the fission process, which results in the formation of more neutrons, takes place throughout the whole of the material and its rate is, therefore, dependent upon the mass."[4] Critical mass, as it is called, marks the point in the ratio when the chain reaction can become self-sustaining. Bomb designers must fashion "reflectors," which inhibit the escape of the neutrons, in order to reduce the size of the critical mass.[5]

Another key task for the designer involves timing. Stray neutrons exist outside the bomb itself or can be generated in a variety of unintended ways. Such random neutrons might, for example, initiate a premature reaction in a fissile mass. Thus, "it is necessary . . . that before detonation, a nuclear weapon should contain no piece of fissionable material that is as large as the critical mass for the given conditions." The designer must fashion a weapon that can be made supercritical (larger than the critical mass) "in a time so short as to preclude a subexplosive change in the configuration, such as by melting."[6]

Early in their programs, weapons specialists in the United States and the Soviet Union proved two general methods for converting a subcritical system into a supercritical system. One, the gun-assembly method, brings two subcritical pieces of material together in a "gun-barrel" mechanism by employing a chemical explosive to drive one piece into another. The other, an implosion method, "makes use of the fact that when a subcritical quantity of an appropriate isotope of uranium (or plutonium) is strongly compressed, it can become critical or supercritical." This compression can be caused "by means of a spherical arrangement of specially fabricated shapes (lenses) of ordinary high explosive."[7]

Soviet (and American) designers deemed the implosion system necessary for a plutonium weapon, and Chinese scientists, gleaning clues from American and European writings on the bomb, largely worked out the theoretical design of such a system on their own.[8] Persuaded of the general advantages of that system, the Chinese decided to design an implosion-type device for their first weapon, even though it contained U^{235}. When, in 1949, the British began to duplicate the Nagasaki-type implosion device, they compiled a working manual based on their memories of Los Alamos, much as the Chinese were later to collect all available data on the theory of implosion.[9] Unlike their British counterparts, the Chinese bomb makers had no hands-on experience in foreign weapons laboratories and had to translate a rough grasp of the theory of implosion into a working model.

The task before them was an enormous one: to master the implosion technologies, then design and manufacture all bomb components. Among the components to be designed were, first (working from the outside to the center of the spherically shaped weapon), detonators that "operated by impulse from a firing device and involved other auxiliaries like safety switches and arming circuits." These detonators had to create a uniform inward-oriented explosion to squeeze the entire surface of the fissile material uniformly. Thus, "the lenses themselves were carefully calculated shapes, containing a combination of fast and slow [burning] explosive so that transit from the detonator to every point on the inner spherical surface of the lens was simultaneous. The detonation from the lenses then reached a spherical shell of homogeneous high explosive called the supercharge. Within the supercharge was the tamper, which converted the convergent detonation wave into a convergent shock wave, reflected some of the neutrons back into the fissile material and generally increased the efficiency of the explosion." Within the tamper of the Chinese bomb was the sphere of U^{235}, and within that a golf-ball-sized initiator that would create an intense neutron source in the core of the weapon in the microseconds of maximum compression.[10]

The engineering of fission weapons posed substantial challenges for China's scientists because weapons makers had to explore in minute detail alternative and complex design paths. Many of the country's most promising physical scientists, including a substantial number of theorists, were enlisted to help chart the course for that engineering exploration. Their efforts were aided by the continuing American and Soviet test programs. As new nuclear weapons were exploded, the Chinese were able to glean small design clues from careful analysis of sample test debris in the atmosphere and other intelligence. The Chinese knew or could roughly estimate developments in the Soviet and American nuclear arsenals, and in that sense can be said to have profited from these foreign nuclear programs.

The evidence of imitation and innovation that can be found in the Soviet nuclear weapons program emerges, though in different form, in the Chinese program.[11] The Chinese received varying kinds and levels of assistance from the Soviet Union in all phases of their program, and assistance, at a quite general theoretical level, was received as well in the weapons design phase. But as we have seen, the promised prototype atomic bomb, removed from a Soviet train at the last moment, never reached China, nor did the Chinese realize their early hope of drawing on the most advanced Soviet knowledge and technologies. This chapter tells the story of how the Chinese designed their first weapons; in doing so, it also attempts to assess the balance between imitation and innovation.

The Northwest Nuclear Weapons Research and Design Academy (Ninth Academy)

The decision handed down by the Politburo in 1955 set in motion the urgent execution of a nuclear weapons program. Less than a year after the formation of Song Renqiong's nuclear ministry in November 1956, work had begun in many domains, including geology, mining, uranium processing, plutonium production, and uranium enrichment. Perhaps the most critical of these domains was the design group.

In the summer of 1957, Song interviewed Li Jue to head the ministry's planned top-secret Nuclear Weapons Bureau, known only as the Ninth Bureau, and, on January 8, 1958, he assigned him to directly oversee the preparatory work for a design group that he was also to head. The preparatory work completed, Li, first in temporary quarters in Beijing and then in Qinghai, assumed the directorship of that group—the Northwest Nuclear Weapons Research and Design Academy (Xibei hewuqi yanjiu sheji yuan).

The construction of the academy began in 1958, and the design teams could not occupy it for several years. In the meantime, Li created a transitional research facility in Beijing—the Beijing Nuclear Weapons Research Institute (Beijing hewuqi yanjiusuo)—to house the initial work on the bomb.[12] The ministry anticipated that the Beijing facility would concentrate on mastering the Soviet nuclear weapons data to be supplied under the New Defense Technical Accord, but that data and the prototype bomb never came. The infamous storeroom for the promised prototype was there. The institute's second mission would be training, and in the end this and the inaugural weapons research became its primary tasks. Operating in these makeshift quarters only added to the pioneering spirit of the start-up academy, and many of the initial breakthroughs were achieved there.

Few outsiders ever knew of the academy, and those who did, if they spoke of it at all, referred to it only by its code name, the Ninth Academy. Its birth came in the tumultuous year of the Great Leap Forward, and the auguries for advanced science and engineering seemed poor. In the high tide of a national anti-Rightist campaign targeted on intellectuals, China undertook to build the Beijing institute and a special weapons laboratory on the scale of America's Los Alamos and to staff it with the country's best scientists and engineers.

In the Los Alamos case, the physicist who suggested the need for such a bomb design facility, Robert Oppenheimer, also laid down the criteria for choosing its site. "He stressed the necessity for free internal communication and conceded this meant tight controls to prevent leaks to the outside. . . . Inaccessibility was most important in selecting a site. There had to be some rail and road facilities, of course, but since the weapons work was not expected to require a large installation, convenience could be sacrificed for the benefits of isolation." In the secluded Los Alamos hideaway east of the rugged Jemez Mountains and within sight of the more majestic Sangre de Cristo range, Oppenheimer assembled some of the best minds in the world to build a machine of mass destruction.[13] Change some of the names, and the picture of the Los Alamos terrain would fit that of Li Jue's academy in China's remote northwest.

Li, at age forty-three, was no stranger to the frontier regions. At the time of his appointment to the nuclear ministry's Ninth Bureau, he was deputy commander and chief of staff of the Xizang (Tibet) Military Region.[14] Illness or injury had interrupted his command in Xizang, however, and in the summer of 1957 he was recuperating in a Beijing hospital. A college student turned warrior in 1936, Li later met Deng Xiaoping, then the deputy head of the Eighth Route Army's political department, who ordered him into clandestine work among the Kuomintang forces. Thereafter, Li's career included defense production and political work in the People's Liberation Army (PLA) prior to

Xizang and the Beijing hospital ward. Song Renqiong considered him the ideal candidate for the Ninth Bureau and so reported to the central leadership. Upon receipt of higher-level approval, Chen Geng, then deputy chief of the PLA's General Staff, went to break the news to Li. The always direct Chen told him: "Lao Li, the army will give you a send-off in a few days." What surprised Li was the advance notice of a change of assignment, the details of which Chen refused to reveal in the hospital room. Since Chen "only replied with a smile, Li Jue thought Chen Geng was joking [about a new post] and didn't take his message seriously."

Out of the hospital, Li realized he had misjudged Chen when he received orders to report to the Third Ministry of Machine Building. There Song told him that his assignment had been decided by the highest authorities and detailed the Politburo's decision to build the atomic bomb. Song communicated Mao Zedong's instruction to Li: "We are stronger than before and will be still stronger in the future. We will have not only more planes and artillery but atomic bombs as well. If we are not to be bullied in this present-day world, we cannot do without the bomb." Song continued by paraphrasing Mao: "It is really time for China to develop an atomic energy program. We possess the [necessary] natural and human resources. We also have laid a certain foundation of scientific research. We will undoubtedly achieve success in speeding the atomic energy program if we pay close attention to it." In short, China could not do without the bomb. At this point Li, overcome by a wave of emotion at the honor his nation had bestowed on him, accepted the assignment. Song promptly gave the new head of the Ninth Bureau the duty of surveying possible locations for "the nuclear weapons academy, the batch process plants, the warehouse, and the nuclear test base" within the overall program.[15]

The search for the spot to build the nuclear weapons academy took Li and his colleagues over "thousands of li" throughout northwestern China (a li is 0.5 kilometer). With the survey data from this search in hand and studiously analyzed, they finally settled on a locale in Qinghai Province and submitted their recommendation to the "central leadership" (presumably meaning the nuclear ministry and the Central Military Commission). In July 1958, these authorities approved the team's proposal and launched what proved to be a protracted construction effort on the northwestern high-desert plateau.

In 1958, within three months of this approval, "a vast contingent of construction workers" set out for Li's remote spot. They went by rail to the Gansu city of Lanzhou, then by truck over the dusty mountain roads to the Qinghai provincial capital of Xining and beyond, to a wilderness area called "gold and silver sand" lying east of Qinghai Hu (Lake Qinghai), in Haiyan County (see Map 3, p. 110).[16] There they pitched tents and began hacking out building sites for the academy. In this initial "contingent" were over 2,000 soldiers, more than 7,000 peasants conscripted from across the land, and some 2,000 seasoned construction workers.[17] Furthermore, the coming years were to see "thousands of experts as well as scholars who had just returned to China from abroad, and students who had graduated from universities and colleges" trekking west to the academy. To help sustain this steady flow of people, and of materiel, the PLA Railway Corps and highway engineers from the Ministry of Communications rushed to build or repair the network of rail lines and highways into the Qinghai base.[18]

The trauma of the "three hard years," 1960-62, struck directly at the academy during the construction period. As in the rest of China, the most serious lack was food. At a time of particularly severe shortages throughout the country, Nie Rongzhen requested that the navy, the Beijing Military Region Command, and the Guangzhou Military Region Command dig into their supplies and deliver 25,000 kilograms (50,000 catties) of soybeans and canned goods to the Qinghai encampment. Nie in desperation ordered Mongolian gazelles to be hunted to help stave off the famine at the academy's units still in Beijing. Further, such national program leaders as Zhang Aiping, Liu Jie, and Liu Xiyao made a special plea to Qinghai provincial officials to assist, and these authorities complied by sending goats and cattle. On the base itself, Director Li Jue led cadres and logistics personnel in planting potatoes, collecting mushrooms, and fishing.

Yet gradually and painfully, their labors succeeded. At the end of 1962, the academy, laid out according to a Soviet design, had occupied "hundreds of square li" on "a plateau where the air is thin and people breathe with difficulty [and] feel dizzy and asthmatic after they have taken a few steps."[19] Numerous laboratories were completed that year, and still more were under construction. The academy could boast a power plant, machine shops, an explosives workship and test area, dormitories, recreation facilities, and guards' barracks.

The Academy's Leading Group

Li Jue picked three principal colleagues to run his Ninth Academy, first at the temporary institute headquarters in Beijing and then in Qinghai. They were Wu Jilin, Zhu Guangya, and Guo Yinghui. The first, Wu, remains unknown to many "even within China's nuclear industrial community, though he made an outstanding contribution to the manufacture of China's first atomic bomb."[20] Unlike Li, who had left school in 1936 for a life of revolutionary underground work, Wu, from the great city of Chengdu in Sichuan Province, stuck with his studies until his graduation in 1937 as a chemistry major. Living in Kuomintang-controlled territory, Wu Jilin had joined Chiang Kai-shek's armies and moved on to Shanxi Province to teach "anti-chemical warfare popular knowledge" to soldiers on the front line. There he met an underground Communist Party member, who persuaded him to defect to the Communist base areas and to sign up with the Eighth Route Army.

Wu's new military commanders quickly recognized his special scientific talents and transferred him to Yan'an's war industry. After the defeat of Japan, Chen Yi, then commander of the Shandong Military Region, ordered Wu to take charge of the same type of industrial work in southern Shandong, where he remained after 1949 as the director of the Shandong Aluminum Refinery. His managerial decisions at the refinery, especially those to "boldly utilize pre-Liberation engineers and technicians," raised the plant's production but also provoked disgruntled well-placed Party cadres to accuse him of "placing bourgeois intellectuals in important positions." These cadres invented charges of corruption against Wu and ordered his arrest.

When the provincial authorities uncovered this miscarriage of justice, Wu was promptly rehabilitated. Even more fortunately for him, by then his case, as required by bureaucratic procedures, had come to the attention of Chen Yun, head of the Ministry of Heavy Industry. Chen took an interest in Wu and transferred him to the ministry. In the latter half of 1957, Chen (who had been a member of the Politburo's Three-Member Group in charge of the nuclear weapons program) remembered Wu's talents and recommended him to Song Renqiong for the academy post.[21] At first Wu took on responsibility for the academy's organizational work as Li's principal deputy. In late 1962, he assumed the concurrent job of director of a leading technological committee.

The second member of Li Jue's leadership group, Zhu Guangya, went to work for the academy on July 1, 1959.[22] A graduate of the Southwest China Associated University (the name applied to the wartime association of refugee colleges), Zhu stayed on at his alma mater as a teaching assistant. There he also worked with Professor Wu Dayou. After the Second World War, Wu was one of a trio of leading scientists picked to tour the United States, and he invited Zhu to accompany him. That invitation became Zhu's ticket to America and a doctorate in nuclear physics at the University of Michigan. When he returned to China four years later, the new Communist government instructed him to organize nuclear physics courses in selected universities. During the Korean War, he was sent to Korea as an interpreter with the Chinese People's Volunteers at the armistice talks. When Zhu joined the academy in 1959, Li Jue assigned him as one of its leading cadres to assist Wu Jilin in supervising the management of scientific research, a job akin to his development of physics curricula in Chinese universities.

As chief of scientific research at the academy, Zhu sought to combine theoretical and applied studies and paid special attention to training in technical quality control. He is remembered for enforcing high scientific standards and for fostering an academy-wide working style that endured into the 1980s. On a state occasion held to pay homage to special achievements in the program, Premier Zhou Enlai singled out Zhu "for his careful and meticulous spirit of work." Thereafter, Zhu rose steadily in the Party hierarchy through the 1960s and 1970s, to become an alternate member and then a member of the Central Committee.

Like Wu Jilin, Guo Yinghui, the third member of the Ninth Academy's leading group, was a close assistant of Li Jue's, but Chinese sources provide only minimal information on his assignments.[23] From the scattered sources on him, we do know that he, along with Wu and Li, accompanied Soviet experts on a survey trip of the northwest to identify possible locations for the Ninth Academy and for the uranium enrichment plant. He is remembered principally for his contributions to organizational work at the academy, particularly in its earliest stages of development.

While the four members of the leading group could point to important credentials in the industrial and scientific fields, only Zhu Guangya had achieved a reputation as a major scientist. But from what we know of Zhu's many contributions to nuclear weapons development at the academy, these had little to do with basic research. Rather he seems to have served primarily as the link between industrial and scientific projects, providing the knowledge of managing large-scale scientific-technical undertakings. In speaking of this four-man directorate, however, we may be drawing too circumscribed a circle around the academy's organizational and technical leadership, because the directorate drew heavily in its decisions on three senior scientists assigned to the academy: Wang Ganchang, Peng Huanwu, and Guo

Yonghuai. Along with Zhu Guangya, these three men were to become deputy directors of the Beijing institute and then of the academy. A clear delineation between decision makers and staff becomes more justifiable only when we come to Chen Nengkuan, Deng Jiaxian, and Zhou Guangzhao or a less renowned coterie of academy scientists, which would include such specialists as the theoretical physicist Yu Min, Professor Fang Zhengzhi, and Associate Professors Qian Jin and Yu Daguang.

Wang Ganchang, a graduate of Qinghua University, had worked on radioactivity with Lise Meitner, a colleague of Otto Hahn (and a later Nobel laureate), and earned his doctorate in physics from the University of Berlin in 1934.[24] Under his mentor, Wang also studied cosmic rays, an interest to which he returned after the war. Stories that Wang engaged in nuclear weapons research for the prewar German government are untrue.[25] Wang came home before the outbreak of the anti-Japanese war, early enough in 1937 in fact to teach physics in Shandong and Zhejiang provinces until the war drove him to Southwest China. His war effort, his biographers record, consisted of "donating all his jewelry and writing a piece for teachers entitled 'Military Physics.'"[26] After the war, the American government brought him to the University of California as a research scientist in nuclear physics, and just prior to the Communist takeover Wang sailed for home.

In 1952, Beijing reportedly sent Wang to Korea on a secret mission to "collect radioactive material." On his return, he resumed research on elementary particles. Both his politics and his scientific achievements appear to have made him an attractive candidate for the positions to which he was appointed in the next years: deputy director of the Institute of Modern Physics and of its later incarnations as the institutes of Physics and of Atomic Energy. He went to the Joint Institute for Nuclear Research at Dubna in 1956 and served as its deputy director from 1959 until his return to Beijing in the spring of 1960.[27] Although he had intended to concentrate on fundamental particle physics, the government ordered him to engage in research on nuclear weapons and gave him three days to accomplish this change in direction.

Two years after joining the transitional Beijing institute in 1960, Wang was appointed director of a second technological committee of the Ninth Academy.[28] He assumed overall direction of both the experiments on the fundamentals of nuclear detonations and the design and manufacture of the bomb's neutron initiator. Li Jue assigned him to oversee the design and manufacture of China's first atomic bomb.

Peng Huanwu checked into the Beijing Nuclear Weapons Research Institute at about the same time as Wang. We reviewed Peng's biography in Chapter 3. By the time he reached the institute, the man with the two doctorates from Edinburgh had gained a reputation as an eccentric genius as well as the first Chinese to be appointed a professor in Great Britain. "When doing mathematics problems," the anecdotal history records, "he never needed the help of assistants [and] because of his devotion to his work, he didn't marry until very late. He was indifferent about his appearance. Even in Wang Ganchang's eyes, Peng was a legendary figure."[29] The Second Ministry also gave Peng the customary three days' notice to report to the academy's institute in 1960, and two years later, he made the transition from aloof genius to the head of a third technological committee.[30] Qian Sanqiang, Peng's friend and schoolmate, personally vouched for Peng to Nie Rongzhen.

In 1960, another famous Qian, Qian Xuesen in the missile program, recommended the physicist Guo Yonghuai to join the academy's institute and later to head a fourth (and final) technological committee.[31] Guo, a native of Shandong, had shifted his university major from physics to aerodynamics after going to the United States for graduate study. Like Qian, a Ph.D. from the California Institute of Technology, Guo had taught at Cornell University and "was well known for the Poincaré-Lighthill-Kuo (PLK) singular-perturbation method in applied mathematics."[32] Although the U.S. government's role in preventing Guo's return to China after 1949 did not gain the same notoriety as had similar efforts in the case of Qian (who had worked on jet propulsion and rockets), he was not permitted to leave until 1956, following an agreement between the two countries on the repatriation of their civilian nationals.[33] Qian promptly made him a deputy director in his own research organization, the Institute of Mechanics, and recommended him for secret work. Guo, who had refused to participate in classified U.S. projects, leapt at the chance to work for his homeland. He continued working at the Ninth Academy until his death in a plane crash in 1968.[34]

At the time of his death, Guo and an engineer named Long Wenguang (later to become the director of the Design Department of the Ninth Academy) were credited with "great contributions" to the design and testing of the bomb.[35] Most of Guo's work delved into the mechanics of nuclear weapons: structural strength, distribution of pressures, and vibration. On the last, for example, he organized scientific and technical personnel to complete tests on shock, centrifugal forces, and temperature and noise-level variations. Wearing two hats as expert and administrator, the American-trained engineer helped develop the overall research-and-development plan for the nuclear weapons program and taught newcomers to the program the fundamentals of detonation mechanics and warhead design.[36]

We know the histories of only a few of the other scientists who worked at the academy. While most of the credit for designing the bomb has gone to Wang Ganchang, Peng Huanwu, and Guo Yonghuai, certain other scientists, as we shall see, made critical contributions along the way. One was Chen Nengkuan, who had received an American doctorate and published important research papers on the physics of metals prior to joining the academy.[37] In bomb work, he had to shift his professional interests to becoming an expert on chemical explosives and detonating caps, in which role he helped to complete and test the atomic-bomb explosive assembly. Chen was joined by Fang Zhengzhi in designing the initiator for the first atomic bomb, and the two scientists conducted one of the critical experiments on the fundamentals of nuclear detonation.[38] Chen and Fang recruited promising graduates from China's top university science and engineering departments to work as their research assistants.

Deng Jiaxian, who came to the academy as an associate research fellow, was a graduate of the Southwest China Associated University. He too had studied in America, earning a Ph.D. at Purdue University.[39] Deng returned to China a year after the establishment of the People's Republic and helped pioneer the creation of the Chinese Academy of Sciences' Institute of Modern Physics. In 1958, when the Ninth Academy's Nuclear Weapons Research Institute was just beginning to take shape in Beijing (and before the academy's permanent facilities were completed in Qinghai), Deng, who was destined to become a leader of the entire academy in the 1970s, participated in its start-up. His first task was to select the top graduates from China's leading universities, and after a three-month selection process, 28 of the best joined him at the institute in Beijing's western suburbs. Even before Peng Huanwu, Wang Ganchang, and Guo Yonghuai arrived at the institute, the Second Ministry assigned Deng to head a so-called Theoretical Forum to begin explorations on the theoretical design of the bomb.[40]

From then on, Deng displayed a special flair for design work. He became the leader of the theoretical section and spent half a year at the academy's institute "directing four youths in achieving a breakthrough in the solving of an equation concerning the state of materials under conditions of high temperature and high pressure." In this connection he once noted: "If the atomic bomb project is likened to a dragon, then the theoretical design of the atomic bomb should be regarded as the head of the dragon."[41] Later, Deng mobilized over ten engineers to complete the calculations on the weapon's mechanics. The average age of these technicians was twenty-three. Deng's style brought the classroom atmosphere into the design group, and Peng Huanwu encouraged open give-and-take and a minimum amount of protocol or bureaucracy within all groups under his command. What Peng wanted were colleagues who were "scrupulous about every detail" on the job, and Deng Jiaxian as section leader was exemplary in this respect. Years later, Deng's associates would claim that the "orientation" he gave them in the 1950s "remains valid up to the present and is regarded as the standard for us to solve key scientific problems."[42]

Another young scientist brought to the academy was Zhou Guangzhao, a protégé of Peng Huanwu. Zhou had studied at the Joint Institute for Nuclear Research in Dubna and, in 1960, had co authored the letter (described in Chapter 5) that brought him and his Dubna colleagues back to China. He enters the story of the nuclear weapons program in association with Deng Jiaxian. When Deng's team got stuck on the calculation of a particular parameter in the bomb's configuration, Peng asked Zhou to go carefully through their figures. After Zhou did so, he judged the problem to be one of presentation, not invalidity. Although he himself thought it "unnecessary to nitpick the results," he felt they should be verified "so as to convince others." Zhou thereupon "gave full play to his special skill and boldly advanced a new theory to verify the results of the calculation." This in turn brought agreement from Peng and other senior scientists.[43] As this episode suggests, the weight of loftier scientific reputation could color the perceptions of scientific validity on the part of higher authorities.

Scientific verification was a role that Zhou was to play again. On October 15, 1964, the day before the first detonation, an official at the Lop Nur Nuclear Weapons Test Base cabled the Second Ministry to request that the theoretical physicists at the academy's group in Beijing rerun the "last calculation to guarantee the probability for the test's success as above 99 percent."[44] The minister assigned Zhou and two mathematicians to what became an all-night marathon. With the results in hand, the three signed a report to the central leadership certifying that the test would succeed.

From the day in 1958 when the Second Ministry chose the first personnel for the academy to the explosion of the first fission bomb, Li Jue's team worked nonstop to design and assemble China's nuclear weapons.[45] In early 1960, in the course of interviewing Wang Ganchang, Peng Huanwu, and Guo Yonghuai, Vice-Ministers Liu Jie and Qian Sanqiang passed on Zhou Enlai's personal message: "Our sophisticated defense program, which has just started, requires sophisticated expertise and leading scientists. It is a political task [for Liu Jie to assemble them]."[46] Shortly after the three arrived to begin work, several senior leaders, including Chen Yi, visited them and thanked them for reporting to the academy on such short notice. Chen exclaimed: "It is easier for me to hold the post of foreign minister with the support of scientists such as you." To spur themselves on, they later dubbed the bomb that they would design "596" as a "reminder of Khrushchev's perfidious actions" in halting delivery of the prototype atomic weapon.[47] The reference was to the year and the month in which Moscow formally communicated its decision to Beijing—June 1959.

Reorganization of the Academy Complex and Initial Experiments

As the cadre responsible for the academy's overall organization, Wu Jilin had to form a managerial system that would maximize the application of advanced technologies from many branches of knowledge, none of which was familiar to him.[48] Nevertheless, it took Wu only a few days to develop a workable plan. His detailed proposal focused on the primary mission of scientific research, and he organized the academy's main divisions according to the needs of that research.

At the outset, in early 1960, Wu recommended to Li Jue that the Beijing institute organize 13 research sections under four departments: the Theoretical Department, the Experimental Department, the Design Department, and the Production Department. In October, Li abandoned this organizational chart in favor of six research sections for Theoretical Physics, Detonation Physics, Neutron Physics and Radiochemistry, Metal Physics, Automatic Controls, and Warhead Ballistics. To carry out his construction responsibilities for the Qinghai academy, he also created two design sections, one for architectural engineering and one for equipment.

While still in Beijing, work on the bomb gradually progressed from theory to engineering design and trial manufacturing. By the end of 1962, Li decided to reestablish the original four departments and abolish the six research sections. The central leadership in October had approved the transfer of 126 senior scientists and engineers (including Professors Zhang Xingqian and Fang Zhengzhi) to do research on the bomb, and their arrival triggered Li's decision.

Li Jue and Wu Jilin believed that the resurrected departments required additional help because of the rudimentary state and small scale of the research capabilities at the Beijing institute. He proposed the formation of an integrated technological commission to oversee four high-level technological committees, each in charge of specific bomb production tasks. The first (under Wu Jilin and Long Wenguang as director and deputy director) had overall responsibility for designing the bomb's technology; the second (Wang Ganchang and Chen Nengkuan) for testing the non-nuclear components; the third (Guo Yonghuai and Cheng Kaijia) for conducting weapons-development experiments ("weaponization"); and the fourth (Peng Huanwu and Zhu Guangya) for work on neutron ignition.

These four committees had direct access to higher officials in the academy and, with Li Jue's great authority as director of the Ninth Bureau, to ministry leaders in Beijing. In this way, ministry, defense academies and institutes, and production organizations could stay in close touch. Moreover, in a program that worked three shifts seven days a week, Li Jue decreed that rank and bureaucratic divisions should be minimized, and he created a close-knit organization based on interdisciplinary and interorganizational scientific seminars. The quest for knowledge and advanced technology overruled tendencies toward excessive compartmentalization and secrecy within the academy itself. This same kind of freedom of discourse within a rigidly protected zone of security had once prevailed at Los Alamos.[49] The dichotomy between impenetrable external secrecy and internal openness seems to have worked in both cases.

Wu planned, finally, to let professional knowledge determine individual assignments. He wrote: "Although most of [the scientists and engineers] are still not Party members, they should be awarded authority commensurate with the posts they hold. The administrative departments should not interfere with their affairs."

As the various reorganizations were occurring, Wu Jilin from 1960 to 1961 tackled the job of setting his own office in order and reviewing the pile of documents, both political and technical, that defined his mission. He announced a system for reporting and record-keeping that would register and analyze daily progress and systematize the directives and technical data from the Central Committee and the ministry. This system catalogued the procedures and provided the authority for internal communications and for maintaining an open channel to Beijing. Finally, Wu used the knowledge from the mass of paper that passed over his desk to formulate a process for regularly canvassing the senior scientists and, through them, the principal staff members.

Wu tested the process by accumulating and analyzing the scientists' projections on the time and resources needed to build the first fission bomb. The scientists had already conducted hours of research on this matter and had prepared reports on their findings. They had drawn extensively on foreign writings and on clues provided by the Soviet design plan (but not technical data) for the Qinghai academy. As we shall see, they also had begun field tests. From their findings and projections, Wu judged that China could produce its first bomb faster than either the United States or the Soviet Union had managed to do.

In early 1961, with the preliminary research and testing stage finished, the academy's leaders told the scientists to master the key technological principles and basically complete the bomb's theoretical design within two years. At this time, Wu called a meeting to convey his prediction about the time needed to build the first atomic weapon to his colleagues, only to be greeted with skepticism and outright hostility. One doubter was especially incensed; he believed that if the academy did not complete the bomb according to this self-imposed deadline, Wu's optimistic projection would be regarded as an attempt to deceive the Party Central Committee.

Silence descended on the meeting. This was a serious charge. Wu, however, was not to be intimidated. He struck the table with his fist and jumped to his feet, saying: "We have taken into account all of the unfavorable factors. To offset them, first of all, our whole Party has been mobilized. Second, an effort with energetic coordination and cooperation has begun throughout the country. Third, we have precious scientific and technical personnel. Fourth, we have made our scientific analyses of the relevant theories and experiments and carefully verified them. Fifth, we have made a decision concerning the direction of our main effort." These five points, Wu continued, made him confident that the fission weapon could be completed in a relatively short period of time.

With this rejoinder as a peroration, the chief research administrator said he was ready to address the question of the type of weapon to be developed. He labeled one A-1 and the other A-2, and from the general description it seems clear he was speaking of an implosion device (A-1) and a gun-barrel device (A-2).* Wu's argument in favor of A-1,

---

* The terms "implosion" (*neizhafa*) and "gun barrel" (*qiangfa*) were unfamiliar to the Chinese bomb builders at the time; they used *yajin* ("squeeze") and *yalong* ("press together") instead. Interviews with Chinese specialists, 1986.

which would "conform to the needs of actual combat," was that it was "more advanced than the A-2 type bomb in both theoretical and structural design." But that did not mean abandoning A-2: "at the same time, we can also do preparatory work on the design of the A-2 type weapon [so as to be in an invincible position]." This determination, he added, was "why we dare to hold ourselves responsible to the Central Committee."

Wu's heroic rhetoric made a strong impact on his boss, Li Jue, who responded that he "was quite certain not only that the A-1 type bomb was more advanced, but also that it would save nuclear fuel for the country." He further agreed that making the A-1 implosion weapon should be the academy's objective even though he understood that it would be more difficult to design and manufacture than A-2. Neither he nor Wu is reported as factoring into this calculation what kinds of design hints had been gleaned from Soviet and Western sources or which fissionable material would be available. It is clear they only grasped the general principles and had no specific designs to copy.

By the words "more difficult to design," Li had in mind two major hurdles in fashioning the implosion system: the explosive assembly and the initiator. As it turned out, the initiator posed the greater challenge. The detonation of the ordinary chemical explosives in the assembly had to be precisely focused and perfectly timed to occur microseconds before the triggering of the initiator to release a flood of neutrons into the fissile core. The very evening that this debate on practical deadlines and alternative bomb types took place, Li and Wu set to work on how to achieve a breakthrough in the theory and practice of nuclear weapons detonations.

Sitting in Wu's dormitory room, the two academy leaders recalled a conversation the year before on how to expose the secrets of the detonation process. They had then recognized the need for some preliminary tests to open the way to resolving a long chain of bomb design problems. After reviewing the academy's human and physical assets, they had concluded that they should enlist army ordnance experts to help them. Wu in the earlier conversation had said: "We had better start the physical experiments as soon as possible at an explosives test range of the Engineering Corps located in the Beijing suburbs." Li agreed, and the experimental program got under way in early 1960.

With a wind-driven powdery snow still on the ground, Chen Nengkuan arrived at the test range located near the town of Donghuayuan in Huailai County, just northwest of Beijing municipality. Driving up to this isolated spot near the Great Wall by jeep, the American Ph.D.

had left far behind the campus life for which he had trained, and was about to conduct the pioneering experiments on explosives to detonate the bomb. In the distance lay an ancient beacon tower, a pillbox left over from the war, and tents and primitive barracks erected by the engineers.[50]

Chen's team members, numbering over 30 men,[51] first tried their hands at casting the lenses for the outer explosive assembly. The chemistry and physical shaping of the different types of explosives presented the first obstacles, and eventually the team would call on a university associate professor, Qian Jin, to refine the techniques for manufacturing the high explosives and the electric spark detonators.[52] Chen had to turn to other academics and professionals, such as Associate Professor Yu Daguang, to design the overall multiple-line synchronous firing mechanisms for the explosive assembly.

For Chen, the first venture into bomb making required more brawn and luck than exotic technologies and scientific understanding. Chen's engineering crew mixed their own chemical brew on the windswept testing grounds at Donghuayuan. They lit a fire and, using an ordinary boiler and some hand-me-down buckets from the army, began melting their explosive mix. "The substance, unpleasant smelling and toxic, melted at a high temperature. Despite these unfavorable conditions, engineers Zhang and Sun and their colleagues persevered and alternated stirring." Though hardly qualifying as high science, this was the first step; and Wang Ganchang and Guo Yonghuai, as they were to do so often, first from Beijing and then as commuters from Qinghai, trekked to the explosives test range to supervise Chen's work. Wang used the occasion of their first visit to "give a personal explanation of the fundamental theories relating to the production of explosives" and also took his turn stirring the bucket.

The Chen team set about their work by devising systematic experiments using different combinations of high explosives and cast shapes— probably of TNT (trinitrotoluene), PETN (pentaerythritol tetranitrate), and RDX (cyclotrimethylenetrinitramine).[53] Chen drew on his own "solid theoretical knowledge" and on "relevant technical data published abroad" in selecting specific chemical explosives and drawing alternative models for molding and assembling the explosive lenses. Helping him, Zhou Guangzhao calculated the maximum force of the explosives, and scientists at a computing institute ran the numbers on the subcritical and ultracritical energy releases. In working out the design of the lenses, the team did not have access to computers sophisticated enough to make reliable calculations, so in the end Guo Yonghuai and Chen had to draw on the general principles of mechanics rather than on rigorous, detailed calculations to make a model for the explosive assembly.[54] After Chen's teammates had "resorted to a unique innovative process of carrying out experiments while calculating and designing," their colleagues accepted their somewhat crude calculations as a rough basis for proceeding.

The actual testing of alternative designs took months. Wang Ganchang personally guided the making of the first experimental components, and Chen double-checked all the numbers by hand. On April 21, 1960, the Chinese made the first of 1,000 prototypes for a test series, and by early the next year, Li Jue and Wu Jilin had enough data to give the go-ahead on the implosion device.[55] For each test, team technicians placed the experimental model on a sand dune near the pillbox, which had become a network of cables and housed control equipment, an oscilloscope, and high-speed remote cameras. The pace was frantic: "Right after the detonation of the first experimental component, heedless of the smoke of explosives filling the air around the site, Wang Ganchang hurried to the sand dune carrying the second experimental component. . . . They worked together piling up another sand dune, connecting cables, and plugging in new detonating caps. Often more than ten experiments were carried out in a day."

After more than a year at this grueling pace, the team achieved a breakthrough. By early September 1962, the team's design perfected, Chen could reliably predict that his unit would produce the explosive assembly for China's first bomb on deadline. Chen and his associates had molded about 2,000 pounds of high explosives into shaped lenses and determined the placement for the more than two dozen ultrasensitive detonators around the assembly.[56] They felt confident that they had found a workable model assembly.

The Initiator

During the years that Chen's team trudged between the pillbox and the sand dune, other groups were at work on the bomb initiator. In the first few microseconds following the detonation of the implosion assembly encasing a nuclear weapon, the fissile core must be flooded with neutrons to initiate and maximize the chain reaction at the precise moment the core achieves critical mass. Most nuclear reactions that lead to a high output of neutrons involve the bombardment of light atomic nuclei such as lithium or beryllium with incident protons, deuterons, or alpha particles. We have already noted that the Chinese were major suppliers of both lithium and beryllium to the Soviet

Union, so the academy team could have experimented with either or both.[57]

In publicly known methods employed to initiate nuclear detonations, alpha particles emitted by naturally radioactive elements such as radium or polonium induce high neutron yields in beryllium. The most stable isotope of radium, for example, is an alpha emitter with a half-life of 1,620 years; one gram of radium-226 produces about 36 billion alpha particles per second. Polonium has a far shorter half-life (138.4 days) than radium, but, like radium, occurs in nature as a product of three natural radioactive decay series: thorium, uranium, and actinium. It can also be produced artificially.[58] The Curies in Paris discovered polonium and then radium while investigating the radio-activity of a certain pitchblende, a uranium ore. Qian Sanqiang, from his work with Irène Joliot-Curie at the Curie Institute, would have fully understood the principles of neutron production. Moreover, as we shall see, when he returned from France, he brought a key element for the initiator with him.

The most probable metal used in the initiator for the first Chinese nuclear reaction experiments (and the first bomb) was polonium. This silvery-gray or black element is formed spontaneously as a result of the decay of radioactive lead (Radium D). It is difficult to work with, however: for a metal, it has a relatively low melting point (254°C) but is intensely radioactive and must be handled with extreme care. Because of its short half-life, it would be unattractive for a large, mature weapons program but not for the first devices.[59] By contrast to the combination of polonium and beryllium, that of radium and beryllium produces only a moderately intense source of neutrons. Radium's much longer half-life means that it produces far fewer alpha particles per second per unit mass than polonium does. In practical terms, its rate of alpha production rules out its use in an initiator.

As the implosion wave in an exploding nuclear weapon hits a sub-critical sphere of U^{235}, the fissionable material becomes compressed and, if the compression is correct, supercritical.[60] The initiator produces neutrons that in the initial reaction with the U^{235} split the heavy nucleus into two reasonably massive intermediate-weight fragments and release a large amount of energy, producing "fast" neutrons (meaning those with energies in the range of 0.5 to 10 MeV).[61] All succeeding fission generations split the atoms of U^{235} and, since some of the fission neutrons are fast, of the heavier isotope U^{238} as well. If the bomb is configured properly, the pace of fissioning within a given mass will outpace the mechanical expansion as the mass blows apart.

If the material as it expands becomes subcritical prior to going through all the requisite number of generations for maximum energy release, the "bomb" will fizzle.[62] More than 99.9 percent of the energy in, say, a 100-kiloton explosion is released in the last seven generations of chain reactions, or a period of roughly 0.07 microsecond.[63] The initiator plays a critical part in injecting a large enough number of neutrons at the beginning so that the number of resultant neutrons is sufficient to beat the deadline for a fizzle.

In 1960, Qian Sanqiang transferred one of his engineers, Wang Fangding, to head the Uranium Chemistry Technology Group, under a section of the Institute of Atomic Energy. Engineer Wang's mission was to undertake research on the raw materials for the initiator. Qian told him: "We have been requested to give assistance on the design and manufacture of the initiator for the atomic bomb." Wang observed that Qian exuded great enthusiasm about the weapons project and later recalled that Qian had already sent Wang Ganchang, Peng Huanwu, and Zhu Guangya from his institute to work under the academy. Qian made a speech that advocated giving unrestricted assistance to the nuclear weapons project and then, pointing to a container on his desk, told Wang: "In this container, there are some raw materials useful for experiments. These raw materials were put there more than ten years ago when I returned to China from France in the forties. Now I give them to you. Please go all out to succeed [in producing the raw materials for the initiator] as soon as possible."[64] We assume Qian gave Wang a chemical sample related to research on polonium, but the short half-life of the element would mean that the ten-year-old material could hardly have been polonium itself. The Chinese would have had to learn how to manufacture the necessary polonium on their own.

Wang Fangding worked with the sample at the Institute of Atomic Energy in Tuoli, south of Beijing, not in Qinghai.[65] He assembled a team of brilliant young technicians to work on the initiator, but the search for the right neutron emitter went poorly. After more than 200 chemical experiments, Wang's researchers failed to obtain suitable results. They blamed their failure partly on unsatisfactory working conditions. Their laboratory was in an unheated workshop on the Tuoli institute grounds, and during the winter of 1960-61 especially, their work was badly impeded. Some nights it was so cold that the lab technicians had to dismantle the waterpipes, reagent bottles, and distillers and cart them off to a heated building. The whole process would be reversed the next morning. Safety problems also plagued the project.

During one particularly oppressive summer heat wave, as the technicians were heating the dangerous raw materials in a lead container, it began to deform. The technicians were so overcome by the heat of the lab they could barely operate the manipulator to open the misshapen container. One of them suffered severe radiation exposure and had to be hospitalized.

Senior scientists from Ninth Academy units in Beijing learned of these many failures and difficulties but could offer few practical remedies. Wang Ganchang, Peng Huanwu, and Zhu Guangya made several inspection visits to the laboratory and proffered general advice to Wang Fangding and the other young experimenters, but they could not speed up the process either. Wang Ganchang took a special interest in the lab's progress. His role in the academy placed him in overall charge of work on the explosive assembly and the initiator, and he frequently called on Wang Fangding for reports.

Engineer Wang finally found the key to success in materials identified only as E and F. In one experiment, Wang tried to combine them. Starting with E, he then added F, sealed the container, produced a vacuum, and then heated the materials. Careful records were kept. As the chemical compound began to expand, Wang witnessed an unusual phenomenon that no one had noticed before, namely, that "within a fraction of a second a proper proportion produced a stable chemical compound." This compound, Wang realized, would provide the key element in the initiator.

As noted, polonium is a natural decay product of radioactive lead, and Wang's distillation experiments to separate polonium drew on the readily available knowledge that radium E (bismuth-210) and radium F (polonium-210)—presumably the E and F in Wang's two compounds—come about naturally in the decay process. Because the institute did not have plutonium production reactors available to it at this time to produce the polonium by the irradiation of bismuth-209, Wang's only choice was to use natural uranium ores, which contain radioactive lead with minute amounts of polonium per ton. He had faced a formidable series of research tasks before his experiments with E and F.

Paralleling Wang's work on polonium compounds, other research related to the initiator proceeded under the direction of Qian Sanqiang and He Zehui. Their tasks included the measurement of neutron flux, the planning of criticality experiments, and the making of initiator devices. Step by step, the Qian-He research team worked out the theories and checked off experiments on each of these tasks.

By the end of 1962, the technicians in Donghuayuan and Tuoli had tested separately the explosive assembly and the initiator. Their next step was to test the two as a unit, and for this step they had to achieve precision timing. After the commencement of the chemical detonation and as the bomb was imploding, they had to fix the exact moment to break the barrier between the alpha emitter (perhaps polonium) and the neutron emitter (perhaps beryllium). The implosion itself constituted a dynamic wave that could force the two components together. The experiment was conducted in two parts, the first with a half-sized prototype and the second with a full-sized one. For both experiments, of course, an ordinary metal ball would substitute for the eventual fissile core between the implosion assembly and the initiator.

Preparations for the first test began during the winter of 1962-63. The ministry's plan at this point called for shifting the research and development on the atomic bomb from Beijing to Qinghai. Work on some of the Ninth Academy's main buildings (for neutron physics and radiochemistry) in Haiyan, however, remained undone, and the Fifteen-Member Special Commission set in motion plans for accelerating the construction. Some sections of the Beijing institute began to move west in March 1963, and started preparations for the first experiment. The scientific and technical personnel at the nuclear weapons academy pinned their hopes for the successful explosion of China's first atomic bomb on the outcome of this experiment.[66]

The day of the first test had the spirit of a grand carnival. A motorcade was mounted from the academy with all major scientists and engineers aboard. On the morning of November 20, 1963, "Li Jue and Wu Jilin rode in a jeep at the lead of the automobiles in order to control the speed of the entire motorcade. Dr. Chen [Nengkuan] and Deputy Chief Engineer Su rode in a sedan carrying the precious experimental components in their arms. They wrapped these [special] components with blankets so as to avoid any shocks. The major component of the atomic bomb was carried by a special van. Complying with Wu Jilin's order, two purplish-red long sofas were placed under the major component so as to prevent its being jolted." After the cavalcade arrived at the test site, workers put the bomb parts together and inserted the detonators. Everyone was told to retire to the safety of bunkers, but Li Jue refused: "No, I must be the last one to withdraw from the test spot because I am in overall charge of the detonation experiment. . . . Comrades, don't panic! Remain calm and confident." He then ordered the detonators installed and the recording equipment turned on.

The control room stirred with excitement. Some of the younger technicians had worked for more than two years simulating this experiment. They knew the equipment had to detect and record the exact moment the initiator ignited within the explosive assembly. "If we fail to acquire accurate data," one of Wang Ganchang's former students realized, "the whole atomic bomb project would be delayed." They pressed on and hoped. Chen Nengkuan gave the order to fire, and Wang's protégé heaved a sigh of relief as the recordings seemed to match the former simulations. Forty minutes later, they checked both the films and the other preliminary data. The implosion wave and initiator had met the required specifications. Li Jue flashed the news to Beijing that the "first atomic bomb could be assembled soon, provided the nuclear component fabricated by the nuclear fuel plant was delivered on schedule."

On June 6 the following year, the academy conducted the second test, this time of a full-sized model weapon. Though this test was more important than the partial test in November, the drama appears to have been far less intense by comparison. The explosive assembly and initiator again worked flawlessly. By this time, another section had solved the theoretical and design problems for the nuclear component.

The Nuclear Component

As China's relations with the Soviet Union deteriorated and all phases of the nuclear weapons effort took on new urgency, Deng Jiaxian and the theoreticians under the Beijing institute's Theoretical Department felt the added pressure. By 1960, they were ready to begin research on the actual design and configuration of the bomb, having completed two years of preparatory theoretical work. During the years 1958-60, they had amassed data on detonation mechanics, neutron transmission, nuclear reactions, and the properties of materials under conditions of high temperatures and pressures.[67]

In the process, the two Soviet nuclear weapons specialists in the academy had provided no help on the bomb design, and the Chinese derided them as "mute monks who would read but not speak."[68] Whenever the Chinese probed for classified information, the Soviets would clear their throats, and silence would reign. The wall of secrecy remained right to the end, when, as the advisers departed, they shredded all documents not shipped home. But then they made a mistake and left some fragments behind.

The Second Ministry directed Deng Jiaxian to reassemble and analyze the pieces, and among them he found crucial data, including "eyebrow-shaped arcs" and numbers, that held clues to implosion. Piece by piece the scraps came together, and Deng prepared a report on the results. Zhu Guangya explained the meaning of this "number-one secret document" to his associates in the academy and pointed out its important clues for their research strategy. Quite unwittingly, the Soviets by their haste had contributed to the bomb's design.

In the months that followed, the academy's officials, still temporarily operating in Beijing until the completion of the facilities in Qinghai, assigned Deng to make the initial calculations on the effects of extremely high temperatures and pressures on materials. Deng, as noted early, had assembled some of the country's brightest young graduates in hydromechanics, aerodynamics, and other engineering sciences at a small house in the Beijing suburbs to run the numbers on a few antiquated hand calculators. They first drew charts composed of empty boxes on large pieces of paper and gradually filled in the cells with "tens of thousands of pieces of data." When the chart had been filled, they solved equations for the effects on materials under varying conditions but kept getting unsatisfactory results, because "the error tolerance resulting from their calculations was still double that of a generally accepted figure." Persevering, they worked in round-the-clock shifts "calculating, drawing, and analyzing." [69]

One unidentified parameter of the design configuration proved particularly intractable even after four rounds of computation. Deng spoke to Peng, who in turn invited the leaders of the Second Ministry to analyze the problem. Minister Song Renqiong told the young scientists to express their views on possible reasons for the difficulty, and they organized a series of seminars to continue the discussions with groups of experts, scholars, and technicians. Guo Yonghuai also attended from time to time, and "on the basis of his erudite knowledge, . . . raised several proposals to be considered." By the end of 1960, armed with new ideas, the young men in Deng's group returned to their calculators and added new factors one by one to each succeeding calculation. It was at the end of this nine-month-long series of iterations that Zhou Guangzhao stepped in to verify the ninth calculation. Deng's engineers believed they had corrected a major mistake made by the Soviet nuclear experts. [70]

The growing demands on the design units at this time quickly strained the existing capabilities, and this growth in turn prompted Deng Jiaxian to suggest reorganizing the theorists into three groups under his leadership. The ministry approved the suggestion, and seven prominent scientists joined Deng's new directorate. It was this directorate that set the direction for research that Chinese sources now describe as the turning point in the bomb design effort.

The first of the three theorists' groups, directly overseen by Deng, continued research on the effects of extremely high temperatures and pressures on materials. Zhou Guangzhao headed the second group, made up of specialists in mechanics; it focused on solving all the technical problems relating to bomb mechanics and followed the direction set by the successful ninth calculation. The third group, subdivided into teams of specialists in hydromechanics and mathematics, each led by a deputy director, had the job of calculating the nondirectional hydromechanics of alternative bomb designs. By the end of 1962, the three groups had mastered the theoretical principles of implosion, including the use of enriched uranium in the core and the mechanics of high explosives in the bomb assembly. And, in September 1963, they completed a draft design for the bomb and performed the key static tests to verify it. They immediately dispatched their approved design to the facilities for manufacturing and machining the fissile core and assembling the bomb.

These facilities consisted of a plant to turn the enriched uranium hexafluoride into weapons-grade uranium metal, a nuclear component manufacturing plant for casting and machining the fissile core, and an assembly workshop to put the entire bomb together.[71] The first of these, the Nuclear Fuel Processing Plant in the Jiuquan Atomic Energy Complex at Subei, received the special containers of enriched uranium hexafluoride from Lanzhou in early 1964. It converted the hexafluoride back to tetrafluoride. In the spring the plant produced a ball of enriched uranium metal from the tetrafluoride and shipped it to the Nuclear Component Manufacturing Plant, also in Subei.[72]

Security was the watchword in the complex. One engineer who had to leave Beijing just before the birth of his first child could not tell his wife what work he was doing, "and his wife never inquired about the work to which he was assigned."[73] When family members wished to visit or there were emergencies that brought outsiders to the Jiuquan complex, security cadres had to escort the visitors, and few were ever allowed to know any details about its work.

One of the few recent visitors to the component plant has given his description of this extraordinary facility. He reports traveling along the ancient Silk Road to a location "at the foot of the Qilian mountains [dividing Gansu and Qinghai provinces] heading for the immense Gobi Desert."[74] He writes: "We got off the train, looked around, but saw no plant. A Japanese-built tourist bus drove us toward the depths of the Gobi Desert, and a green belt, with green trees, meadows, brooks, flowers, . . . gradually appeared before us. Tall buildings towered above the green woods. Our hosts told us that we had come in the best season for the Gobi Desert. Had we come in the windy season, we might have seen . . . crushed stones as big as basketballs rolling everywhere. . . . More than 10,000 residents live in this oasis, which has been turned into a new city. . . . An enormous modern integrated nuclear industrial complex located 10 km from the residential quarters operates day and night."[75]

The author, a journalist, records seeing workers in protective clothing, waste treatment areas, fully automated workshops, medical facilities, and computerized control rooms. He also records a conversation with Zhu Linfang, who stressed that the windy, dry, and remote location ensured that radioactive particles emitted into the atmosphere would be broadly dispersed and not endanger populated regions.

The man responsible for manufacturing the fissile component for the first bomb was Zhu Linfang. Zhu, then aged thirty, was dispatched to Jiuquan Prefecture from the city of Chongqing in Sichuan Province in 1959.[76] A college major in machinery manufacturing, he had worked in a factory as head of a technology production section and as a secretary of the local Communist Youth League committee before arriving in the Subei area to take charge of the preparatory work for the construction of "China's sole nuclear component manufacturing workshop."

Joining the atomic energy complex as a complete novice in nuclear matters, Zhu began by querying the Soviet experts assigned to assist in the manufacture of the nuclear component, only to get back insulting responses: "The experts replied to his first question saying: 'It is unnecessary for you to know the answer to this question.' As for his second question, the experts answered: 'Young fellow, don't be impatient.' Mustering his courage, Zhu asked another question: 'Would you please tell me the shape of the product [that is, the nuclear component]?' The answer came back: 'It is similar to a watermelon rind.' Finally: 'How would you wrap the product?' Answer: 'Just as you would wrap an apple.'"[77]

Although the component plant was not to be completed until long after the Soviets left, research had begun on the fabrication process well before the final brick was laid. In the spring of 1961, Zhu Linfang and several others from the plant traveled to Beijing to discuss the smelting and casting of the nuclear component.[78] There they consulted specialists for expert assistance and outfitted several one-story buildings in the Beijing institute to conduct preliminary tests.[79]

Zhu's group began by studying how to obtain a vacuum in a chemical reaction chamber in order to "determine some of the parameters that would be useful in smelting and casting the nuclear fuel."[80] The group technicians installed some makeshift equipment to make the vacuum but ran into problems manufacturing the stack of the experimental reaction chamber. Zhu personally took charge. In search of a factory that could do the job, he trudged "the length and breadth" of the city in the summer of 1961 with a stainless steel plate weighing over ten kilograms. At each plant, his plea for assistance was rebuffed because, for security reasons, he was forbidden to reveal his identity or the purpose of the equipment to be made from the steel plate. Finally, on the twenty-first try, the manager of a hardware factory agreed to fabricate Zhu's chamber stack. But this factory could make only the stack, and impediments to the completion of the vacuum chamber continued to plague Zhu after his return to the Jiuquan complex near Subei.[81] Some time later, the Second Ministry allocated the new plant a reaction chamber from East Germany, but this chamber was designed for refining general alloy steel and not nuclear metals.[82] It was to take many months to modify the imported chamber.

The time was the latter half of 1962, and Zhu was feeling the pressure from Beijing to forge ahead. By the late summer, specialists at the Ninth Academy had achieved major breakthroughs on the bomb's theoretical design, explosive assembly, and initiator; and workers at the Lanzhou Gaseous Diffusion Plant had finally put their jigsaw puzzle of equipment together. In September, the chief of the General Staff, Luo Ruiqing, called a meeting to accelerate the nuclear program. That same month, the Party leading group of the Second Ministry issued a "Report on the Development of the Atomic Energy Program Taking the Road of Self Reliance." This report, originally prepared in August and usually referred to as the Two-Year Plan (September 11, 1962), set the date for the first test sometime between the last half of 1964 and the first half of 1965. Mao himself read the report and, on November 3, 1962, noted: "Very good. Act accordingly. [All departments concerned] should vigorously carry out coordination and cooperation so as to complete this [strategic] program." Two weeks later, the Politburo created the Fifteen-Member Special Commission under Zhou Enlai, and the following month the commission held its third session in Zhongnanhai. At this meeting, Minister Liu Jie personally assured Zhou Enlai that the two-year time horizon was realistic. Word of Liu's commitment spread quickly to the ministry and to Zhu Linfang.[83]

In the winter of 1962-63, Zhu learned that his deadline for the fissile core—"the product"—was July 1964. To meet the schedule, he proposed to scrap the Soviet-supplied plans for constructing a large, sophisticated workshop to process highly radioactive metals, presumably both plutonium and enriched uranium. The nuclear component plant had neither the manpower nor the resources to build such a workshop and finish the bomb within the shorter time limit. At this point some of the plant's special equipment was still on order from a Shanghai factory, including the ducts for conveying radioactive material, which the cadres at the Jiuquan complex were disturbed to learn could not be produced for at least a year.[84] This lag, if not overcome, would set back all work on the bomb manufacture.

Zhu felt he could solve the problem by substituting equipment. "Suddenly, Zhu Linfang had an inspiration: 'Would it be possible for us to manufacture the nuclear component in a small workshop of the laboratory type that could be built using indigenous methods?'" What Zhu had in mind was using the humble laboratory left behind in Beijing as a model for his workshop. His colleagues disagreed. In an attempt to persuade them, Zhu Linfang gave this explanation: "Highly enriched U^{235} oxidizes more slowly than Pu^{239}. In addition, U^{235} is not so poisonous as Pu^{239}. Thus, I think it is feasible to put the core made of U^{235} in an airtight container and to use a small cart as the means of delivery instead of the automatically controlled delivery ducts [from Shanghai]."[85]

His proposal sparked even more controversy. Someone reminded Zhu: "Lao Zhu, you should remember the regulations stipulated by the ministry that no action can be taken to modify the Soviet design plan."[86] Zhu replied that he was quite aware of the regulations, since during the Great Leap Forward, "some people got so dizzy as to modify the Soviet design plans even though they didn't have a good grasp of the . . . plan or the equipment supplied by the Soviet Union." Zhu said he remembered Mao's warning, "*Xian xie kaishu hou xie caoshu*" (First write in regular script and then write in cursive script).

Obviously, Zhu said, the ministry and Mao were correct. "However," Zhu continued, "at present, we have had enough experience acquired in tackling technical problems during the past one or two years." Thus, his workshop technicians could now "write in cursive script"—that is, proceed to conduct modifications of the Soviet plan.

Zhu acknowledged that any modified plans would have to be submitted to the ministry for approval. Zhu got his approval in due course, and by the time the Lanzhou Gaseous Diffusion Plant succeeded in making weapons-grade uranium hexafluoride (late December 1963), the technicians at the Nuclear Component Manufacturing Plant had tested his plan. The workshop building (code name number 18) had been completed six months earlier, in July, and its equipment installed by October.[87]

The metallurgists, now under ever greater pressure from Beijing to fabricate the core of enriched uranium, almost immediately encountered a variety of technical unknowns in the casting process. One of the most intractable problems was the development of air bubbles in the casting. In Beijing, Zhu recalled, "they had taken into account the problems they would confront in changing the raw material [of the core from impure to highly pure metal]. While tackling the technical problems in Beijing, the problem of the emergence of air bubbles in the casting had not been very obvious because there were many impurities in the raw material."[88] At that time, Zhu's technicians could not eliminate air bubbles in sample castings and, after laboring through more than 1,000 experiments, had realized that their process was too complicated. Zhu Linfang, surprised that the bubbles were now showing up in the higher-grade uranium, put the problem at the top of his priority list.

Now that they had moved to their new Jiuquan plant with somewhat more modern equipment, Zhu remembered the years in the primitive facilities in Beijing and of the fight against hunger. Perhaps, he thought, this was the moment to return to some of the earlier methods and the working style tested in Beijing.[89] He assumed, based on past experiments, that the fault lay in impurities in the metal, but his analysis made him less sure. According to the published history, Zhu's team took a long time to ferret out the reasons for the appearance of the bubbles in the uranium. Zhu himself reportedly "stayed by the side of the vacuum reaction chamber day and night."[90] For him, this was what Mao would term the "key link" in the entire process.

When the casting team could report no real progress, Vice-Minister Yuan Chenglong and one of the ministry's officials, Chief Engineer Zhang Peilin, headed for Jiuquan Prefecture. There they formed a group of experts that included the chief engineer of the Jiuquan Atomic Energy Complex, Jiang Shengjie, a leading nuclear chemist who had been trained at Columbia University in the United States. These experts had the authority to give direct orders to Zhu's team, which had stepped up the rate of its experiments after 50 tries without success. At a technical meeting called by the ministry, one of the experts suggested that the team abandon the line of inquiry it had been pursuing. A nuclear metallurgist on the team replied: "If we strike out on a new path, all our previous efforts will have been wasted." Zhang Peilin, reputed to be adept at "summing up the experiences obtained by his subordinates," then pronounced judgment. Because China had not been able to obtain sufficient useful information from abroad on refining and casting the uranium component, he thought "the only feasible way to solve the technical problems [was] to resort to repeated experiments." He added: "I think the technical direction chosen by us is correct. We have acquired a huge amount of technical data that can be used like gravel to pave the path of success."[91]

This settled the matter, and the experiments continued through the spring festival of 1964. One after another, alternative plans and technical innovations were devised, and for each the casting team collected and annotated the relevant data throughout the more than 100 experiments. They then embarked on testing one of seven promising approaches, and this time they succeeded in removing the air bubbles and produced a nuclear component to specification. On April 30, 1964, technicians at the plant began machining the bomb's uranium core.[92]

Machining the Nuclear Component and the Final Assembly

Preparing for this event, the lathe operators at the plant had begun to carry out simulations and practice exercises six months earlier. One master worker, Yuan Gongfu, trained so hard in these sessions that he lost more than 30 catties (15 kilograms). But eventually he became skilled in operating the specially designed lathes and other machine tools under the rigorous and cumbersome conditions imposed by the use of highly enriched uranium. He learned to machine the practice material within extremely close tolerances and without error, "wearing a special gauze mask and a pair of gloves of two-ply thickness." Zhu Linfang, acknowledging that Yuan was the best person for the job, now authorized him to machine the ball of uranium.[93]

As happened in other critical dramas during the bomb-design program, this one, on April 30, drew a crowd of distinguished kibitzers. Vice-Minister Yuan Chenglong and leading cadres from the atomic energy complex joined the entire workshop crew. They nodded in a gesture of respect to the master lathe operator and waited for him to speak. Yuan, temporarily full of confidence, told them: "You can rest assured that we will without question succeed in machining the nuclear component to the required standard."

All was fine until the uranium ball was put in the vise. Faced with this unfamiliar reality, Yuan lost his poise and became visibly upset. He suddenly appreciated the high risks involved, for he was machining lethal uranium, not ordinary steel. His work would determine the success or failure of a decade of labor by tens of thousands. "He worried about the consequences of a serious accident. He couldn't help trembling when he considered the possibilities." A colleague, sensing Yuan's stage fright, reminded him to start machining. That only made matters worse, and his increased trembling caused the uranium ball to slip in the vise. This, of course, made Yuan even more panicky; he became drenched with sweat.

The leaders thereupon adjourned the exercise and reassembled the crew outside the machine shop. Director Zhou Zhi of the atomic energy complex showed sympathy and asked Yuan: "Xiao Yuan, what has happened to you? Are you certain you can complete the job?" By this time, Yuan appeared to have totally lost his confidence, and one of his shopmates suggested that the machining be postponed. The complex director then asked Zhu Linfang's opinion. Zhu replied that Yuan should go right back to the lathe. Soothingly, Zhu told Yuan: "It will be no problem for you. The mishap didn't result from your lack of technical skill but from your nervousness." He urged Yuan to relax for a few minutes, have a glass of milk, and then return. Even though it was now late at night, Yuan agreed and, back at the lathe, began again. "This time he was more cautious and halted once in a while so as to measure the core accurately." Step by step he moved from machining, to measuring, to adjusting, and then back to machining. By early morning, on May 1, 1964, the nuclear core for the bomb was ready.

For the Jiuquan Atomic Energy Complex, the last step in making bomb 596 came that August.* All the parts—explosive assembly, tamper, uranium core, and initiator plus electrical assembly—had been moved to the Assembly Workshop under the direction of Engineer Li, and in June, the final safety experiments on the component's criticality had succeeded.[94] At the last stage in the assembly, once again senior

---

* The procedures for the assembly took place between July 20 and August 19, 1964. By August 20, the bomb was completely assembled and ready for delivery to the Lop Nur Nuclear Weapons Test Base. "Riddle of Research and Development," p. 6; Li Jue et al., *Dangdai Zhongguo*, pp. 55, 273-74.

leaders, including Zhang Aiping from the Lop Nur test base, came to watch the action.[95] The final assembly took 72 hours.

At the workshop, pairs of white and velvet curtains were draped over the windows to shut out the sunlight. Workers had installed electrostatic copper wires at the doors to ground the static electricity of anyone entering the assembly hall. Those in attendance went first to a changing room to don white coveralls and cloth slippers. The room was a model clean room, and the sign on one wall read, "*Zhiliang diyi, anquan diyi*" (Quality is number one, safety is number one). Everything had been inspected, but Engineer Li ordered everything reinspected one last time. He took a phone call from the Ninth Academy's Wu Jilin who, on being reassured that all was ready, said: "Please pass along [to the assembly technicians and workers] the directives of Chairman Mao and Premier Zhou: 'Be bold but cautious.'[96] Forget everything except the assembly work. Carefully perform self-inspection and mutual-inspection before commencing. You must achieve complete success in the assembly work this one time." Wu informed Li that Zhang Aiping had come to watch the assembly and then gave the order to proceed as soon as Zhang and Vice-Minister Liu Xiyao arrived. When the two leaders had taken their seats outside the designated security area, the assembling began.

This time the presence of dignitaries seems to have inspired confidence, not nervousness. The assembly work proceeded rapidly and smoothly over the next two days and down to the last item on the checklist. Li then invited Zhang and Liu to approach the security line to supervise the final assembly stage, completing the last point on the list. They complied. Li and his master workers "inserted the nuclear component [*hexin bujian*] into the shell case," and the plant's work on bomb 596 was done. Zhang and Liu burst into applause.

SEVEN

The Final Countdown

The sites for China's nuclear weapons tests and the impact zones for most of its missile tests lie hidden in the deserts of the Xinjiang Uygur Autonomous Region. Here Chinese archeologists even now unearth the relics of long-buried kingdoms and silk-laden caravans in digs close by the paraphernalia of strategic death. Korla, Minfeng (historically called Niya), Yanqi (Agni and, later, Karashahr), and Lop Nur, all locations in a network of weapons test bases, evoke memories of the Western Han dynasty's warriors fighting the Huns, of the oasis kingdom of Loulan on the western edge of the Lop Nur marshes, and of the adventures of Marco Polo along the Silk Road. In the Mongol Empire of the thirteenth and fourteenth centuries, Chagatai, an obscure fief of Genghis Khan's second son, occupied much of present-day Xinjiang. Some physical evidence of this grand legacy dots the halls of the British Museum in London and the Hermitage in Leningrad even as modern-day explorers search the military zones to discover what of the past remains.[1]

The Land of Lop Nur

The Chinese name for this northwestern province, Xinjiang, now an autonomous region, means "new domain." With its vast Tarim and Junggar basins and majestic Tian Shan and Kunlun Shan ranges, Xinjiang accounts for one-sixth of all China.[2] An admixture of ancient and modern heightens the mystery that envelops this land, which long ago served as the meeting ground for Chinese from the east, Mongols and tribal bands from the north, Slavs and other peoples from across the Pamirs, Europeans from the west, and Indians and Tibetans from the south. As the Silk Road headed east into Xinjiang from the Pamirs, it split into northern and southern routes, one passing via the oases south of the Tian Shan and the other along the watering holes north of the Kunlun Shan. The two branches rejoined in Gansu, at Anxi, where travelers from Xinjiang's remoteness still recount their adventures and vow to return.

Traces of the region's bygone times endure and enrich the story of China's nuclear quest. Religions beckoned some to this frontier land, but the hold of most faiths that competed in Xinjiang withered in the desert fastnesses. During the fifth century, the Buddhist Faxian ventured northwest to the fortress at Dunhuang, later famed for its art and literary treasures, and then beyond to the country of Agni, a rocky oasis at modern-day Yanqi on the western edge of the Lop Nur Nuclear Weapons Test Base. He met more than 4,000 monks of the Hinayana school at Agni, but remembered the people there as "lacking in courtesy" and as having "treated their guests rather coolly." Two hundred years after Faxian, the Tang Buddhist master Xuanzang made it to Agni and visited the Arghai-bulak, or "teacher's fountain," where a legendary monk had brought forth water from a dry rock before he entered Nirvana.[3] In the end, Buddhism lost out to the Muslim faith, and the Uygur, Kazakh, Uzbek, and Hui peoples thereafter rooted their religion in Islam.

In the thirteenth century, Marco Polo headed to the Middle Kingdom along the southern route of the Silk Road, which took him through the city of Lop (now Ruoqiang) and then, down the Gansu corridor, to the town of Etzina. At Lop, Marco Polo recalled the "well-known fact that this desert is the abode of many evil spirits, which amuse travelers to their destruction with most extraordinary illusions." Among the mirages seen by some on the Lop desert was that of "a body of armed men advancing toward them." These "spirits of the desert," he wrote, lure men to their deaths as they "fill the air with the sounds of all kinds of musical instruments, and also of drums and the clash of arms." At Etzina, now a ruined "Black Town" (Kharakhoto) that lies within the present-day missile test base, Marco Polo judged the inhabitants to be idolaters and their land so self-sufficient that "they do not concern themselves with trade."[4]

The accompanying map centers on one of the most forbidding and least-known regions in all China, the basin nestled between the Tian Shan and the Kuruktag mountains. Reginald Schomberg, one of the few Western visitors to this region, recorded his impressions of the area in the 1920s. He remembered hearing that the Kuruktag (Quruq Tagh) mountains northwest of the salt lakes of Lop Nur were the victim of a desiccation bad enough to "[play] a part in the destruction of Loulan" in the Han period.[5] But in fact Schomberg, like Sir Aurel Stein before him, found the hills of the Kuruktag less dry than he had anticipated.[6] "The country was pleasant," he recalled, "and certainly the name Quruq or Dry was a misnomer, as the people at Kurla [Korla] had already told us." He noted: "Brushwood and grass were ample, the toghraqs [*populus balsamifera*] abundant and well grown, and vegetation generally good [to the point where the] Quruq Tagh is the winter pasture of the Khoshut Mongols from Qara Shahr [the modern-day Yanqi area]."[7]

Stein too was pleasantly surprised and wrote of how the sight of "the reddish crest line of the barren Kuruktagh . . . raised the spirits of the men." In this reportedly sterile terrain, Stein's Loplik laborers found water and grazing grounds for their camels and met "traces of occupation in early historical times."[8] Stein's assistant surveyed the land north of the Kuruktag to the "only permanent homestead in that vast area of barren plateaus and hills"—the settlement of Qinggir (Singer)—a few kilometers from today's nuclear ground zero.[9] This forbidding desert hosts China's secret nuclear weapons test range and an associated science city.

The test facility now traverses the boundaries of Turpan Prefecture and Bayingolin Mongol Autonomous Prefecture southeast of the provincial capital of Ürümqi. We know little of the people in this sparsely populated region of the Xinjiang Uygur Autonomous Region, though the nineteenth-century Russian explorer Mikhail Pevtsov has conveyed some of the local history.[10] Before the turn of the century, by his count, there were only about 800 people living in the whole of the Lop Nur basin. Most of those people, like those to the north of the Kuruktag, had Turkic and Mongol ancestors. The Turkic groups had come to the area some 400 years earlier, when they moved into largely Mongol settlements.[11] Gradually the nomadic Mongols had drifted elsewhere, leaving the territory to more sedentary hunting, herding, fishing, and foraging communities.[12] The immigrants grouped into small settlements of ten to twenty households and lived in poor cane huts (*satmy*). More oriental appearing than their distant and far more numerous Uygur cousins, the newcomers retained a variety of Mongol marriage and burial customs, which they still practiced in Pevtsov's time.[13] The only outsiders in that period were pockets of Chinese soldiers assigned to frontier forts or to guard exiled political offenders in isolated fortress prisons.[14] The descendants of this mixed group of Turkic-Mongols and Chinese were to bear the brunt of the changes wrought by the test base and nuclear fallout on their lands.

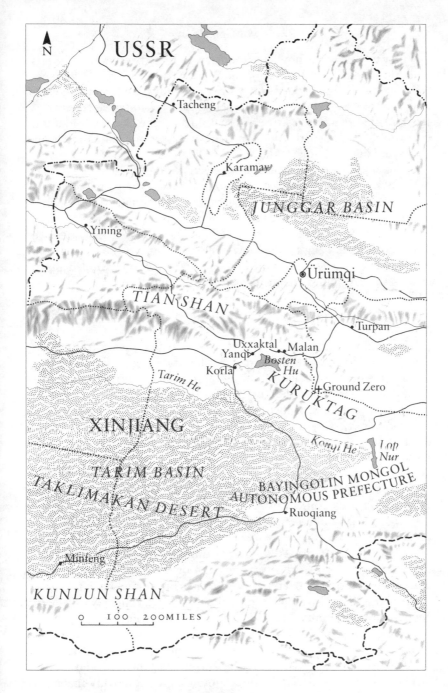

Map 4. Xinjiang

Still, we must stress how few and isolated these people were. In this land, the Turki farewell, "*Yol bolsun*" (May there be a road), has special meaning. Wind-eroded clay ridges (*yardangs*) and, running east-northeast to west-southwest, "an unending succession of steep terraces separated by trenches and troughs" cut through the clay soil and taunt the would-be road maker.[15] Han dynasty historians called the fearsome yardangs at Loulan "the White Dragon Mounds."[16]

In picking this spot of long history, fierce climate, and hostile topography to carve out the nuclear weapons test base, the Chinese leaders were returning to an area they first came to as revolutionary victors. In May 1949, as the Nationalist-Communist civil war was ending, units under the command of Peng Dehuai launched a major offensive in the northwest. Peng's force, once the Northwest People's Liberation Army and now redesignated the First Field Army, barely avoided a trap thanks to last-minute reinforcements from Nie Rongzhen's armies, and the combined forces subsequently fought some of the most costly engagements of the war as they battled for Gansu, Ningxia, and Qinghai provinces.[17]

In the autumn of 1949, Mao Zedong ordered his agent in Moscow, Deng Liqun, to go to Xinjiang to see if he could negotiate the surrender of Chinese Nationalist forces there. In September, the Nationalist commander agreed to the surrender terms. The following month, Wang Zhen, the commander of a corps under Peng's First Field Army, sent forces to occupy the province. Up to this moment, the Nationalists had been locked in a fight with troops of the Soviet-sponsored Eastern Turkestan People's Army. After some weeks of intrigue, that force was ordered by Moscow to join Peng's field army. With the desertion of the Nationalist forces to the Communist side at about the same time, Peng's armies finished their occupation of Xinjiang.[18]

By 1954, many units of the First Field Army, including former Nationalist troops, had been deactivated or redesignated as Production and Construction Corps (PC Corps) units to remain in Xinjiang.[19] From the end of 1954 to 1957, the PC Corps' ranks increased from more than 200,000 to over 300,000.[20] The corps dated back to 1950, when Wang Zhen, by then a top political as well as military official in Xinjiang, had responded to Mao's directive of December 5, 1949, to turn the army into a working force.[21] Demobilized soldiers of the People's Liberation Army (PLA) by the tens of thousands were ordered to construction jobs in the oases of the Tarim and Junggar basins. They tackled the desert wastelands to reclaim fields for new army-managed farms, to open up communications, to plant trees, and to finish water conservation and irrigation projects. Farming, mining, small manufacturing, and urban housing all fitted into the corps' broad job description.

Some of the demobilized units also provided backup to the border defense, public security, and local garrison units, and carried small arms. They worked alongside Party cadres to enforce the regime's still shaky authority in the region and to engage in political indoctrination and mobilization campaigns. These mostly Han Chinese ex-soldiers constituted Beijing's first wave of immigrants into Northwest China, thus perpetuating an imperial policy from the Qing era.[22]

In 1956, Wang Zhen, who only recently had moved to Beijing as commander of the PLA's Railway Corps, assumed responsibility for the new Ministry of State Farms and Land Reclamation and central control of the Xinjiang PC Corps, which was under the political leadership of his former subordinate Wang Enmao. Later, the two Wangs were to oversee part of the construction for the system of strategic weapons test bases and their peripheral communications. This provincial PC Corps furnished the heavy machinery and backbone work battalions that were to construct the peripheral facilities for the nuclear test grounds, as well as the associated missile test facilities.

The Search for the Test Site

On August 10, 1958, a handpicked unit of PLA officers and men from a garrison in Shangqiu, Henan Province, boarded a train and began rolling west.[23] All they knew, from their predeparture briefings, was that Beijing had ordered them to scour the western lands and find suitable locations for a secret facility. The briefings revealed neither their ultimate destination nor the length of their assignment. Speculation and rumors rumbled through the train, but none of the soldiers guessed the unit's actual assignment.

The train's first stop was the historic town of Dunhuang in northwestern Gansu. There, ten days after leaving Henan, the unit disembarked and set out to reconnoiter nearby stretches of the Gobi Desert to get a feel for the more intensive field surveys that lay ahead. A week later, at the end of August, the leaders and Party members of the PLA unit recalled the survey teams, and in the Dunhuang movie theater, the men learned why they had come to the northwest: their mission was to select the location of the nuclear test base and to operate as the core unit to build it. This was heady news: "The atmosphere of the meeting was immediately enlivened. Several comrades were even moved to tears. All determined to work with all their might to make contributions to China's atomic and hydrogen bomb development and to China's defense modernization."

The spirit of high adventure quickly subsided, however, as the teams reentered the vast Gobi Desert. Somewhere in this wretched land, far from all they had known, they would find and build their permanent home. And the requirements for that home severely constrained their choices. Yet they purportedly set off on their mission undaunted by the challenge, and for the next three months, they scoured the desert for possible ground zeros, living quarters, and command posts.

In October, the Central Military Commission designated Zhang Yunyu to be the future commander of the projected test base, and in November, with several potential base locations settled on, he flew from Beijing to Dunhuang for an on-site inspection. The Soviet advisers assigned to the survey teams had especially recommended a location some 140 kilometers northwest of Dunhuang. They had devised a plan specifying a site for a maximum explosion of 20 kilotons. This was sufficient, they said, for China's nuclear weapons program.

After making "a careful analysis of the topography, geology, highways, and water resources" and comparing the various alternatives, Zhang concluded that none would do. Noting that the United States and the Soviet Union had tested megaton weapons, he rejected the 20-kiloton limit as falling far short of China's needs. He also concluded that the whole region lacked sufficient water and transportation. All possible sites for the base lay too close to Dunhuang, which might be in jeopardy from radioactive fallout. Even the living quarters and on-site command post were downwind of ground zero. Reluctantly but firmly, Zhang vetoed all the possible choices and reported his decision to Beijing.

In December, the leadership in Beijing accepted Zhang's judgment and ordered the search for suitable base locations to move farther west. Zhang established his headquarters in Dunhuang and sent the Henan unit's survey teams on their way. Soon afterward, the unit's aerial survey teams set up temporary operations in the Xinjiang county of Turpan about 550 kilometers northwest of Dunhuang. Few adequate charts then existed for this largely unexplored territory north of the Taklimakan Desert, and Zhang ordered survey aircraft to conduct the initial reconnaissance missions. The planes flew from the airport at Hami, a town to the east, with the goal of covering as much of the territory north and northwest of the Lop Nur marshes as possible. By mid-December, the pilots had spotted some likely candidates for base locations, and on December 22, Zhang once more dispatched his survey teams into the desert. All the ground reconnaissance teams, each consisting of some 20 soldiers, could communicate with each other and with their airborne comrades by radio. They covered this territory of the ancient kingdom of Loulan by jeep, basically following the routes of the ancient Silk Road between Yanqi and Turpan and Yanqi and Lop Nur.

By the time one of the survey parties arrived at the oasis of Huang-yanggou in the winter of 1958-59, the surveyors had a good understanding of the local topography, including terrain, water resources, and soil conditions. They liked what they found at Huangyanggou. The area around this oasis formed a large desert valley with the high Tian Shan range to the north. The valley extended more than 100 kilometers west to east and 60 kilometers north to south. The surrounding country looked promising, and in March Zhang told the entire headquarters unit to leave Dunhuang for Turpan.

Meanwhile, the surveyors had done some preliminary work on siting a future scientific center and living quarters, and had begun to look for a ground zero on the valley floor. Determining the right places for these vital locations took several months—from sometime in the winter of 1958-59 through the following spring. The technical requirements were to choose a detonation point and an "on-the-spot command post" close enough to each other to permit direct observation without jeopardizing the safety of the observers. The most suitable ground zero according to one set of criteria proved to be only 50 kilometers from the command post—too close to avoid dangerous levels of radioactive fallout. A reconnaissance unit discovered a place where the separation would be 90 kilometers, but direct observation appeared to be a problem: "They then delivered two trucks of firewood to this location, poured gasoline on it, and set it ablaze. However, those who remained at the command post could not find the flame even with the aid of a theodolite." Finally a place was found and settled on where the distance was only 70 kilometers. It had taken more than 1,000 men two years to fix the future ground zero. This was the general region of 41.5° north, 88.5° east.

The Construction of the Base

On June 13, 1959, while the search for ground zero was proceeding and four months before the formal establishment of the Lop Nur Nuclear Weapons Test Base on October 16, the Party committee of the proposed base convened its first enlarged meeting.[24] It met in a cellar on the base's still-makeshift center to form the base headquarters under Zhang Yunyu. Because Zhang Aiping lived in the cellar when staying at the center (presumably on inspection trips as deputy head of Nie Rongzhen's commission), the basement became known as the General's Residence, as well as the place where base officials made all their initial decisions.[25] Here Zhang Aiping later held the post of commander of the First Atomic Bomb Test On-Site Headquarters (Yuanzidan shouci shiyan xianchang zhihuibu), which was the next higher level in the chain of command above the test base command and maintained liaison with the Ninth Academy (see Fig. 1, p. 56). Zhang was also to serve as the director of the First Atomic Bomb Test Commission (Yuanzidan shouci shiyan weiyuanhui), a body still higher in the command chain that reported directly to Mao Zedong's Central Military Commission and to Zhou Enlai as head of the Fifteen-Member Special Commission.[26]

A tract of more than 100,000 square kilometers was set aside for the test base—an area as large as China's eastern province of Zhejiang.[27] Over the years, as the testing program unfolded, the region close to the detonation zone was to become filled with the remnants of former tests conducted on or near the desert surface—from explosions in towers, in the air (delivered both by plane and missile), and in both horizontal and vertical shafts. According to a former commander of the test base, "more than 2,000 kilometers of highways have been built on the base [by 1984]. At each test site there is a command center, a communications hub, a control center, and a permanent survey station. At the air testing grounds, there also are some simple houses, airports, and underground water pipes. In the distance there is an airport and a factory to assemble test items." Some of these facilities remain in use to this day, standing witness to the various pockmarked ground zeros of past tests, littered with scarred debris and abandoned. Two reporters who visited the area have written of the "fantastic picture" before them: "Broken cars on dispersed rocks; piles of scrap iron that had originally been armored personnel carriers; the wreckage of planes; destroyed cement buildings some of which had a surface of melted glass." In early 1960, Zhang Yunyu's units aimed toward the tests that would yield this landscape of nuclear refuse.

On the Lop Nur base, too, the "three hard years" exacted their human toll.[28] In 1960, hunger and malnutrition struck the builders of the base, and the army construction workers soon consumed all available vegetables. To avoid starvation, most base personnel thereupon resorted to eating elm leaves and foraging for wild plants. Solving the food shortages proved particularly severe in this area because the lack of rainfall ruled out any possibility of the base workers' growing their own crops. Picked for its extreme isolation and desolation, this desert valley north of the Tarim Basin averages only 16 millimeters of annual rainfall, against an annual evaporation rate of about 3,600 millimeters. Men drawing water for washing or cooking would watch it evaporate in minutes, and any water found had to be used for both purposes.

For a time, much of the construction work concentrated on road building, water conservation projects, and utilities. The Central Military Commission handpicked units of the Engineering Corps and the Railway Corps to complete the main construction work on the base and assigned several units under the Production and Construction Corps to build the peripheral infrastructure for base communications. We have interviewed former soldiers now living in the United States who served in one of the ten regiments under the Agricultural Second Division of the Xinjiang PC Corps.[29] The division provided some of the manpower to build the communications network for the test base and for the missile impact centers. Each construction regiment maintained its own farms, factories, medical clinics, and logistical facilities, and operated special mobile engineering units for construction. Each regiment contained several thousand men and an equal number of family members.

Of the ten to twenty companies in each regiment, only four or five were composed of youth who had joined the army as volunteers and carried arms. Many, if not most, of the other companies were manned by soldier-prisoners.[30] The former soldiers we have interviewed have estimated that about two-thirds of the 100,000 men in the division were current or former prisoners. These men did not bear arms; their duty, especially for former Kuomintang soldiers, was considered a somewhat less onerous type of reform-through-labor. In the 1950s, many companies contained both ex-Kuomintang troops and ordinary prisoners and were labeled *xinsheng liandui* (newborn companies).[31] Those men classified as prisoners had lost their political rights and had to perform especially strenuous labor under the supervision of armed soldiers. The armed soldiers in the companies enjoyed higher salaries than the other groups, received special welfare benefits, and were exempted from the heaviest work.

The mobile engineering companies assigned to the nuclear test base and missile impact areas were organized under a divisional engineering detachment. They had an unusual composition, because the detachment enlisted a number of minority nationalities as a result of the

Yita incident in May 1962. By this time, tensions had flared along the
Sino-Soviet border, and the Chinese allege that Soviet agents had instigated tens of thousands of local minority nationality peoples to flee
to the Soviet Union from Yining Municipality and Tacheng County
(hence the abbreviation Yita). In the aftermath of this incident, Chinese authorities moved almost all the remaining minority nationality
peoples from the Yita areas to the interior of the autonomous region,
and some who were able to do physical labor wound up as a reinforced company in the engineering detachment. These Yita construction workers received favored treatment, which included the higher
pay and privileges accorded to troops stationed on the border.

Units of the engineering detachment plus others built towns as well
as communication systems on the periphery of the test base.[32] One
such town, Malan, has yet to appear on Chinese maps; the engineers
created it at a road junction between the settlements of Uxxaktal and
Yushugou. In addition to civilian houses and barracks, the town
boasts an airport, a hospital, a bank, a department store, several res-
taurants, a post office, and a reception center. It serves as a recreation
and tourist center for nearby residents. Malan is the headquarters city
(the "capital") of the Lop Nur Nuclear Weapons Test Base.[33]

The public security forces rarely permit visitors to proceed north
beyond the city proper to a heavily guarded restricted area. In this
Scientific Research District (Kexue yanjiu qu), special construction
companies from the Engineering Corps (not the PC Corps) assigned
to the units originally dispatched from Henan did all or most of the
construction. Located in the foothills of the Tian Shan some tens of
kilometers northwest of Malan city center and about 200 kilometers
northwest of ground zero, the district has turned a high valley into a
modern scientific hub for nuclear weapons research and testing. Its
middle school and kindergarten carry inscriptions by Nie Rongzhen,
but apart from that and the fact that entrance to its inner zone is
restricted, there is not much to distinguish it from any recently constructed Chinese urban area. This restricted inner zone constitutes a
Research Institute, with buildings devoted to hydromechanics, solid
mechanics, optics, physics, radiation chemistry, computing, and data
management. Established in 1963, the institute maintains a large ar-
chive on nuclear explosions, anti-nuclear warfare, nuclear weapons
designs, and so on. "Almost one thousand pieces of technical materials
concerning the scientific results of [previous] nuclear test explosions
are classified 'top secret' and preserved in the archives." The scientific

and technical personnel in the zone believe they have "succeeded in developing a school of their own concerning nuclear weapons development performed by all the relevant countries of the world."[34]

According to one of the former PC Corps workers we interviewed, outsiders only knew that there was a very special top-secret district somewhere to the north of Malan. Vehicles coming from the outside had to change to specially screened military drivers, who often would be replaced three times en route to the district. Outsiders had also learned that nuclear scientists lived there, and that monkeys, dogs, mice, and goats were raised there for use in the nuclear test experiments. According to those we interviewed, "stories about the district spread far and wide among the local people," despite its secrecy.

By the mid-1960s, the Central Military Commission had approved the sending of additional construction units from the Engineering Corps and the Railway Corps to expand the base facilities and overall infrastructure. By early 1967, these units had built an airbase for transport aircraft and medium-range bombers and a railway system for the base.[35]

While the number of people participating in any given test would "range from several thousand to nearly ten thousand," the community centered on the science district appears to have remained reasonably stable.[36] Chinese sources mention members of the cable or transport company and the Yangpingli weather station who worked on test after test in the fierce cold or heat of the Xinjiang desert.[37] Over the years, as the base community grew and matured, the personal and family aspects of base life became almost as important as scientific work and preparations for next tests. One unexpected problem was that well-educated females—scientists, technicians, nurses, and clerical staff—quickly outnumbered intellectually acceptable males. Each year, at graduation time, a new batch of young male scientists and technicians would arrive at the base, and this annual event reportedly blossomed into a community marriage contest. According to one account: "It was really a sight. It seemed that the girls and their parents were all trying their best to win over a young man."[38]

Thus over the years did this special frontier base expand its horizons beyond nuclear weapons testing. Like the forts on China's ancient borders, the science district gradually became a hub of local activity and an integrated community. Its unique mission brought together Chinese science and engineering and created bonds among them that remain strong and politically relevant to this day.

The Arrival of the First Weapon and Final Preparations

In 1964, planning for the detonation of the atomic bomb dominated everything else in the science district and throughout the base.[39] With the work finished on the bomb assembly at the plant in the Jiuquan complex, the word was flashed to the Lop Nur test base that the bomb and its assemblers would begin to arrive. As head of the First Atomic Bomb Test Commission, Zhang Aiping intensified his contact with Zhou Enlai's Fifteen-Member Special Commission and leaders of the Ninth Academy in Qinghai.

For safety, the engineers disassembled the bomb and repackaged it in two parts, one to be transported by train, the other by air. Zhou Enlai had instructed: "Guarantee the security of the assembly, delivery, and detonation of the first atomic bomb," and Wu Jilin, one of the senior leaders of the Ninth Academy, went to extraordinary lengths to comply. A special train manned by China's best engineers was provided. It would carry the nonnuclear explosive assembly. A senior cadre had to assure Wu "that all the coal allocated for the No. 1 Special Train had been screened so as to pick out detonating caps and any other explosives that might have been mixed into the coal at the mines." This meant that workers had to screen over 100 tons of coal. Along the route from Gansu to Xinjiang, "all passenger trains were ordered sidetracked for the special train, and the power supply was ordered cut to the high-tension wires that crossed the tracks." Wu also noticed that railway workers sent to inspect the special train had been issued special copper hammers that would not produce sparks. By the time the train set off with its explosive cargo, all the railroad and security personnel had gone through a number of rehearsals and had reviewed all possible contingencies of what might go wrong.

Authorities in Beijing were especially concerned about the physical security of the train that left the Jiuquan assembly plant in late August.[40] One rail car carried both the bomb component and the Ninth Academy specialists who were to participate in the first test. On the passage through Gansu and Xinjiang, the heads of the provincial Public Security Bureaus joined the train for the trip through their respective jurisdictions. They provided Wu Jilin, who with Wang Ganchang journeyed to the Lop Nur Nuclear Weapons Test Base, with a detailed itinerary and an itemization of what security measures had been taken. At the provincial border, Wu and Wang were "attracted by a moving sight on the platform." As the train pulled out they noticed lines of railway policemen and security officers standing "straight as ramrods just like a group of statues." The public security chief commented: "They have been there since last night. The same security measures have been adopted at all the major stations along this railroad line."

At each stop in Gansu and Xinjiang, Wu Jilin joined the public security officers in personally inspecting the carriage containing the explosive assembly. Security officers and technicians watched both the device and a special control panel for the compartment. Each time the train began to move, Wu asked for continuous reports on temperature and vibration conditions, and at each "dangerous" curve in the rail line, he inspected the stability of the container for the assembly. Wu had direct communications with the engineer and could order the train to proceed at lowered speeds to reduce vibration. In this way, the nonnuclear assembly made its way on the long journey by rail to the test base and then by truck to ground zero.

Several days after this part of the bomb assembly had been delivered to the Lop Nur test base, Nie Rongzhen authorized the air shipment of the nuclear core. Assistant Research Fellow Hu and Deputy Chief Gao of the Second Ministry's security department were assigned to escort it. En route to the west, the security officer, upon discovering the nature of the cargo he was shepherding, became concerned about the possibilities of a premature chain reaction and asked Hu: "What influence can be exerted by cosmic rays on the device?" Hu assured him there was no possibility of an accidental chain reaction, but immediately went to the monitoring panel. This made the security officer even more anxious, and throughout the trip he repeatedly instructed Hu to measure the level of cosmic rays. Gao expressed great relief when the plane landed at the Lop Nur base, and he could hand the nuclear component over to Li Jue, head of the Ninth Academy, and Wu Jilin.

Security and formalism reigned supreme even at the base landing strip. The pilot and his passengers unloaded the container with the uranium core and placed it, still sealed and locked, in a special vehicle. Only after they arrived at the test site did Hu break the seal and remove the lock in the presence of the on-site leading cadres, technicians, and security officers. Once these people had verified the contents, they signed for custody of the nuclear component on behalf of the assembly team.

Under Zhang Aiping's overall authority, two men—Zhang Yunyu, commander of the Lop Nur Nuclear Weapons Test Base, and a scientist named Cheng Kaijia—now put the final preparations for the full test in high gear.[41] In the summer, before the delivery of the components, crews had put up an iron tower with an elevator that would lift the bomb 120 meters above the desert. Moving out from the tower, workers now began to lay out a radial pattern of monitors, cables, live animals, and military equipment (including aircraft, tanks, artillery, and the superstructures of ships). Various military officials and observers began to arrive, and leaders from the Ninth Academy—Li Jue, Wu Jilin, Zhu Guangya, and Wang Ganchang—undertook the meticulous inspection of the general test zone and the tower and other ground-zero facilities.

An aerial inspection of the test zone produced the only real surprise during the preparations. Photos revealed the presence of human inhabitants, and the safety officer issued an order for experienced desert trackers to conduct a search. After several days, a team discovered signs of an encampment. The search continued until one night the base trackers stumbled across a ragged group of Kuomintang fugitives, holdouts from the surrender of 1949: "Several men, haggard and shabbily dressed, were staggering along. . . . A baby uttered a cry, and the adults boorishly shouted curses." The male fugitives fled on horseback, and it took three more days to round up the more than 200 hideaways living near ground zero.[42]

Everything else went smoothly and by the book. Zhang Aiping had the prime responsibility for the on-site test operations as head of the First Atomic Bomb Test Commission and of the First Atomic Bomb Test On-Site Headquarters.[43] He and Vice-Minister Liu Xiyao (concurrently deputy commander of the on-site headquarters) personally inspected the tower. On one ascent to the top of the tower with Li Jue, Zhang asked: "Can you assure the security measures for the delivery of the atomic bomb from the ground to the top of the tower?" Li answered in the affirmative and volunteered that those "who would take part in the nuclear test had even climbed barehanded to the top of the tower several times."

With his confidence thus bolstered, Zhang directed all the predetonation procedures to be rehearsed one last time with a simulation bomb; such simulations had already been conducted in August. As the exercise got under way, however, a violent storm suddenly struck ground zero. Gale-force winds, not uncommon in that part of the world, began to batter the tower before all of the technicians could descend, but Li Jue ordered the simulation to continue. In a tent at the foot of the tower, Li, Wu Jilin, and others kept in touch with the remaining technicians by phone and ordered them to take readings from all their instruments throughout the stormy night. In one severe gust, the relentless wind put the elevator out of commission, and Li commanded specially chosen soldiers of the Technological Team to climb the tower with food and water for those at the top. In the early dawn, the simulation and the storm ended, and with the exception of the elevator's failure, Zhang declared the risky exercise a success.

Only two more steps in the preparations remained to be done: checking the calculations and setting the zero hour. The first step took place the day before the detonation. Still concerned about the possibility of firing a dud, Zhang Aiping sent an "urgent telegram" to the Second Ministry to ask for one last round of calculations. He wanted the theoretical physicists of the Ninth Academy who still worked in Beijing to "guarantee [by their calculations] that the probability of success for the test would be above 99 percent." Minister Liu Jie immediately assigned Zhou Guangzhao and two other mathematicians to the task even though most of the technical data had been shipped to the test base. The gaps in the data made it necessary for Zhou's team to reconstruct much of their earlier work from memory and forced them to work in relays to complete their calculations. Early the next morning, the date set for the explosion, Zhou and his colleagues signed their names to a memorandum for the Central Committee certifying that the probability of success indeed exceeded the required 99 percent.

By then, zero hour had already been determined. Chinese sources report that a young technician named Yang prophesied the time from his dream on China's National Day, October 1, 1964. Yang awoke and rushed from his tent, yelling, "The zero hour has been approved by the Party Central Committee."

Engineer Cai and his colleagues were awakened by his cry. "What is the matter?" "Technician Yang had a dream." Yang cried out with excitement: "I dreamed that the Party Central Committee has approved the zero hour. It consists of three fifteens!" "Three fifteens?" It suddenly dawned on his colleagues: "The first fifteen is the fifteenth anniversary of the founding of the People's Republic of China; the second means that today is October 1, and it will be on October 16, fifteen days from now; and the third means the atomic bomb will be detonated at 1500 hours that day."

Several days later, the Central Committee scanned the long-range weather forecasts and issued the zero-hour order: 1500, October 16. No one at the base voiced surprise, and whether true or not, the story of Yang's dream has become part of the lore of all who witnessed the first atomic explosion.

Detonation

The initial code-word command, "shoot the basket" (*tou lan*), was given on the fifteenth.[44] The base basketball players had suggested it. The order meant to insert the uranium component and initiator into the explosive assembly. The components from the Jiuquan complex already lay housed in the test-site assembly workshop, and only five people were permitted in this underground shop as the assembly commenced. Li Jue felt he had to be one of the five "so as to set at ease the minds of those engaged in the dangerous work." Two days before, technicians had put together the explosive assembly. Four of them now inserted the nuclear component and bolted the entire device together. Li Jue congratulated them on a flawless job.

The bomb was now partially armed. The Chinese say that, having reached this point, they were not worried.[45] They had fully prepared for the test, and according to their own account, had taken measures to protect personnel. This was true in respect to ensuring maximum protection from a premature bomb explosion. When technicians partially armed the device, all unnecessary people withdrew to shelters. Of the several thousand individuals involved in the test, only a few remained at ground zero.[46] Before midnight on the fifteenth, the whole nuclear device was assembled, and safety people from the ministry's Security and Protection Bureau determined that no electrical power was turned on and that no detonators were installed.

Fewer precautions had apparently been taken against the post-detonation dangers, though Zhou Enlai ordered personnel from the Anti-Chemical Corps and medical units to measure the fallout (using sample blood tests) and to assess its consequences on exposed civilians.[47] In this and subsequent tests, the safety specialists came to appreciate the seriousness and extent of the many problems associated with atmospheric testing.[48] Chinese sources state that some scientists and technicians on the base subsequently suffered from fallout radiation during the early test program and fault their own safety measures. The initial tests within the ground-zero "military restricted zone" produced serious radiation contamination throughout the zone, and cars now cross it only with tightly sealed doors and at high speed.[49] Nie Rongzhen reportedly became ever more worried about radiation health hazards at the base and in January 1967 made an inspection of safety procedures in Xinjiang. He discovered that many scientists and technicians on the Lop Nur base had been exposed to high levels of nuclear fallout contamination. "Youths were becoming prematurely bald."[50]

But on the fifteenth, as zero hour approached, few if any of the participants at ground zero gave a moment's reflection to any such threat. Calmly, Master Worker Cao supervised the moving of the bomb in its cylindrical container (about 2 meters high and 1.5 meters in diameter) from the silo workshop to the surface. At ground level, his men placed it on a handcart, and two technicians delicately pushed it to the tower. Several days before, Premier Zhou Enlai had spoken to leaders of the base by phone to discuss their roles at this penultimate point in the test and "even inquired about several details that were apt to be neglected by others."[51] Both technicians reportedly recalled Zhou's words as they shoved the cart to the tower elevator.

There, two other specialists took charge and, as had happened at each minute step along the way, formally and in writing transferred custody of the bomb to the tower crew. Li Jue, Wu Jilin, Wang Ganchang, and Zhu Guangya stood by in silence as the transfer ceremony proceeded. Now with the bomb theirs, the tower crew's chief, Chen Nengkuan, issued the order to the bomb escorts: "Hoist up!" The film cameras of the August 1 Film Studio began to roll.[52]

One-hundred-and-twenty meters up, the final technical crew received the bomb and began to install it on the tower. Over the next hours, they took readings from all the tower's on-board instruments and then inserted the detonating caps in the implosion assembly. For days Li Jue, Zhu Guangya, and their crew had worked around the clock. Now Li, Zhang Yunyu, and two technicians ascended the tower one last time to fix the electrical connections and make the final checks. When they reached the foot of the tower at T minus 50 minutes and counting, those on the ground asked them why they had descended four or five minutes later than scheduled. Li replied: "I myself had to be quite certain that there was no danger of anything going wrong."[53]

The tower team then moved to the test-site control room 23 kilometers away. Li Jue handed over the key to the tower electrical controls to the control room chief. This step had been introduced as a security measure so that the bomb could not be exploded with anyone near the tower. This was the last checklist item before final countdown. In the seconds before 1500, the ever-familiar numbers were called out: ten, nine, . . . At the exact time, the command was issued: "Fire!" (*Qibao*).[54]

In the seconds that followed the explosion, on-site commanders barked additional orders.[55] In the skies above the test, a special plane flew directly through the rising cloud, and another, piloted by a woman, began a 36-hour flight to collect air samples for later assessments of fallout.[56] Seventy kilometers away, technicians in the observation and command post peered at the explosion and towering funnel through protective glasses. The artillery unit launched rockets to collect samples from the mushroom cloud. Some took greater chances: "Wearing antichemical clothing, those assigned responsibility for chemical analysis drove vehicles into the test site to collect technical data relating to radiation and the [effects of] the shock wave." Special armored units headed directly to ground zero to prove the combat capability of their vehicles under nuclear conditions.[57] Watching the "moving scene," Li Jue and Wu Jilin remained speechless. Wang Ganchang, Peng Huanwu, Guo Yonghuai, Zhu Guangya, and Chen Nengkuan were overcome with the emotions released after long years of trial. They wept.

A mood of joy swept the command post.[58] Zhang Aiping, Liu Xiyao, and Zhang Yunyu all "trembled" with excitement. Zhang Aiping demanded scientific confirmation of the spectacle rising before them. "Is it a nuclear explosion?" Wang Ganchang said yes, and Zhang then placed a call to the Second Ministry in Beijing. This is recorded:

"This is Zhang Aiping. Please put me through to comrade Liu Jie [the Minister]."
At the Atomic Bomb Test Office of the Second Ministry, . . . Liu Jie and several cadres were waiting. . . . When the phone rang, the cadre [who answered] was so nervous that he dropped the receiver to the floor. Liu Jie picked it up.
"Please report to Chairman Mao and Premier Zhou that our first atomic bomb has been detonated."
"Repeat, please!"
"The atomic bomb actually detonated. We have seen the mushroom cloud!"
"I will report it immediately!" After this, Liu Jie picked up the special-use phone: "This is Liu Jie. Please put me through to Premier Zhou."
"This is Zhou Enlai."
"Comrade Zhang Aiping has phoned us from the test base. He has told us that the atomic bomb has been detonated and the mushroom cloud seen!"
"Good. I will report it immediately to Chairman Mao."

Several minutes later, Zhou rang Liu Jie back: "Chairman Mao has directed us to verify once again whether it really was a nuclear explosion so as to convince the foreigners!" Dutifully, Liu transmitted the Chairman's order to Zhang, who replied that there had indeed been a nuclear explosion, and that it had been verified as such. By this time, Liu had begun to shake noticeably, but he again placed a call to the premier. "Our first atomic bomb really has been detonated," he told Zhou. "It was a successful nuclear test! The Central Committee and Chairman Mao can rest easy." For the next few minutes, Liu felt too exhausted to speak. The burden of the ceaseless toil that began with his first plan at the Ninth Academy had lifted.

That afternoon joy erupted in Beijing.[59] By chance, thousands of actors and actresses had gathered in the Great Hall of the People after a performance of the musical extravaganza "The East Is Red" to meet the nation's central leaders. At 4:00 in the afternoon, they were greeted by Zhou Enlai. The premier gestured for silence and announced: "Comrades, Chairman Mao has asked me to tell you the good news. Our country's first atomic bomb has been successfully detonated!" At first the crowd remained silent, even stunned. Then the cheering began. Zhou responded: "You may celebrate all you like. But mind that you don't damage the floor." A few hours later, Radio Beijing began releasing the news to the world.[60]

EIGHT

Strategic Doctrines and the Hydrogen Bomb

Mao Zedong seems never to have entertained the notion that nuclear weapons had changed basic military and political realities or undermined his own preconceptions about war. He regarded everything he had said or written on international security and warfare as enduring and relevant even as his philosophy stressed change and the need for rethinking. What Albert Einstein once said about nuclear weapons altering everything save our ways of thinking applied with special force to Mao.[1] The Chinese leader would have tolerated no tinkering with his ideas on people's war and grand strategy in the name of nuclear reality.

On matters of war, Mao took the same dogmatic approach to doctrines that Stalin did. As the political scientist Herbert Dinerstein has observed: "Stalin's dogma merely provided a general framework for the discussion of war. The true criterion of its usefulness as a basis of discussion was its ability to solve the problems raised by the existence of nuclear weapons. This capacity was lacking in the [Stalinist] doctrine of the permanently operating factors, which debarred any discussion of the real issues."[2] Mao, of course, would have rejected in principle the idea of "permanently operating factors," but his rigid adherence to revolutionary precepts from the past amounted to the same thing.

This rigidity, however, undergirded and protected the nuclear weapons program once the Chairman placed his cachet on it. Thus supported, the program advanced quickly, and by 1963, forecasts about the date of a first atomic test were becoming firmer. Operational plans soon had to anticipate approaching deployment requirements, and publicly declared doctrines and military practice began to diverge.

We now examine the competing pressures on China's nuclear weapons policies in the months leading up to the October 1964 test and into the climactic years of the hydrogen bomb and ballistic missile programs. As the weapons effort proceeded, the policies associated with Mao Zedong became embroiled and reaffirmed in a fierce polemic with the Soviet Union and, after 1966, in the high drama of the Cultural Revolution. Yet within the planning and operations centers of the emerging strategic rocket forces—the Second Artillery Corps—military commanders sought guidance on where and how to deploy the new weaponry, on who the presumed enemy was, and on who could give the orders to launch. For these decisions, they needed more than Mao's "little red book" of revolutionary maxims.

Chinese Nuclear Policies, 1963-1964

Beijing celebrated the opening of 1963 by thrashing over and expanding its doctrinal arguments with Moscow.[3] To be sure, Chinese disquisitions on strategy and conflict defiantly adhered to Mao's core ideas, but in the year leading up to China's own bomb, Beijing added significant details to the well-known themes. These additions coincided with the final rush to the first nuclear detonation.

In January, the Central Committee's organ *Hongqi* once again invoked the name of Lenin to refute the "modern revisionism" of the Soviet Union and Moscow's counterrevolutionary opposition to Mao's theses on imperialism and atomic weapons. It labeled Soviet doctrines "music for the consolation of slaves."[4] Then, in March, *Hongqi* published a long tirade that, inter alia, rehashed what were by then familiar Maoist positions on nuclear war and strategy.[5] It quoted the Chairman at length on imperialism's impending political and military decline and then adapted this forecast to Mao's line of despising the enemy strategically while taking it seriously tactically. The perceived urgency for intensified struggle, *Hongqi* said, mirrored a tactical assessment, for "imperialists and reactionaries will not automatically tumble from their thrones."[6] In the coming showdown, China's enemies could be expected to resist and might be tempted to use nuclear weapons. Ahead, the Chinese could foresee the coming fight in Indochina.

Hongqi prefaced its review of strategy and tactics with a general essay on war and peace. A main theme in the historical part of the review was that wars stem from the aggressive nature of imperialist states and the inherent contradictions among them, not from popular resistance or the policies of socialist states. This fact rendered foolish any unilateral or premature disarmament as paths to peace. Despite the contention of Soviet authors that revolutionary struggle would result in global war, such struggle, not disarmament, was the only viable means of averting that war. Further, atomic weapons had not altered "the laws of social development" or caused Marxism-Leninism to become outmoded. Indeed, by brandishing nuclear weapons, the imperialists had merely uncloaked "their evil ambition" and their suppressed fears, and had exposed themselves to greater isolation within the international community. If used by the imperialists, these weapons, as Mao had argued in the past, would destroy the very wealth and populations they were intended to conquer. Finally, the *Hongqi* article declared, "the secret of nuclear weapons has long since ceased to be a monopoly." In socialists' hands, these weapons would have "a purely defensive purpose," and there would be no danger of war "with nuclear superiority in their hands."[7]

Three months later, the Chinese press expanded on these refrains in the first of Beijing's major "polemics on the general line of the international Communist movement."[8] This document reiterated the argument that disarmament proposals had the twin virtues of exposing the sinister designs of the imperialists and fostering political combat with them. But disarmament itself would be attained only after "combined struggle." In such a struggle, the demand for "the complete banning and destruction of nuclear weapons" alone would be appropriate; the Chinese countenanced arms negotiations between the socialist and imperialist countries solely when the negotiations stayed confined to this uncompromising end.

As noted, the ideas expressed in these 1963 statements mostly echoed fundamental Maoist sentiments reaching back to the 1950s or earlier, but the constancy of Beijing's position appears to have altered somewhat in the face of the limited test ban treaty signed by the United States, Great Britain, and the Soviet Union on August 5.[9] By this act, the Chinese asserted, the Soviet leaders had switched sides. They had joined the imperialists "to consolidate their nuclear monopoly and bind the hands of all peace-loving countries subjected to the nuclear threat." The test ban treaty had met the demand for a "genuine peace" by appeasement, a "fake peace."[10]

What is relevant to this study is the tripartite treaty's impact on Beijing's thinking about its adversaries and on its strategic outlook in light of its own nuclear program. In this respect, the Politburo judged "the central purpose" of the treaty to be the prevention of "all the threatened peace-loving countries, including China, from increasing their defence capability."[11] The Chinese leaders argued that "every socialist country has to rely in the first place on its own defence capability" and not on Moscow for its survival.[12] For socialist countries other than the Soviet Union to have weapons would promote peace: "the greater the strength on our side, the better."[13]

From the Chinese point of view, the limited test ban treaty signaled a reversal by the Soviet Union—its repudiation of a coordinated strategic doctrine for all socialist states. From now on, the members of the Soviet Politburo would be considered renegade leaders who aimed "only to preserve themselves and would leave other people to sink or swim." While proclaiming the traditional links to the "Soviet people," the Chinese assailed the collusion of "Soviet leaders" with Washington in their "attempt to manacle China" and repeated a theme from memoranda sent to Moscow since September 1962: "It was a matter for the Soviet Government whether it committed itself to refrain from transferring nuclear weapons and technical information concerning their manufacture to China; but . . . the Chinese Government hoped the Soviet Government would not infringe on China's sovereign rights and act for China in assuming an obligation to refrain from manufacturing nuclear weapons." From the Chinese perspective, the test ban treaty was an undisguised Soviet-American attempt to "bring pressure to bear on China and force her into commitments."[14] A year away from conducting its own tests, Beijing could only regard the accord as a means of exerting hostile influence on China to put brakes on its nuclear program.

In a follow-up attack on a Soviet rebuttal to their accusations, the Chinese responded to Moscow's question, "Would the Chinese leaders feel more secure, even if they sat on their own atom bomb?"[15] The answer, Beijing replied, was obviously yes. A newly created nuclear arsenal, to be sure, could cause the United States to "aim many more atom bombs at China," and the People's Republic would have to remain vigilant to this unpleasant possibility. But since the United States had already targeted China with nuclear weapons, "a few more" would not add to the danger. On the contrary, China's nuclear forces would make the country safer.[16]

The Chinese Politburo used the occasion of this response to refine still further what it deemed appropriate actions for socialist countries to take in the face of nuclear threats. It held that in the Soviet-American Cuban missile crisis of 1962, Moscow had departed from doctrinally acceptable policies by practicing "both the error of adventurism and the error of capitulationism." A proper socialist policy would have avoided the dangers of head-on clashes with the United

States in the first place. But once committed to the Cuban deployments, the Soviet Union should not have yielded to Washington's "humiliating terms" and withdrawn the missiles from Cuba. This retreat could only inflate "the aggressiveness and arrogance of the imperialists." The Soviets had made the error of reversing the formula for a correct line of strategy and tactics: "If one does not take full account of the enemy tactically and is heedless and reckless, while strategically one dares not despise the enemy, it is inevitable that one will commit the error of adventurism in tactics and that of capitulationism in strategy." Any principled policy would have to combine realistic caution and rational resolve.[17]

At the time of this comment, the detonation of bomb 596 was still more than a year away; the final work on uranium enrichment and on the bomb design was progressing in Lanzhou and Qinghai. Within weeks of the comment, Foreign Minister Chen Yi told a group of Japanese correspondents: "Atomic bombs, missiles and supersonic aircraft are reflections of the technical level of a nation's industry. China will have to resolve this issue within the next several years; otherwise, it will degenerate into a second-class or third-class nation."[18] This statement seemed to confirm Soviet accusations that, as the strategic specialist Alice Langley Hsieh once noted, "the reason why the Chinese want to develop their own nuclear weapons is in order to pursue special aims and interests that cannot be supported by the military forces of the socialist camp."[19]

On November 19, 1963, the Chinese enunciated just such a reason—a reason far more ominous for Moscow than the one given by Chen Yi.[20] Reaffirming their long-standing support for no-first-use policies and their opposition to the employment of nuclear weapons in revolutions, they called on socialist countries to "achieve and maintain nuclear superiority." But which countries deserved to be called socialist? Not the Soviet Union, the Chinese implied: "The leaders of the CPSU [Communist Party of the Soviet Union] direct the edge of their struggle . . . at the socialist camp. . . . They use nuclear blackmail to intimidate the people of the socialist countries. . . . They use nuclear blackmail to intimidate the oppressed peoples and nations." The Chinese were on the verge of proclaiming that a nuclear arsenal of their own might defend the nation against that Soviet threat.

For the next several months, in late 1963 and early 1964, Chinese military writings turned to other issues—shifting the emphases to "man as the primary factor," to studying the works of Mao Zedong, and to "political and ideological work in the army"[21]—and the Sino-

Soviet debates merely replayed old rhetoric on matters of war and peace. Still, some small signs that a serious doctrinal review was under way did surface. In April 1964, for example, the Chinese published a rare article on Western defense doctrines and examined the U.S. strategy of flexible response. The writer, Hung Fan-ti, recalled President John F. Kennedy's remarks that the United States had not solved the problem of dealing "with a whole arsenal of missiles coming at us at maximum speed with decoys." Hung noted the passing of the strategy of massive retaliation associated with Eisenhower's New Look, a strategy that had been based on a "willingness to commit suicide." He stressed that the United States had been forced to forsake the Eisenhower strategy because of the growing Soviet nuclear arsenal, a change that showed that nuclear weapons in socialist hands could produce positive results, and that the "enemy" was in fact not nuclear weapons but "U.S. imperialism."[22]

Several months later, the Chinese used the occasion of the Tenth World Conference Against Atomic and Hydrogen Bombs to underscore this last point. As China was readying its own weapon, Mao's Politburo wanted to deflect the international peace movement away from opposing the bomb itself and toward condemning "U.S. imperialism, the most ferocious enemy of world peace." In a lead speech to the conferees, Liu Ningyi, head of the Chinese delegation, ridiculed a Soviet-backed line that saw "nuclear weapons rather than imperialism headed by the United States as the source of nuclear war." Liu again denounced the limited test ban treaty, an agreement that was "favorable to nuclear monopoly and nuclear blackmail by nuclear powers and to U.S. imperialism." He repeated the allegation that the Soviet Union had used the treaty to "oppose socialist China." While some "kind-hearted and well-intentioned" people might think that "a partial prohibition [of nuclear weapons was] better than no prohibition," they were simply the victims of a Soviet-American trick.[23]

Liu's speech was to be China's last major statement on the question of nuclear weapons prior to the explosion of October 16, 1964. Most of the motifs found in the earlier statements were to be repeated throughout the rest of the 1960s. However, these motifs bore increasingly little relationship to the Chinese military's most pressing concerns as the nation divided and became entangled in the initial disputes preceding the Great Proletarian Cultural Revolution.[24]

Rather than spurring the leadership to explore alternative doctrines on the role of nuclear weapons now that they were becoming part of China's arsenal, the October atomic test was used mainly to reinforce

China's prevailing international line. The exploration of new strategic doctrines ceased. Beijing's open writings on strategic weapons served important polemical and tactical purposes but skirted fundamental security issues. For the leaders at the top, nuclear weapons were soon to become a sideshow as the grand struggle for power commenced.

The Development of the Hydrogen Bomb

The assembling of the first atomic bomb marked a moment of high drama, but it did not mean a slackening in the Second Ministry's work load. In the first place, the atomic energy complex in Jiuquan would need to prepare for the serial production of fission weapons and to finish the facilities for plutonium, and the uranium program would have to press ahead from the mines to Lanzhou. And then there was the matter of the hydrogen bomb.

The basis for the transition from the work on one type of weapon to the next can be traced to 1959 and China's "place in line" as a follow-on nuclear state. At that time, according to Vice-Minister Liu Xiyao, the central leadership in Beijing began to "draw up an important plan and to concentrate in time our manpower and material resources to develop the necessary new materials, equipment, and instruments concerning whose secrets we could not acquire from foreign countries." This plan included both atomic and hydrogen weapons. By the next year, the Ninth Academy's Beijing Nuclear Weapons Research Institute was preoccupied with the atomic bomb, and the Second Ministry therefore directed the Institute of Atomic Energy to form a "leading group" for research on thermonuclear materials and reactions. This group, composed of Qian Sanqiang, Huang Zuqia, Yu Min, and others, began "to accumulate continuously several basic elements [*jiben canshu*] related to the process of thermonuclear fusion."[25] Knowledge of these elements, Liu Xiyao says, "gradually gave us command of the relevant basic laws."

Immediately after the atomic bomb design group at the Ninth Academy had finished its long work on weapon 596 in September 1963, Nie Rongzhen ordered its members to stay together and shift to fashioning a thermonuclear device.[26] In Qinghai, the Theoretical Department (led by Deng Jiaxian) assumed the principal burden of designing China's first thermonuclear weapons, as it had in the last stages of designing the atomic bomb. While we know few details of this work, one Chinese source asserts that Beijing's decision to proceed along multiple lines to the fusion device without delay greatly shortened the interval between the fission and fusion weapons.[27]

Storm hitting a meteorological station at Lop Nur

Living quarters at Lop Nur in the 1960s

Soldiers of the Chemical Defense Corps training at Lop Nur

Zhang Yunyu checking plans at Lop Nur, 1962

The cylinder containing China's first atomic bomb, on its way to the tower at Lop Nur, October 15, 1964

The bomb being hoisted to the tower, Lop Nur, October 15, 1964. The men in the photograph are seated on the cylinder containing the bomb.

Zhang Aiping inspecting tank to be driven across area adjoining ground zero after explosion of the first bomb

ABOVE: The tower on which China's first atomic bomb was detonated on October 16, 1964

RIGHT: The mushroom cloud of China's first atomic bomb

Soldiers of the Chemical Defense Corps advancing toward contaminated areas after nuclear explosion

Zhang Aiping phoning Zhou Enlai to report the successful explosion, October 16, 1964

Zhou Enlai in Beijing, announcing the successful explosion

Women technicians at ground zero of China's first nuclear exposion

Lift-off of the nuclear-armed
DF-2 missile, October 27, 1966

A Qiang-5 attack aircraft, of the type
configured for launching an atomic
bomb on January 7, 1972; the explosion is shown on the cover of this
book

Marshal Nie (center) receiving Zhang Yunyu (left) and Zhang Zhishan (right),
commander and deputy commander of the Lop Nur base, November 1966

China's first hydrogen bomb dropping from a Hong-6 bomber, June
17, 1967

Deng Jiaxian (left) and Yu Min

Wang Ganchang

The explosion of China's first hydrogen bomb

Embedded in any H-bomb effort lay a fundamental choice. The issue, loosely conceptualized, centered on the size of the weapon. The explosive force of the bomb dropped on Hiroshima was equivalent to about 12,500 tons of TNT (12.50 kilotons), a force that the Chinese matched in their first atomic test on October 16, 1964. Nie Rongzhen's designers might have settled for a next-generation weapon ten times the size of the bombs dropped on Japan and their own first atomic bomb. However, PRC strategists knew that megaton-sized warheads deployed on a planned force of ballistic missiles could compensate partially for the initial gross inaccuracies of the delivery systems. That calculation was simple. More significantly, they believed that a megaton arsenal would add universal credibility to Chinese military power, even though they recognized that the attempt to develop such an arsenal would demand great creativity and might fail.

The real value of megaton weapons was that the bigger bang might disproportionately influence the perceptions and, ultimately, the decisions of China's foes. These weapons could have decisive strategic significance in preventing a nuclear attack. Beijing's foreign policy specialists and military planners understood that technological attainments—for example, the making of high-yield warheads or the mastery of missile engineering—could send important political messages to worst-case planners in Washington and Moscow. In the aggregate, such messages might serve in place of declared new strategic doctrines; they could supersede people's war and paper tigers.

At the outset, then, Chinese decision makers had to agree on which type of hydrogen weapons to build. To appreciate the choice facing them, we must discuss briefly the nature of the technologies involved. In thermonuclear weapons, the fusion process exploits the extreme temperatures produced by a fission explosion. These temperatures recreate conditions similar to those at the center of the sun, where hydrogen or one of its isotopes, deuterium (H^2) and tritium (H^3), become fused into helium. "Several different fusion reactions have been observed between the nuclei of the three hydrogen isotopes, involving either two similar or two different nuclei."[28] When the helium nucleus forms in this thermonuclear process, energy is released as fast neutrons, fast light nuclei, and other forms.

Fusion weapons are of two general types: fusion-boosted fission weapons and multistage thermonuclear weapons.[29] In the first type, the bomb makers place fusion material in the core of an implosion fission weapon. The fusion material is surrounded by fissionable uranium or plutonium encased in high explosives. "When [this core] is compressed by the chemical explosion, an uncontrolled chain reaction begins. [As the temperatures rapidly rise, the] fusionable material inside the device . . . ignites." The resulting fusion reactions release large numbers of fast neutrons that speed up the fission chain reaction; the consequent higher temperatures in this "synergistic process" create a more efficient fission bomb. Its explosive power could reach several hundred kilotons.

In the true thermonuclear bomb, a fission device within the weapon acts as a primer or trigger. The detonation of the trigger creates enormous temperatures within the bomb casing, which initially acts like a bottle and keeps the energy confined. "Because the casing is massive, the large expansive forces exerted on it do not cause it to disassemble before the compressive forces can act on the secondary [fusion material]." The high temperatures and compressive forces cause the fusion materials to ignite, releasing large amounts of energy and fast neutrons. "If the massive casing consists of natural uranium, the fusion neutrons will cause the uranium nuclei to undergo fission." A device of this sort is sometimes called a three-stage fission-fusion-fission bomb. Its power has no theoretical upward limit (though engineering feasibilities would impose constraints). In the early 1950s, American and Soviet weapons specialists had demonstrated the design principles of such a multistage bomb; as theory, these principles were known to a small group of Chinese scientists as well.

In May 1964, Mao Zedong urged these scientists to speed up the hydrogen bomb project. The initial issue for them, Liu Xiyao states, was whether to settle for a boosted weapon or press on to a full-fledged multistage weapon. In July 1964, when the Second Ministry received a directive from Premier Zhou Enlai that called for marrying a nuclear weapon to a guided missile, the officials understood Zhou's message as a directive to produce a multistage missile warhead. Liu explains that China chose the high-level, or multistage, thermonuclear path because the United States posed a high-level threat: "Since we had to overcome nuclear blackmail [with the acquisition of fusion weapons], we thought we had to develop hydrogen bombs with deterrent force. Facts have proved that we made a correct decision." Liu adds: "As for the high-level goal, we had to be able to produce thermonuclear warheads that would be fixed to our intermediate- and long-range missiles."

The July directive, presumably endorsed by the Fifteen-Member Special Commission, apparently was the basis for Nie Rongzhen's directive to the Ninth Academy. Three months later, within hours of the first detonation at the Lop Nur test grounds, Zhou underscored the directive and issued another implementing order: "With the weapons of [Mao's] articles 'On Practice' and 'On Contradiction' and of the experiences acquired in the Daqing oil field, you should depend on a leading body composed of cadres, specialists, and workers to make a breakthrough as soon as possible in the development of the hydrogen bomb project." For those at the Ninth Academy, the arrival of Zhou's edict intensified the high emotion of the first test and forced the momentum to continue.

One reason why it was so much easier to build the thermonuclear weapon than a fission one was that a staff with the basic scientific and technical expertise had already been assembled and had gone through the experience of building a nuclear weapon. According to Liu Xiyao, "The situation was entirely in our favor when we decided to start the development of our hydrogen bomb project." Here, the relationship between theoretical and practical design was much closer and required far less intermediate testing. Moreover, Liu states, "We knew what kinds of materials should be used to carry out thermonuclear fusion. We also commanded the necessary fundamentals [that] had been used to make hydrogen bombs in the United States, the Soviet Union, and Great Britain." Some of the key materials, including those in the fission device itself, were in production. By this time, too, more sophisticated Chinese computing systems had become available: "We had calculators and rudimentary computers sufficient for calculating and verifying the theoretical design" of the thermonuclear reactions.[30] Chinese theoretical scientists had started work on "key elements" as early as 1960, but most of all, perhaps, officials in Beijing had confidence in their scientists. Liu declares: "We had a large number of nuclear scientists who, compared with foreign nuclear scientists, were second to none. What the foreigners had accomplished, we could accomplish."

Yet all did not depend on self-reliance, as Liu acknowledges: "Without Soviet industrial help, . . . it would have been impossible for us to have achieved such a rapid success in making the atomic and hydrogen bombs." Without giving details, Liu adds that China could not get "any secret scientific or technical data concerning hydrogen bomb development," but that it did benefit from analyzing the relevant reports published abroad. These reports especially "contributed to the unification of our ideas and to the determination of our goals."

What the Chinese learned from this reading of the foreign literature made them aware of the materials they needed for the thermonuclear reaction, and by the winter of 1964-65, they were completing the pro-

duction line for lithium-6 deuteride, one of the principal thermonuclear materials.* They "knew nothing, however, about the performance of those materials . . . or about the concrete conditions under
which a hydrogen bomb would be detonated." They had yet to learn
how to employ an atomic bomb to create those conditions, and their
task was complicated by having to use enriched uranium instead of
plutonium in the fission trigger. These fundamental issues preoccupied
the design group that, under the Ninth Academy, contained specialists
from both the Beijing Nuclear Weapons Research Institute and the
Institute of Atomic Energy.

In February 1965, the ministry directed the latter institute to work
on measuring the lithium deuteride molecule and analyzing its reactions. Qian Sanqiang assigned 50 of his top scientists and engineers to
the task and allocated four accelerators and other scarce equipment
for their use. He placed the woman scientist He Zehui in charge of
coordinating this delicate research.

The design group's principal work on the H-bomb took fourteen
months. At first, the research calculations and experiments at the
Qinghai Ninth Academy and elsewhere were centered on the possible
designs for the hydrogen bomb's igniter as a first step in understanding
the entire detonation process. In round-the-clock shifts, scientists under Deng Jiaxian worked out the theoretical basis for the most promising plans by the end of 1965, when the Chinese discovered "some
key points relating to the internal causes and external requirements
for burning thermonuclear materials."

This discovery hinged on the successful completion of extensive calculations and required the use of China's best computing facilities.
Some time in late September 1965, the academy sent Yu Min, the
deputy director of the Theoretical Department, to Shanghai to make
these calculations, and within about two months he cabled Qinghai

---

*Lithium-6 deuteride is a chemical compound. During the fusion reaction in the
weapon, tritium is produced by the neutron bombardment of the Li^6. Using Soviet plans,
the Second Ministry in 1959 began building the Lithium-6 Deuteride Workshop at the
Nuclear Fuel Component Plant (No. 202) in Baotou, Nei Mongol, to complete the
isotope separation, produce the deuterium, and synthesize the compound. The withdrawal of Soviet assistance denied the Chinese some of the critical equipment and scientific data for the workshop, and the Chinese responded by transferring scientists from
the Institute of Atomic Energy to a new Lithium Isotope Research Section of Plant 202.
The Ministry of Chemical Industry provided heavy water for the deuterium, and the
Ministry of Metallurgical Industry supplied the minerals containing lithium. The first
batch of lithium-6 was separated on September 17, 1964, and a week later the workshop
had its first lithium-6 deuteride. During these same years, the Chinese also worked on
the production of tritium in a self-designed plant, which went into full operation in May
1968. Li Jue et al., *Dangdai Zhongguo*, pp. 245-55 passim, 470.

that he had "found a shortcut" to the super weapon.[31] This optimistic report prompted Deng Jiaxian to join Yu in Shanghai, where he confirmed Yu's findings. The two then discussed their results with ministry officials and convinced Vice-Minister Liu Xiyao (who was in charge of the ministry's daily affairs) to order a battery of "cold tests."[*] It was these tests, probably carried out in early 1966, that proved the "key points . . . for burning thermonuclear materials" and led to the first test explosions of a fusion-boosted fission weapon.

The "booster" proved feasible in a test on May 9, 1966, when a Hong 6 medium bomber dropped a 200–300-kiloton uranium device that contained lithium-6. This test focused on the performance of the thermonuclear materials, and on December 28, 1966, a follow-on test examined the "fundamentals of a thermonuclear explosion" in a 300–500-kiloton uranium-lithium device. The two detonations proved the validity of the Deng-Yu "theoretical plan," and with their success, the Fifteen-Member Special Commission decided to proceed directly to the testing of a multistage thermonuclear bomb, a three-megaton device, on June 17, 1967.

While the theoretical and experimental work proceeded at the academy and the test base, Party radicals, moving toward the Cultural Revolution, instigated a diverting series of political movements.[32] Time had begun to run out for the national consensus on the strategic program that had been forged at the end of 1962. In the winter of 1965-66, the political-military leadership in Beijing began to fragment, and the following spring the unleashing of ultra-Leftist forces started. In April, the army announced the launching of the Cultural Revolution.[33] The upheaval that followed over the next decade, a major story in itself, steadily undermined the strategic weapons program despite Mao's initial inclination to protect it.[34]

The hydrogen bomb project did not escape the effects of the Cultural Revolution. As the violence of the mass movements spread during the summer of 1966, the Ninth Academy no longer seemed so remote: "All of a sudden, riots erupted in China. Several well-known scientists had to suspend their research. [At home in Beijing,] Deng Jiaxian's wife was publicly denounced, and later his son and daughter were compelled to relocate to the countryside."[35] Despite what could not help being a personally upsetting turn of events, Deng judged that no routine or administrative response to the growing unrest in Qinghai would stem the tide of mass criticism. He therefore attempted to

---

[*] Chinese normally use the term "cold tests" (*leng shiyan*) to mean the physical testing of components without the inclusion of fissile material.

make use of the mass enthusiasm by organizing and enlisting it in support of the thermonuclear project. A contemporary report notes that the theoretical design unit put forward the slogan: "Carry out the hydrogen bomb test ahead of schedule in tribute to the Great Proletarian Cultural Revolution." The political campaign reportedly spurred the members of the unit to work overtime, with the result that they finished the design a month ahead of schedule.[36]

Another tactic was to promote patriotic competition with the French, who at that time "were speeding up their hydrogen bomb development program. Learning of this information, Chinese scientists . . . suggested that 'we have to succeed in detonating our first hydrogen bomb ahead of the French.'" Deng Jiaxian supported this suggestion and won approval from the ministry to use the scientists' words as a slogan to inspire a mood of competition among all involved. As banners bearing the slogan went up in the Ninth Academy, the mood of hostility and the personal conflicts temporarily ebbed: "The scientists and technicians who had been caught up in factionalism began to discuss all possible ways to reach a final solution [for the theoretical design of the hydrogen bomb]." For the time being, a spirit of teamwork returned to the Theoretical Department.

The Cultural Revolution and the H-Bomb Test

That spirit could not long remain insulated from the enervating influence of the Cultural Revolution. In the country at large, where the tempo of events was accelerating, the mass movement had become fully identified with the testing of nuclear weapons by mid-1966. It was as if a bizarre compact had been arranged between man's frenzy and the unleashed atom.

As the banners of the rebellious Red Guards were unfurling in Beijing during 1966, for example, soldiers of the Second Artillery Corps at Shuangchengzi, near Jiuquan, had launched a DF-2 missile armed with an atomic warhead against a ground zero in Xinjiang 800 kilometers to the west. When announced, this test on October 27 prompted the young fanatics to take to the Beijing streets with dancing. They sang the praises of Mao's thought and pasted up "big character" wall posters that acclaimed the bond between the creative force of Mao's radicalism and the force of the bomb. The banners in the streets displayed the Chinese characters for "double happiness" as well as condemnations of U.S. imperialism.[37]

Nie Rongzhen, fully recognizing the extraordinary risk involved in the October 27 test, had flown to Shuangchengzi two days before to take personal charge.[38] Nie understood that "if by any chance the nuclear warhead exploded prematurely, fell after it was launched, or went beyond the designated target area, the consequences would be too ghastly to contemplate." He ordered the missile commanders to brief him in detail about the safety procedures and, after the firing, quickly flew to the impact zone at the Xinjiang test base to assess the missile's accuracy. Nie determined that the DF-2 was "deadly accurate."

Meanwhile, back in Beijing, all the staff members of Nie Rongzhen's Defense Science and Technology Commission now concentrated on the study of Mao's writings and acclaimed "the great pioneering effort whereby the spiritual atomic bomb [had] led to the explosion of the material atomic bomb." Party leaders enjoined Nie's weapon makers to "go to the toughest places to temper and remould themselves." What if they did have to "live in tents, drink brackish water, and always work under the scorching sun or in sandstorms?" In doing so, they were answering Mao's call for scientific and technical personnel to "plunge themselves into struggle and practice. . . . By getting down into the dirt and mud, they have become imbued with the same thoughts and feelings as the workers, peasants and soldiers. . . . They work out all data and trace every line on a blueprint in the spirit of being responsible to the people of all China and of the world." Enduring through hardship had proved their mettle as staunch loyalists in the face of those who would betray or desert Chairman Mao in his hour of trial.[39]

In January 1967, Nie's strategy of allying with the radicals to fend off radical penetration of the strategic program began to unravel. Contending factions had already begun to fragment the Second Ministry's Spartan order, and then, suddenly, the system collapsed. Three main factional groupings rallied their warriors and rushed forth to battle each other. The largest, the Revolutionary Rebellion General Headquarters (Geming zaofan zongbu) directed its attack against the ministry's leadership and staff, who organized themselves into the Red Flag General Headquarters (Hongqi zongbu). A small extremist faction, the Extract Poisonous Nails General Headquarters (Ba duding zongbu), seized the opportunity for maneuver and for a time greatly intensified the conflict. When the battle was over, the Geming zaofan zongbu had prevailed.[40]

With discipline collapsing at the ministry in Beijing, staff members of the Nuclear Weapons Test Base defied standing orders and invited units of the Red Rebellious Corps (Hongse zaofan tuan) of the Harbin

Military Engineering Institute (HMEI) to Xinjiang to "exchange experiences." The Defense Science and Technology Commission had assigned the institute to train personnel for the strategic weapons program, so that there were many Harbin alumni on the test base's staff who had maintained close ties with their alma mater. Mao Zedong had incited Mao Yuanxin, his nephew and a recent HMEI graduate, to attack higher authorities, and he and his band of institute fanatics set off on the 4,000-kilometer journey to the Lop Nur base. A week before their arrival, Nie Rongzhen had gone to the base; his immediate reaction to the impending "revolutionary rebellion," as such Red Guards' exchanges were called, was outrage.[41]

Nie promptly ordered the Xinjiang Military Region Command to secure the base as an emergency measure and in accordance with standing security regulations.[42] Base security forces arrested the visitors from Harbin as they stormed toward the test grounds, and the confrontation grew as rival groups mobilized their factions. To add to the crisis, one young rebel phoned Beijing and reached Chen Boda, head of the Cultural Revolution Group under the Central Committee. Chen immediately phoned Nie. The high-riding Chen reached Nie at the base command center, and a tense conversation ensued. Chen: "Who issued the order [to detain the rebels]?" Nie: "It was I." In the argument, Nie reminded Chen that the Central Military Commission had made him, Nie, responsible for the success of the test program; then he rudely hung up.

This response of arrests and defiance momentarily brought the situation under control, but Nie apparently recognized the precariousness of his position. He ordered a scheduled experiment to proceed on the morning of January 10, 1967, and shortly thereafter returned to Beijing for a key meeting of the Central Military Commission on January 19. At this meeting, the military command, jointly with the Central Committee and the State Council, decided that "troops must be dispatched at once to exercise military control over all important granaries and warehouses, prisons, and other important units that must be protected and watched according to the provisions of the Central Committee."[43] "Other important units" included the strategic weapons bases. Thus reaffirmed, the shaky union of high science and crude politics continued, though not without major disruptions, into the spring of 1967.[44]

In early May, the Nuclear Component Manufacturing Plant in the Jiuquan complex completed the assembly of the bomb, and on May 9 the Fifteen-Member Special Commission ordered the Lop Nur base to finish preparations for the test by June 20. Nie Rongzhen's commission then instructed the base to conduct a trial run with a prototype weapon without its nuclear component.[45]

Though the nuclear program was momentarily spared the worst of the climate of chaos, political intrigue and professional indifference disrupted the preparations for the first multistage thermonuclear detonation, now scheduled for June 17. At the Lop Nur Nuclear Weapons Test Base, some of the workers and staff members had become embroiled in the Cultural Revolution's compulsory rituals: "speaking out freely, airing views fully, holding great debates, and writing big character posters." Furthermore, they had moved to "seize power" despite official prohibitions against such actions by workers in the strategic weapons program, and their preparations for the test had stalled.[46] Politics had "taken command."

A few days before the test, Wang Ganchang from the Ninth Academy came to the dormitory room of a workshop director at the base and ordered him to put his men back on the job. The director replied: "The regular methods [on the test base] just don't work because authority in the workshop has been seized." He noted that the difficulties even affected assembling and feeding his personnel, and told Wang that there was no money to pay the cooks. Wang in dismay paid the cooking staff from his own pocket and turned to organizing the truck drivers in the motor pool to deliver equipment and the bomb to the test site.

At about this same time, Premier Zhou learned in a national briefing that the rehearsals for dropping the bomb by parachute had revealed three tears in the parachute.[47] The air force had been conducting training drops of model bombs since April. He ordered the implementation of new safety procedures, including additional tests of the detonation control system, and sent Nie Rongzhen to supervise the June 17 test.

On that day, technicians at the test site, under the personal leadership of Nie Rongzhen, posted a large portrait of Mao Zedong alongside clusters of red flags and placards inscribed with Mao's sayings.[48] In the preparations for the test, young workers at the nuclear test base had purportedly broken with official procedure and started on their assignments long before the design of the weapons had even been completed at the Ninth Academy.[49] Base workers, it seems, had manufactured their own monitoring instruments rather than obtaining them from factories in order to prove their faith in Mao's principle of self-reliance. The technicians at the control center for the test "carried on their designing, testing, and trial manufacturing while condemning and repudiating the towering crimes committed by the top party persons in authority taking the capitalist road."

A few hours before the test, the bomb was hoisted onto a Hong 6 medium-range bomber for delivery to the assigned drop point over the Lop Nur test base. On board the plane, He Xianjue, a 1964 graduate of the Northwest Industrial University, had the responsibility of completing the final check of the bomb. He performed the last-minute adjustments in the instrument settings "so as to make it possible to increase the precision [timing] of the detonation" after release.[50]

Once airborne, however, some crewmen on bomber 726 suffered a case of nerves. For his part, the pilot found it difficult to operate the aircraft normally, and the navigator once over the target failed to release the bomb at 8:00 A.M., as scheduled. Word of the delay was flashed to Beijing, and Zhou Enlai personally radioed the crew members and encouraged them "to remain calm and act resolutely." The pilot, Xu Kejiang, then made a second run, and the bomb was released at 8:20.

The explosion was devastating. Steel plate 400 meters from ground zero melted, as did concrete blocks whose surfaces turned to glass. The shock wave struck a 54-ton locomotive 3 kilometers from ground zero and shoved it 18 meters. Semi-buried fortifications were sliced apart, and brick houses 14 kilometers away collapsed. The spectacular fireball rapidly became a massive mushroom cloud.

On the ground, politics took command. The military press credited a combination of technical competence and political zeal for the flawless working of the control center's equipment in the "split second of the H-bomb explosion." The operators of this equipment reportedly had just arrived "fresh from college," and their ideological orientation symbolized the heroic mood surrounding the test and politicized the event.[51]

Moments after the order to release the bomb was given, soldiers from the base guards unit leaped from their protective trenches and shouted "Long live Chairman Mao!" "The flash was an order; the explosion was the signal to charge; and the smoke [of the mushroom cloud] was the target. They pressed ahead toward the explosion center." The media equated their mass charge to the surging forward of the Chinese nation toward new peaks of greatness. The H-bomb had "shocked the whole world." The Chinese people had demonstrated their "will power and ability to catch up with all the countries and surpass them."

From Test Devices to Weapons

By the time the Hong 6 bomber released the thermonuclear device in June 1967, plans for China's nuclear arsenal were already far advanced. As early as 1960, the Central Military Commission had ordered the Ninth Bureau to begin studying how to "weaponize" (*wuqihua*) the nuclear program, how to move beyond test devices to deliverable bombs and warheads.[52]

Almost immediately, the Beijing Nuclear Weapons Research Institute added weapons development to its planning priorities. It began investigations into the aerodynamic properties of different shapes, the structures of warheads, and detonation control systems. The Second Ministry also responded. It made contact with the ministry-level bodies responsible for aircraft, electronics, conventional weapons, and missiles in order to start research and development on air-dropped nuclear weapons. Such weapons, the Chinese soon concluded, would require the building of greatly improved data control and test facilities, and the order went out to begin their construction.

Starting in April 1960, the design teams fashioned a number of weapons models with different aerodynamic configurations and began wind-tunnel testing. By the end of 1961, the tests had proceeded to the point that units could begin experimenting with small-scale models at the Artillery Corps test grounds. The testing of full-sized models started a year later at air force bases, while other units worked on the detonation system composed of batteries, fuses, safety locks, and the detonator itself. The system was ready for flight testing by the end of 1962.

In 1963, the air force assigned its Independent Fourth Regiment to carry out this mission. Working with the regiment's air crews, specialists from Malan at the Lop Nur test base used both optical and radio-telemetry instruments for data retrieval, and they passed the data affecting designs to the bomb manufacturers immediately after each test. This phase of the development work was stamped completed before the first atomic test, but afterwards, on the basis of data from that test, they had to introduce changes to improve performance reliability.

During the flight testing program, the air force also discovered the need to modify its new medium-range bomber (the Tu-16, first supplied by the Soviet Union and then copied as the Hong 6). Air force and Lop Nur test base engineers gave the aircraft designers at the Third Ministry (for aircraft production) the necessary specifications, and in August 1964, Plant 172 completed modifications on the plane.

Additionally, the Second Ministry needed to know how transporting and air launching would affect the bomb itself. It ordered remote-controlled simulation tests, first with models and then with real atomic bombs, in trucks, trains, and aircraft. One critical target of concern was the effect of environmental change and movement on the detonation system. It took the Chinese three months to devise a detector for this one problem.

Up to this point, the weaponization program had been proceeding according to a directive issued by Zhou Enlai as head of the Fifteen-Member Special Commission. The directive, written in 1963 or 1964, stated: "Finalize the design [of the air-dropped weapon] while conducting tests; start producing [the weapon] in small numbers after finalizing good designs; guarantee the quality [of the weapons] and then equip the troops to enhance combat readiness." With so much of the research-and-development timetable in the Zhou directive now behind it, the Second Ministry, on February 3, 1965, submitted a "Report on Problems in Speeding Up the Development of Nuclear Weapons" to the commission. The ministry's report directly echoed Zhou's call for the rapid deployment of nuclear weapons. It recommended the development of one-megaton thermonuclear weapons weighing about one ton each. Zhou convened a meeting of the commission, which adopted the report and designated specific atomic tests from then through 1967 for weaponization.

Shortly after the spring festival in 1965, preparations began for the test of an air-dropped atomic bomb, and the Central Military Commission ordered the air force to pick a crew for the mission. Zhang Aiping assumed overall command of this atomic test, as he had in the test the previous October. He started by ordering a series of drills with concrete dummies and, later, with TNT. The target of white chalk formed a spot 200 meters in diameter. In one drill, the pilots of the Independent Fourth Regiment missed their target by 700 meters and knocked out a bridge built for the actual test. Zhang, calmly we are informed, told the crew not to worry but to continue practicing, and when the genuine three-ton article was released on May 14, 1965, it detonated at an altitude of 500 meters, within 40 meters of the target center.

The next step in the weaponization program focused on missile warheads. Even while work was proceeding on the air-dropped system in the early 1960s, the Beijing Nuclear Weapons Research Institute undertook preliminary investigations on their design, but it was not until August 1963 that the actual research effort began. The following spring, the Fifteen-Member Special Commission directed the Ninth Academy to accelerate that effort, and within weeks it laid out a plan of attack.

The missile-warhead program roughly followed the same two design paths that had characterized the air-delivered-bomb program: weapon configuration and weapon detonation. In April, the Beijing institute, which now shared responsibility for weapons development with its parent academy in Qinghai, assumed the principal burden for designing the warheads and conducted countless experiments on the technologies, detonation systems, and environmental impact. Structural design on a prototype warhead began in early 1965. At the end of the year, the Ninth Academy reviewed the progress of the ongoing warhead tests and the development of the hydrogen bomb and, with the approval of higher authorities, decided to give highest priority to the design of a missile-carried thermonuclear weapon.

The institute struggled hardest on the detonation system. In the same two years that its scientists were working on the warhead configuration, others at the institute and elsewhere studied alternative designs for detonators, telemetry systems, and self-destruct and other safety devices. In June 1965, almost as soon as testing began on the detonation system, the institute started a series of large-scale experiments. The first consisted of transporting the warhead or its components by truck over long distances. During the runs, scientists measured the effects of changes in acceleration, vibration, shock, temperature, and humidity. Next came missile tests of the warhead without the nuclear component. In the third round the crews at Shuangchengzi launched a missile to test the self-destruct system. The fourth round verified the reliability of the detonation system during flight. On personal orders from Zhou Enlai, experiments in the last two rounds concentrated on the integrity of the system under emergency or crash conditions.

In our discussion of the Cultural Revolution, we have described the last event in the weaponization program on which we have information—the launch of a missile armed with an atomic warhead on October 27, 1966. The experimental team had assembled in Shuangchengzi the previous month, and Nie Rongzhen had assumed overall on-site command. This was the first and last time China conducted such a risky test, but it prompted Mao to exclaim: "Who holds that we Chinese can't make missile-carried nuclear weapons? Now we have succeeded!" By the end of the decade nuclear weapons had entered the force.

Missiles and Military Doctrine

The communique on the H-bomb test in 1967 lauded Mao for his "brilliant prediction" in 1958 that it would be "entirely possible for some atom bombs and hydrogen bombs to be made in ten years' time." Though the announcement asserted that the "success of this hydrogen bomb test represents another leap in the development of China's nuclear weapons" and "marks the entry of the development of China's nuclear weapons into an entirely new stage," the accent of the message fell on Mao the leader and on the Cultural Revolution. The success was said to heighten "the morale of the revolutionary people throughout the world" and to strike a blow against the nuclear monopoly of the United States and the Soviet Union. Yet, as in past test announcements, the explicit goal of the new-stage weapons was held to be "entirely for the purpose of defense, with the ultimate aim of abolishing nuclear weapons."[53]

The communique also stressed that "man is the factor that decides victory and defeat in war," to which the army newspaper, in its own article at the time, added: "atom bombs, guided missiles, and hydrogen bombs, all in all, are nothing much to speak of."[54] Work on "China's first self-designed, self-made hydrogen bomb" had coincided with forced progress in the development of ballistic missile delivery systems during the first half of the 1960s, but Beijing took the occasion of the H-bomb test to reaffirm Mao's theses on people's war. Thus the contradictions abounded: the weapons that would thwart imperialism were nothing much to speak of; the weapons of a new stage would simply reaffirm older ideas; and the weapons made by experts were embraced by extremists making war on experts. Caught between such countervailing strains, few were prepared to tackle the intellectual tasks of refining or enunciating any new doctrines for China.

Nevertheless, a change occurred, though it remained unspoken. China had no clearly articulated nuclear doctrine that would shape its early nuclear weapons procurement and deployment policies. Yet, originating with the decisions to build the multistage hydrogen bomb and then to complete an ambitious missile program, technological imperatives began to drive the army's actual policy decisions, even though these decisions were handled with some delicacy, and their full meaning was mostly left unexpressed.

In the final analysis, the delivery systems selected for the nuclear warheads constituted the military's fundamental recasting of the emphases on revolutionary struggle, stern warnings, and visible military movements that had prevailed since the 1950s.[55] The emerging view decisively endorsed the military-technical side of doctrine and a posture to deter a nuclear attack. One might say the weapons, once deployed, spoke for themselves. In their silent vigils, megaton warheads could proclaim a powerful new doctrine of assured retaliation.

Upon reflection, moreover, the leaders of the strategic program could trace the origins of their new posture to conclusions they had reached during the Korean War and the era of American nuclear threats. Most of those manning China's strategic arsenal had long felt, as Nie Rongzhen did, that "since the Korean War, we have often been disturbed by the point [that] we lagged far behind the then enemy in military technologies."[56] They could recall with pride mixed with some bitterness the arduous quest for security begun over a decade before. Their shared goal had been to catch up, and by the mid-1960s, that goal was becoming a reality. Based on the military's guideline of giving first place to military hardware and quality, the strategic policy was becoming "based on a state of war—being prepared for 'an outbreak of an early war, an all-out war, and a nuclear war' [*zaoda, dada, da hezhanzhen*]."[57] The actual doctrine of constant military preparedness represented a logical extrapolation from the nuclear achievement.

Still, Chinese analysts did give Mao Zedong credit for providing some of the underlying principles for the new doctrine. They heralded his speech "On the Ten Major Relationships" of April 1956, in which he pointed out the fixed relationship between economic construction and defense construction.[58] In defining the nuclear weapons program, Mao had limited its scale—China's nuclear weapons "won't be numerous even if we succeed"—and he had argued that the program's success would "boost our courage and scare others."[59] In 1981, the Party Central Committee concluded, in its grand evaluation of Mao's historic role: "After the founding of the People's Republic, he put forward the important guideline that we must strengthen our national defence . . . and develop modern defence technology (including the making of nuclear weapons for self-defence)."[60] The ballistic missile program, naturally enough, followed from these broad statements attributed to Mao.

In early 1963, on behalf of the Fifteen-Member Special Commission, Zhao Erlu, whose background was in defense production, visited the missile organ, the Fifth Academy. There, he called on the missile specialists to devise a *banian sidan* plan; that is, to build "four types of missiles in eight years." As he envisaged it, the plan would lay out a schedule to complete a *dongfeng* (DF)—or East Wind—series, DF-2 through DF-5, each with a different range based on a specific imaginary target. The imaginary targets in the draft plan as originally formulated in 1964 were Japan (DF-2), the Philippines (DF-3), Guam (DF-4), and the continental United States (DF-5).[61] A brief review of the details of the DF plan illuminates how emerging technologies circumscribed and yet reflected the broader strategic goals and common strategic assumptions of the Fifth Academy's missileers.

As formally endorsed by the newly created Seventh Ministry of Machine Building (replacing the Fifth Academy in January 1965), the Eight-Year Plan for the Development of Rocket Technology (1965-72) summarized the suggestions of some 3,000 missile workers and engineers on the original draft. When approved by the Fifteen-Member Special Commission, the plan, adopted in March 1965, set the guidelines for the full-scale pursuit of the *dongfeng* missile program, as well as other missile technologies.[62]

Even before the 1965 guidelines were finalized, work on the missiles themselves had begun. As a result of negotiations concluded in October 1957, Moscow had shipped two R-2 missiles (U.S. code name SS-2) to China in January 1958.[63] By 1960, the Chinese had already followed Soviet blueprints of the R-2 rocket and had built, for research and educational purposes, an experimental liquid-fueled missile (Chinese code name 1059) with an approximate range of 590 kilometers. Special units of Chinese artillery had first test-launched the missile in November 1960. The Fifth Academy had also tested the engine of a follow-on surface-to-surface missile (the DF-1) between 1960 and 1963, but then temporarily abandoned it.

Well before the 1965 decision, the missile academy agreed on a modified Soviet R-5 (U.S. designation SS-3)—on which they had only notes and drawings—as the basis of the prototype for China's first operational missile. The 1,450-kilometer-range missile, designated DF-2, had a single stage fueled by nonstorable alcohol and liquid oxygen and was road-mobile.[64] The DF-2 was designed to carry a 20-kiloton warhead, and did so during the nuclear weapon test on October 27, 1966. Numerous problems, resulting from elastic vibration and the failure of the propellant supply pumps, plagued the DF-2. Its first successful test-launch came in June 1964,[65] and the deployment schedule, arranged to coincide with the first atomic test, slipped two years. Upon deployment of the DF-2 in 1966, the Central Military Commission ordered it sited in Northeast China and targeted on cities and U.S. military bases in Japan.

The next two missiles to come on line were the DF-3 and DF-4.* In 1963, the Chinese designers scrapped the DF-1, predecessor of the DF-3, because each of its four engines was capable of only 16 metric tons of thrust. They then basically remade the missile, with more thrust (increased from 64 to 104 metric tons) and a more powerful storable fuel (unsymmetrical dimethylhydrazine [UDMH], and a liquid oxidizer [AK-27], nitric acid and nitrogen tetroxide). They designed this missile, the new DF-3, for an "intermediate range" of 2,800 kilometers and a yet-to-be-completed thermonuclear warhead, and test-launched it in December 1966. As a military vehicle, the DF-3 could launch warheads against U.S. bases in the Philippines. It also served as the first stage of the DF-4 and of the Changzheng-1 (CZ-1), the launcher of China's first satellite, in April 1970. By the end of the 1970s, the DF-2 and DF-3 had been deployed in permanent sites "at over 100 launchers, all of which probably possess a refire capability." [66]

China began research and development on the DF-4 missile in 1965 and managed to begin the initial deployment in 1971. The limited-range intercontinental DF-4 could carry a megaton warhead to a maximum range of 4,800 kilometers, [67] and the Central Military Commission made the U.S. Strategic Air Command base on Guam the theoretical target of this very capable system. The first stage, which was the DF-3, had four engines, and the second stage had a single engine. Over the next decade or so, the Second Artillery Academy successfully conducted range-extending tests on the first stage. [68] The greater range of the DF-3 and DF-4 missiles permitted their deployment in more protected locations deep inside China.

In 1969, the Sino-Soviet conflict flared into fierce armed conflicts, and Beijing ordered its strategic rocket units to implement the *banian sidan* plan with a new adversary in mind. Soviet cities now became the designated targets, and starting in 1971 the first DF-4 units moved to Qinghai (Xiao Qaidam and Da Qaidam) and other sites in Northwest China, closer to key Soviet targets.

In 1965, the Chinese began development of the DF-5, a two-stage,

---

* According to Zhang Jun (*Dangdai Zhongguo de Hangtian Shiye*, p. 567), the DF-3 (*zhongcheng huojian*) and the DF-4 (*zhongyuancheng huojian*) were first successfully flight-tested in December 1966 and January 1970, respectively. We should note that the terms *zhongjincheng* (medium-short-range, a classification for the DF-2), *zhongcheng* (medium-range), and *zhongyuangcheng* (intermediate-range) imply ranges that do not precisely coincide with the terms "medium-range," "intermediate-range," and "limited-range intercontinental" as used in the West. The U.S. Department of Defense categorized the DF-2 as medium-range, the DF-3 as intermediate-range, and the DF-4 as limited-range intercontinental missiles. See Brown, *United States Military Posture for FY 1976*, pp. 48-49. In this discussion, we have chosen to follow the Western usage.

full-range (13,000-kilometer) intercontinental ballistic missile capable of delivering a multimegaton warhead to Hawaii and the continental United States. The Chinese formed two competing teams to design a large missile that would be based in an underground silo, one team to analyze the advantages of a nonstorable, liquid-fueled system, the other the advantages of a storable one. Over the months, the teams debated the pros and cons of each system, and in the debate, Chinese perceptions of the 1962 Cuban missile crisis between the United States and the Soviet Union loomed large. To most, the crisis demonstrated that missiles with nonstorable fuels were ineffective in times of great international tension. Such missiles simply took too long to launch in a rapidly escalating showdown and could not be held in readiness over extended periods. Storable propellants, by contrast, would offer the advantage of high readiness for the DF-5 and could be maintained in a launch mode for prolonged periods.[69]

The argument settled, the Chinese designed the DF-5 to use storable liquid fuels, similar to but not the same as those used in the DF-3 and DF-4.[70] It took the developers until May 1980 to test-launch the DF-5 missile over a range near its maximum, though deployments began earlier, after partial tests of the missile in the 1970s. The Cultural Revolution took a major toll on both the DF-4 and the DF-5 program, causing serious delays.

The guidance systems for the DF-2 through DF-4 series provided the strategic units with weapons of limited accuracies. In all three, graphite control vanes provided attitude control.[71] Not until the initial deployment of the DF-5 against Soviet targets in the late 1970s* did the Chinese install stabilized-platform inertial guidance systems, gimballed thrust chambers, vernier engines, and swivelling main engines for attitude and thrust-vector control.[72] On the matter of guidance and accuracy, Zhang Aiping had deemphasized precision: "It is unnecessary for us to achieve tremendous accuracy. If a nuclear war breaks out between China and the Soviet Union, I don't think there is too much difference between the results, provided China's ICBM misses its predetermined target, the Kremlin, and instead hits the Bolshoi Theater."[73] What mattered, Zhang implied, was the ability to destroy urban areas or "soft" military targets in a retaliatory strike.

While downgrading the importance of precise accuracy of the nation's missile forces, the Central Military Commission stressed the

---

* Even though China did not test the DF-5 over the full Pacific range until 1980, the missile had been tested over shorter ranges in the 1970s and was initially deployed late in the decade, probably starting in 1979.

need for absolute command and control. It deemed such authority essential under all conditions, but especially because of China's minimal warning capabilities and overall inferiority vis-à-vis the nuclear superpowers. In June 1959, the first strategic missile battalion was formed within the Artillery Corps.[74] Then, in the mid-1960s, this battalion and subsequently established missile units separated from the Artillery Corps, and the Central Military Commission began to organize them as the Second Artillery Corps (Dier paobing). Security was tightened even further, with elements of the original Gonganjun (Security Forces) helping to man the corps' headquarters and command centers. At this early point, according to one Chinese source, China's missile units were "forbidden to launch any missiles unless given orders by China's supreme command," the Central Military Commission.[75] The formal establishment of the Second Artillery in July 1966 reinforced the centralized control system for China's strategic forces.[76]

Within the limits of available technology, the corps devised stratagems to increase the probable survivability of the mobile, medium-range missiles and then of the longer-range missiles, placing them in caves and deep canyons where they were well hidden from Soviet and American satellites. In 1974, the U.S. Secretary of Defense noted, "The Chinese are clearly sensitive to the importance of second-strike capabilities and are making a considerable effort to minimize the vulnerability of their strategic offensive forces."[77] Three years later, the chairman of the Joint Chiefs of Staff added that China's military strategy would "likely continue to be defensive in outlook with deterrence of an attack its primary objective." An effective deterrent, he noted, "does not always require the highest level of sophistication."[78] Chinese missile commanders would have agreed, and their leaders within official councils quietly expressed confidence that they now deployed a formidable and relatively invulnerable retaliatory arsenal.

Although Mao's avowal that nuclear weapons bestowed no special power to deter a nuclear attack or dominate the battlefield still stood as official policy, the strategy inherent in the initial deployments of Chinese strategic forces was manifestly a contradiction of Mao's position.[79] The announcement of the first atomic test had proclaimed that China would not use nuclear weapons in a first strike and would design them "for defence and for protecting the Chinese people from U.S. threats to launch a nuclear war." When the Chinese had deliverable nuclear weapons, the Americans would "not be so haughty, their policy of nuclear blackmail and nuclear threats [would] not be so effective." Beijing understood it could not credibly use its tiny nuclear arsenal to make threats against its enemies, let alone actually launch them without committing suicide, and thus it would not repeat Moscow's error in the Cuban missile crisis of "adventurism" and "capitulationism."

Instead, the Central Military Commission had chosen to create a force of high-yield, moderately accurate weapons that could with reasonable probability survive and retaliate if China suffered a nuclear blow of its own. Given Mao's well-known views about nuclear war, none would doubt that he had the will to order a second-strike retaliation, and it was this belief, coupled with several dozen survivable weapons, that led to Mao's prediction that the nuclear program would "boost our courage and scare others." De facto, Mao had embraced a form of minimum nuclear deterrence. In his memoirs, Nie Rongzhen makes this doctrine explicit. The development of the strategic rocket forces had enabled China "to own the minimal means to stage a counterattack in case our country suffered a surprise nuclear attack by the imperialists."[80]

According to a senior participant in the missile program, "Between 1956 and 1964, Mao Zedong drew the outline for socialist China's nuclear strategic theory." Implemented though not ever articulated, that strategy depended on a secure, comparatively small nuclear delivery system able to "strike against the principal population centers" of a more powerfully nuclear-armed adversary.[81] In connection with the border clashes with the Soviet Union in 1969, the Chinese declared: "Should a handful of war maniacs dare to raid China's strategic sites, . . . that will be war, that will be aggression, and the 700 million Chinese people will rise up in resistance and use revolutionary war to eliminate the war of aggression."[82]

For the Second Artillery, this meant limited nuclear retaliation at a time and against targets of Beijing's own choosing. The commander of one missile-launching brigade expressed that mission in the starkest terms:[83]

I am often lost in thought. Can our country survive a first strike inflicted by our adversaries? We put ourselves in a passive position because of our strategic principle *houfa zhiren* [to gain mastery by striking only after the enemy has struck first]. However, [our adversaries] also are afraid of the ICBMs we have deployed because it is impossible for them to destroy all our missile bases [in a first strike].

In slightly modified form, this retaliatory doctrine continues to be operational.[84]

By examining Chinese thinking about nuclear weapons and their influence on modern warfare and global politics from 1955 to 1967, we have glimpsed Chinese strategic doctrines in theory and practice. During the early years of dependence on the Soviet Union, as articles published in 1955 show, the Chinese echoed Soviet writings on the preemptive use of nuclear weapons and surprise attack. The accent in these writings, of course, fit China's mood of danger in the face of the threatened use of American nuclear weapons. As that mood dissipated, Chinese views returned to Mao's dicta on "paper tigers" and survivability in a nuclear war and parted company with Soviet views. In the years 1955-58, a period of high leadership involvement in the nuclear program, the top leaders seldom forayed into the labyrinthine world of nuclear philosophy and policy. They did not accept the need to provide ideas or explanations for the program or its future impact on military planning.

In the years of increasing strains in Sino-Soviet relations from 1958 to 1960, Mao did articulate his views on nuclear weapons. His debates with Khrushchev introduced more detail than at any point before or after concerning his thinking about the possible causes of nuclear war and its consequences. Mao factored nuclear weapons and long-range missiles into new formulas designed to account for the changing balance of forces between East and West; to him, Sputnik had dramatized the superiority of the Soviet Union in the delivery of nuclear weapons. Strategic forces also became the focal point in the clash of Soviet and Chinese security interests and helped precipitate the broader bloc conflict.

We now know that military leaders below Mao considered the more mundane doctrinal ramifications of introducing nuclear technology at some future date. They certainly worried about the implications of their nation's acute vulnerability to nuclear attack even as Mao sneered at U.S. threats. The army began training its forces to survive a nuclear attack as early as 1957 and, in 1961, initiated high-level discussions on tactical training under "modern conditions." Units at and above the regimental level were directed to "find out how to use our atomic and chemical weapons for sudden attacks."[85] By the mid-1960s, the military leaders had begun readying selected units for the day the army would acquire and deploy nuclear bombs, and, presumably, educational cadres had to teach something practical about the command and control of strategic forces. Release and targeting policies, as we have seen, did carry a broader doctrinal message, though no one could have made the message explicit or systematic until Mao Zedong had spoken. He did not speak, and in his lifetime Mao did not permit the development of open-ended strategic studies. Military research organs to conduct such studies were not set up until several years after his death in 1976.

The introduction of nuclear-armed missiles did communicate a doctrine at odds with declared policies and national dogma on strategy and war. No technology had ever introduced such fundamental contradictions in thinking about international conflict, and Mao's worldview should have led him to search for the "key link" in the opposition between the bomb and his outmoded doctrines. He missed the contradiction and thereby lost his chance to influence China's future nuclear strategy.

NINE

Chinese Lessons and the Global Nuclear Experience

The success of China's nuclear weapons effort does not eclipse the decade of chaos and failure in which it occurred, but it does stand as a major accomplishment, a marked contrast to China's general fate in that decade. That contrast causes us to review possible explanations for the success of the program in this concluding chapter. We will pursue our search for understanding from four separate perspectives: an analysis of the evolution of the program itself, a comparison of the Chinese effort to that of other nations, an evaluation of the program against general criteria for success or failure, and a summary assessment by Nie Rongzhen, the program's leader.

Dependence, Interdependence, and Self-Reliance

The history of the Chinese bomb project provides a basis for looking at the issues raised at the beginning of the book, as well as for pursuing the explanations for the success of the Chinese program. In the introduction, we examined the context and immediate background of the Chinese decision to launch the weapons project and then posed some questions about leadership, organization, and the development of science and technology. We return to those questions here.

Our study has illustrated how the Chinese Communist Party, even as it reactivated revolutionary leadership systems and norms in the Great Leap Forward, simultaneously modified or scrapped revolutionary techniques and prescriptions in the pursuit of nuclear weaponry. This modification occurred over many years; its manifestations differed from institution to institution and from one technological challenge to the next. Because of the part played by individual scientists and engineers, as well as by local-level mining, processing, designing, manufacturing, and testing units, it would be inaccurate to conclude that a single institutional system emerged from the weapons program. In the jargon of social science, there was no single dominant model; there were many. Some of these were appropriate only to mining or enrichment, and some, even though later discarded, provided important lessons and experience for leadership, organization, and technological development in other areas inside the program. Perhaps the most important lesson of all is that no single method or system works equally well in all areas of such a complex, prolonged program.

Still, the microsystems created at each stage in the nuclear acquisition process—from uranium exploration to the final bomb test—clearly operated within a broader nuclear program under the direction of Nie Rongzhen and, ultimately, key Politburo members like Zhou Enlai. That program imposed national-level direction and coordination and fostered selective uniformities, even as it tolerated local flexibility and diversity. It was the power of the program that it promoted an effective synergism among leadership, organization, and technology.

The comparative study of the various national nuclear programs confirms that all of them drew on outside knowledge and help. Exogenous knowledge came directly and indirectly, overtly and covertly. That complicating fact inhibits an understanding of what clearly delineates the indigenous from the foreign. Considered from this perspective, the national management of China's nuclear weapons program passed through several different stages determined by the evolution of the Sino-Soviet alliance. These stages seem similar to the ones found in the Anglo-American alliance. British military nuclear research, development, and production, for example, underwent progressive changes that were connected to the evolution of the alliance beyond Britain's initial dependency on the United States.[1] In China, as in the United Kingdom, the nuclear program's leadership, organization, and technology evolved through stages—from dependency to interdependency and, finally, to self-sufficiency—as defined by the country's alliance relationship. Within the Chinese program, to be sure, each action affected leadership, organization, or technology and had its own domestic programmatic and political rationale, but each also fit within and was strongly related to an internationally defined stage. Thus, when we test the influence of leadership, organization, or technology on the program's success, we must examine not only their mutual interactions but also how that influence shifted from stage to stage.

In the era of nuclear dependency, 1955-58, the Chinese anticipated and aimed toward future self-reliance but brought their immediate decisions and organizations into line with those of the Soviet Union. The Soviet government responded generously, as we have seen, but it also imposed demands and restrictions on Chinese initiative. It treated the Chinese as students and subordinate allies, and the Chinese accorded their Soviet mentors great deference and respect. Beijing's nuclear weapons program paid a price for its dependence on Soviet science and technology, perhaps, but the gains in training, resources, designs, and equipment ensured the program's rapid pace and product quality, and much of its long-range vitality. Dependency also introduced strains and promoted an enduring institutional split within China between the strategic and conventional programs and the political-military leaders who ran them.

The nuclear program in these start-up years did manifest nascent qualities of self-sufficiency and programmatic experimentation, but the overt emphasis fell on managing the high-level politics involved. Leadership carried the main burden. The assessment of the American threat, the evaluation of the country's nuclear resources and capabilities, and periodic negotiations with the Kremlin prompted the top Chinese political and military leaders to work together. These leaders drew up the initial plans, accepted a coalition with the scientists, identified and empowered the operating managers and systems, and set the tone of high adventure.

The conceptual world of Chairman Mao clouded Chinese thinking about the ultimate goal of self-reliance and the intermediate means of depending on Soviet assistance. While China required the skills of scientific and technical specialists, Mao had always found it galling to cooperate with them, let alone to work alongside them as partners. Such reactions channeled his approach to the very scientists upon whom any self-reliant nuclear weapons program must eventually depend. The Soviet factor further compounded Mao's qualms about relying on questionable allies, whether domestic or foreign. In the years of dependency, Mao both demanded and resented Soviet assistance. In the main, he treated potential allies as competitors, not equals, and reacted poorly to changes or ideas that others initiated or that he could not control. As his mind kept returning to the stark clarity of the revolutionary war, Mao only occasionally led the nuclear weapons program in any consistent or significant way. On balance, he most often remained aloof, even indifferent, as his mind fastened on his revolution's ultimate destiny.

The system of organizations the Politburo initially created to launch the program was experimental and did not last long, and the scientific and technical community had yet to develop enough strength or self-confidence to have a serious impact on the leadership's directives. In this stage of dependency, the leadership remained committed to the ultimate objective of national independence and tried alternative ways to balance the competing claims of short-term objectives of dependency and long-term objectives of sovereignty. Nevertheless, dependency caused the Chinese to be flexible, almost philosophical, about the long term and to concentrate on the immediate. At the very time the senior leaders in Beijing enjoyed their maximum potential for influencing the nuclear weapons program, they had the least practical scope. Soviet assistance thus reinforced their desire to break free as soon as possible.

Cooperation with Moscow did make a lasting mark on China's strategic outlook. It allowed Beijing to take some technological shortcuts and left the Chinese with what might be called an abbreviated textbook for future learning and application. In their dependent status, the Chinese principally echoed Soviet military doctrines or worked to modify them in accordance with Mao Zedong's preconceptions. By accepting Soviet assistance in the era of Sino-Soviet nuclear cooperation, the Chinese postponed a serious consideration of their own future status as an independent nuclear power.

Despite the high degree of China's dependency on the Soviet Union, the Chinese and Soviet nuclear programs never achieved binational integration. Even as Moscow's dutiful students, the Chinese resisted Soviet moves to enlarge the arena of their dependency from technological aid toward a common defense strategy or to introduce new patterns of Soviet military penetration. No nuclear dependency has ended in a jointly integrated nuclear weapons program; China's dependency on the Soviet Union was no exception. The Chinese, while leaning to the Soviet side, maintained fiercely independent political-strategic values that have characterized all nuclear weapons states; such independence, it would appear, constitutes the raison d'etre of all atomic programs.

The years of interdependence, 1959-60, differed from the previous stage because China now put more emphasis on its own pathway within the dual-track approach to the bomb. This change greatly aggravated Sino-Soviet tensions and provided grist for the emerging polemics. It induced Moscow to cancel delivery of the prototype atomic bomb and such strategic materials as the precious uranium hexafluoride for Lanzhou.

Interdependency as a stage in the nuclear weapons program also coincided with the Great Leap Forward and its collapse, and the period witnessed major strains throughout the entire national organizational system. That system, at its bureaucratic worst, was one of Mao's principal targets for reform. By this stage, the Soviet-style machine-building ministries and Chinese Academy of Sciences institutes had proved to be organizationally inadequate for carrying out the nuclear program, just at the point when the mass movements and economic-political upheaval created the need for stability and continuity in that program. Here, the Politburo's top leaders failed to manage well the strategic relationship with Moscow and the program's intellectual resources that had been threatened in the anti-Rightist campaign of 1957. It fell to senior program officials such as Nie Rongzhen and Song Renqiong, backed by Zhou Enlai, to mold a strong military-scientific organization of a special kind. Taking advantage of the isolation of the prospecting, design, production, and test facilities, they forged new institutions on the basis of strong personal networks, a need for secrecy, and geographical insularity. Nie Rongzhen provided much of the genius for constructing the organizational system that was to survive numerous extreme tests for the next two decades.

Within the islands of activity overseen by Nie and his lieutenants, the local systems adopted novel organizational techniques to fashion binding relationships between hitherto opposed cultures—military and civilian, scientist and worker, Chinese and Soviet. Those who had never done so before now worked well together. The novelty lay in multiformity. We have noted the diversity of the organizations within Nie's institutional system, and this diversity was important. One might compare the organizational multiformity to the guerrilla bases that existed under the loose but critical authority of Yan'an in the revolutionary war. Driven by protracted struggle and enduring weakness, the Party had to follow Lenin's principles of democratic centralism: active participation and debate prior to a command decision and disciplined implementation thereafter. Under revolutionary conditions, strict uniformity and top-down command would have been impossible to enforce and might have proved suicidal. In the nuclear program during the stage of interdependence, the Chinese again experimented with organizational variety and bottom-up initiative, and these innovations seemed to have served Nie Rongzhen well as he created the Defense Science and Technology Commission and the panoply of organizations for the total program. His was not a system built on textbook efficiency or mindless obedience. That was the secret of its success.

The stage of independence began, not by choice, with the Soviet evacuation in the summer of 1960. The Chinese label 1960-62 "the years of standing firm."[2] The Soviet departure reactivated the somewhat dormant interest of the Chinese Politburo in the future of the strategic weapons programs and kindled a high-level conflict. This conflict temporarily swamped and challenged the institutional processes that had developed under Nie's control. The economic disaster of the "three hard years" (1960-62) made more pressing the matter of privileged organizations and their priority allocations. The collapse of the Sino-Soviet alliance also coincided with the mounting threat of U.S. military intervention in Southeast Asia.

More than a struggle for power or programs, the transition from interdependence to independence was a time for national reaffirmation and reintegration. The Politburo acted to forge a working alliance to oversee the program. As one Chinese specialist has put it: "This was the time we stood together." Beijing temporarily asserted central leadership and organization and struggled toward a coherent policy in the wake of the costly Soviet withdrawal. The problems created by the withdrawal of data, plans, and advisers fell heavily on the installation of equipment for uranium processing and enrichment. Many leaders questioned the rush toward the bomb in a time of national emergency. Mao himself is credited with resolving the dispute and ordering the creation of the Fifteen-Member Special Commission. This body had great symbolic significance as the strategic program adjusted to the trauma of total independence. The commission did not end bureaucratic conflict, though it did produce a sufficient operational consensus to allow the nuclear weapons program to forge ahead.

Thus, national leadership and organization once again and for a short time played a vital role in the program. The upheaval of the new stage made this inevitable and a prerequisite for renewed progress, and here the key figure was Zhou Enlai. His stature, sense of mission, and leadership qualities allowed him to fashion an agreement that gave all relevant leaders a stake in the acquisition of nuclear weapons. The consensus did not last beyond the preliminary bouts of the Cultural Revolution, but by that time China had become a nuclear power.

With the grand compromise of 1962 out of the way, the full implications of independence could ramify throughout the program. Now the specialists would have their day. For the next two years, national leadership and organization, stable and supportive, receded into the background to give China's scientists and technological specialists a dominant voice. These years, 1962-64, are called the "stage of flowering" (*kaihua jieguo*). In this "tense and exciting period," Chinese technicians overcame the two major problems: the production of the enriched uranium and the design of the neutron initiator.[3]

The symbiotic relationship of leadership power and technical expertise in China combined what David Holloway has termed "demand-pull and discovery-push in weapons innovation."[4] The high command in Beijing did place demands and deadlines on the scientists and engineers, and these were understood and respected. However, the record shows that in most instances the nature and level of the programmatic demands emanated from the experts themselves; the central authorities looked to the scientists for realistic targets and for a validation of specific technological paths.

These same experts generated many of the most perplexing challenges for the program's leadership and organization; by the stage of independence in the nuclear weapons program the most important of these involved industrial and engineering applications more than basic science or high-level policy. The necessity of problem solving in engineering required an attitude of discovery and invention, and a number of "scientific" breakthroughs and "technical" innovations entailed making the political and bureaucratic system experiment in novel modes of nationwide cooperation with the science and technology system. When the managerial institutions worked best, innovative concepts and engineering designs either functioned well the first time or were revised through trial and error by specialists at many bureaucratic levels in a rare spirit of mutual help.

The Chinese nuclear weapons program (like its Soviet counterpart) was able to overcome bureaucratic inertia and anti-intellectual Party conservatism. The singular priority accorded the program gave research and production units substantial autonomy and put technical competence on a par with ideological dogma. The years of independence from 1960 to 1964 helped produce somewhat greater decentralization and freed most operational units from excessive bureaucratic red tape. The senior leaders became sufficiently immersed in the substance of the program to try to bridge the gap between their own powerful but ill-equipped realms and the research, production, and test facilities. To an unusual degree for the political system of the 1960s, Beijing avoided clumsy interference or incompetent micromanagement.

China's singular concern with national independence reinforced this developmental process. Yet, for all its preoccupation with untarnished sovereignty, the nation could not escape the inhibitions of nuclear status. In China as elsewhere, the possession of nuclear weapons exposed the country to the nuclear paradox: such power imposes unprecedented risks. Over time, the acquisition of nuclear weapons undermines familiar modes of nationalism and drives nuclear states to a tacit reciprocal tolerance in order to survive. Although China's very insularity made its planners oblivious to their imitativeness, nuclear weapons did impose their implacable logic of caution on Beijing's war preparations. China has responded, though slowly and with considerable resistance, to the global nuclear system, and has acknowledged the special burdens of managing a nuclear arsenal. Step by step, the Chinese have come to plan their national nuclear forces with the unavoidable awareness of constraints found among most nuclear planners throughout the world. The weapons themselves, not the Soviet Union, limited Chinese defense planning. What first liberated Beijing's policy makers soon imprisoned them.

Comparative Nuclear History and the Chinese Case

In the history of discovery, few revolutions match for geographic sweep the sudden insight into the fission of the uranium atom and thence the race for the bomb. Physicists and other scientists fleeing Nazi tyranny teamed up with counterparts from Great Britain, Canada, and the United States, first in research and then in the Manhattan Project. American industry, a beacon of freedom in a dark world, made a decisive practical contribution to the project, but the science bore no single national flag. The genius of many lands shared the excitement and made possible the "great blinding light" in the desert at Alamogordo. China's program thus perpetuated a global transformation, and its evaluation must distinguish its national characteristics from the universal process.

As the mysterious power of fission became known, the same awe that struck Hahn, Meitner, Frisch, Fermi, Peierls, Bohr, Szilard, Kurchatov, Wigner, Wheeler, and Bethe engulfed Qian Sanqiang, Peng Huanwu, and Wang Ganchang. These and a few other Chinese had bonds to the international fraternity of physicists, those whom the Soviet physicist Yakov Frenkel' has called a "narrow caste."[5] The scientific wonder of fission and its potential enraptured and drew to it men and women of China just as much as it did all attached to that fraternity. That this must be said derives more from American parochialism than from anything special about China.

The exultation of Chinese partaking in a global moment also became submerged in power politics, just as occurred in every other nation that pursued fission's military potential. The story of the first Chinese nuclear weapons illustrates once more the internationalism of one of scientific history's strategic moments and the global myopia in grasping that moment's universal significance.

Yet while recognizing the flaws of politics and man, we must still answer questions about the significance of nuclear weaponry and its development for individual countries, and one way to probe such questions for China is to place its experience in comparative perspective. The difficulty of making valid comparisons, however, should not be underestimated. The landmark studies by Margaret Gowing of the British program and by Richard Hewlett and Oscar Anderson or Richard Rhodes of the American program drew on a large data base that simply is not available for China. David Irving similarly had access to rich documentary files to write his history of the aborted German nuclear weapons project. Less-detailed studies are available on the Soviet and French programs, though David Holloway is completing a major work on the former.[6]

If we contrast our fund of knowledge about the British and American programs to what we know about nuclear weapons development in China, we encounter large gaps in the Chinese data and become aware that we will have to be content with much less information on personnel, policy and technical issues, resources, industrial facilities and support infrastructure, and test results. Furthermore, Gowing's close examination of the rise and fall of Anglo-American collaboration in the nuclear field far surpasses the possibilities for a similar treatment of Sino-Soviet cooperation. On the other hand, the Chinese have been far more forthcoming in releasing information than either the Soviet Union or France, two countries for which we have only sparse reliable data. If we acknowledge and keep in mind the limits, we can attempt a preliminary comparative assessment.

Let us begin by comparing the British and Chinese roles in the history of nuclear weapons development. Too often slighted in the studies of the American nuclear weapons program, scientists in Britain, some British and some refugees, were "the first in the Second World War to establish the feasibility of an atomic bomb."[7] A committee of British scientists operating under the code name Maud in the fall of 1940 had established liaison with an American counterpart committee. The Maud Report of July 1941 proved instrumental, even decisive, in getting the U.S. Manhattan Project moving. Although the British wound up as junior and strictly regimented partners in the Manhattan Project, they made vital contributions to it as respected co-workers in, for example, uranium separation and bomb fabrication. Such scientists as

John Cockcroft and William Penney, moreover, brought home invaluable experiences for Britain's own postwar program.[8] U.S.-British cooperation ended abruptly with the American Atomic Energy Act of August 1946, but in January 1947, when the government of the United Kingdom made the decision for an independent nuclear weapons program, it could proceed on the foundation of five years of solid partnership with the United States.

The Chinese could draw on no equivalent experience as participants in the Soviet nuclear weapons program. In the main, Soviet support for the Chinese program constituted limited, one-way scientific-technological-industrial assistance, not an association of co-workers. To be sure, China's programs for defense science and technology, including, as we have seen, the pursuit of nuclear weapons, were significantly benefited by Soviet aid. But the Chinese never penetrated the inner sanctum of Soviet nuclear facilities as had their British counterparts in the U.S. programs. We know of no Chinese nationals intimately involved in the Soviet or Western weapons programs, though a small number of ethnic Chinese reportedly played peripheral roles in the U.S. postwar test program and later returned to China.[9] We have already observed the backgrounds of such Chinese in nonmilitary nuclear physics laboratories in the West, but their role did not resemble the one played by the British in the Manhattan Project.

Washington and Moscow terminated atomic interdependence with their respective British and Chinese allies, causing bitterness and considerable misunderstandings, though both London and Beijing would acknowledge their debts to their more powerful allies. When atomic cooperation ruptured in the Chinese case, the break partly provoked and partly reflected a more fundamental disintegration of the alliance. Nuclear weapons policies and procurement undoubtedly exacerbated the rift in the alliance with Moscow and accelerated the collapse of the relationship. A comparable political rupturing did not occur as the British and Americans parted company and went their separate nuclear ways. To the irritation of other Europeans, the Anglo-American partnership persisted and, in the face of deteriorating postwar Western relations with the Soviet Union, became the core of the NATO alliance. A decision to pursue an independent nuclear program need not precipitate a political crisis or generate uncontrolled hostility toward an ally.

The third and fifth nuclear powers—Britain and China—had concluded very early that an independent nuclear arsenal would add measurable weight to their international position.[10] Their success in acquiring nuclear weapons, they believed, would in itself validate the potency of their defense and technological capabilities and advertise their future military potential. It would bring recognition from friend and foe alike, and when Great Britain and China set off their first weapons, the citizens of those countries hailed the explosions as a signal triumph. As Margaret Gowing comments, in respect to the first British bombs, "few voices were raised to doubt the morality of making them."[11] In China, the populace would have scoffed at Lyndon Johnson's characterization of the October 1964 explosion as a "tragedy for the Chinese people."[12] It is not fortuitous that dozens of China's current leaders once held important positions in the strategic weapons program.

Plainly, the Chinese and British decisions to obtain nuclear weapons sprang from far different motivations. For years throughout the 1950s, the American government had attempted to exploit a calculated nuclear threat against China in Korea, Indochina, and the Taiwan Strait. Washington had issued its threats in ways never dared by Soviet leaders toward the allies of the United States. Those in the United States who fear and oppose nuclear proliferation would do well to remember that it was the United States that made the Chinese program essential to Chinese security. Seen in this way, Beijing's decision rested on a much more genuine military rationale than did the British decision to acquire the bomb.

No nuclear weapons program has succeeded without the formation of a coalition between high politics and grand science. When asked why Nazi Germany had done so little work on the atomic bomb, the former Reichsminister Albert Speer said: "We never got beyond primitive laboratory experiments, and even these were not ready for decision." Unsolved scientific questions caused part of the problem. Nonetheless, David Irving has concluded that the inability of German nuclear scientists to "fire Speer's imagination with the possibilities of atomic fission was their greatest shortcoming." Irving blames this failure on the shyness of academic physicists, and thus "it was not surprising that they failed to establish *rapport* with great men of government or industry."[13]

Soviet scientists had somewhat better relations with officials in the Kremlin after June 1940, when the Presidium of the Soviet Academy of Sciences approved the creation of the Uranium Commission headed by V. G. Khlopin, a radiochemist.[14] Soon thereafter, however, the Soviet scientists themselves became divided on the future possibilities for atomic energy, including its military potential, and lost their persuasive influence. Many of these scientists disagreed with Igor Kurcha-tov's optimism about the military implications of fission and argued that any technological or military applications remained decades off. Carrying the pro-development banner alone, Kurchatov and a few col-leagues, such as Yu. B. Khariton of the Institute of Physical Chemistry, did not enjoy "the authority or reputation that would enable them to approach the political leadership." Stalin's purges of the 1930s had left the Soviet military with unimaginative and incompetent special-ists, and the war stole resources from all research programs, especially those deemed to be marginal. Stalin neither appreciated the potential of an atomic bomb nor, according to Holloway, "would have decided to expand research" even if he had seen the possibilities. Nevertheless, his knowledge that Germany, Britain, and the United States were working on a bomb led Stalin to set up a small project in 1942; this was converted into a crash program after the bombing of Hiroshima. It was Hiroshima rather than Soviet scientists that finally proved per-suasive to Stalin and caused him and his successors to listen, however grudgingly, to selected scientists thereafter.

In China, the Party's history of anti-intellectualism did not prevent politically well-placed scientists like Qian Sanqiang from inspiring the confidence and exciting the imagination of senior officials. Their task at Zhongnanhai was intrinsically easier than Werner Heisenberg's at Speer's Munitions Ministry or Igor Kurchatov's at the Kremlin during the Second World War: Mao Zedong, China's leader, knew that the atomic bomb worked. Besides, given Qian Sanqiang's scientific *and* political standing, he was more likely to be heeded than Kurchatov in Moscow in his role as mediator between the scientific community and the elite. Qian's Janus-faced authority derived primarily from his access to Zhou Enlai and Nie Rongzhen; he was not the scientific equivalent of Igor Kurchatov, for he did not run what were considered central components of the nuclear weapons program itself.

A working rapport of scientists and officials may constitute a sine qua non of program success; it does not ensure it. In no case, including China's, did the coalition between national politics and the bomb proj-ect proceed smoothly. Nevertheless, within each successful program, the polity fostered institutional arrangements that engendered per-sonal commitment and loyalty and overcame a proclivity toward bu-reaucratic inertia.

The British and Chinese programs illustrate that the excitement and commitment found in the American Manhattan Project could be readily and successfully replicated in quite different political and industrial-scientific environments. These attributes arise independently of advanced modernization and scientific-technical development on the Western model. They rest instead on national will. What Gowing says about Britain—"The purse strings had been opened and the priorities had been granted because Ministers and officials alike knew that the atomic bomb had revolutionized war"[15]—could be said of China as well. Were it not for the British comparison, we would be tempted to argue that the reason for the depth and persistence of the Chinese resolve derived from the American nuclear threat alone. Certainly, Chinese sources imply this, but the existence of danger from abroad offers only a partial explanation of that dedication. The leadership's commitment and faith in the nation's scientific prowess or potential over the long run played an important role.

National resolve can also make unbearable the very idea of failure, and its aura need not translate into day-to-day interference. The distancing of politics from direct programmatic decisions seems to increase the polity's operational punch. In China at least, Beijing's directives, especially those from Mao and his Politburo comrades, had the ring of Delphic decrees, and at the ministerial level, much of what passed as decision making involved the bureaucratic mustering of political and economic backing, the fending off of political opposition, and the ensuring of limited guarantees of noninterference during moments of upheaval. Without these, the selfless commitment of lower-level personnel might well have waned, for powerful mentors, protectors, and suppliers with revolutionary backgrounds clearly helped to foster the extraordinary esprit de corps among those personnel in the first place.

The discipline imposed by rigid security measures does not advance the explanation for that commitment much further. Though the military establishment probably played a stronger role in China's early program than in Britain's or America's, military-like security shrouded all the nuclear energy and weapons programs.[16] While Soviet military advice, Holloway concludes, "appears to have played no part in the decision to build the atomic bomb,"[17] the secret police imposed a strict security regime on the Soviet program. In the design of their program, the Chinese selectively adapted Soviet security procedures but circumscribed the impact of those procedures on the program's inner workings. Commitment grew principally from within the program; it was not simply imposed.

Centralization and power characterized the overall Soviet system,[18] and the Chinese program certainly manifested some of these characteristics. But with the possible exception of the building of the Lop Nur Nuclear Weapons Test Base, the Chinese relied almost exclusively on elite military construction units, not prisoners, and promoted nuclear scientists and technicians into the prestigious strategic weapons establishment. China's hitherto parochial organizations had to accept universally defined standards of knowledge and performance to make the program succeed. The nuclear weapons program bore less of a Soviet imprint than almost any other part of the Chinese military-industrial system, a fact that has added to its authority in the post-Mao era.

An adaptation of the revolutionary model was needed to accommodate and empower somewhat suspect intellectuals and to meet industrial-technological requirements of a poor and industrially underdeveloped society in a wartime-like atmosphere. Although Chinese personnel of all backgrounds in the nuclear weapons program moved to the bases, factories, and institutes under the mantle of the military, the inner sanctum in places like Jiuquan Prefecture, Qinghai, and Lop Nur had more the feel of the seminar than of the barracks or the hills of Yan'an. In its anonymous hideaways, the "narrow caste" became a fiercely guarded fraternity. The Party and Central Military Commission commanded what was essentially a military-run effort, to be sure, but this effort altered the military as much as the military dominated it.

Managerial Criteria and Individual Motivation

In China, all involved in the nuclear program realized that, in practice at least, the scientific and technical decision making had to be left to the experts. In the nuclear weapons world, China's real innovation was to tolerate and then mandate a reform of the political system in order to leave the hard choices to the nonpolitical specialists. In their coalition with the military-political command structure, the scientists reciprocated by according the political leadership professional respect and the final, albeit mostly perfunctory, authority on most nontechnical policy matters. Cooperation also developed between scientists and engineers or technicians, for in nuclear weapons matters all the participants were learning through practice and as they went along. The Red and expert—politician and technical specialist—successfully worked together in this case. As the central system crumbled in the Cultural Revolution, that special relationship endured even as some individual members of the nuclear weapons program fell victim to the turmoil around them.

As one result of the bridging of the political and operational domains, China's program exhibited important management features that, in a much later and different context, would be identified as underlying large, successful defense acquisitions,[19] namely, clear command channels; stability; limited reporting requirements; small, high-quality staffs; communications with users; and prototyping and testing. We will briefly examine each of these features in relation to the Chinese nuclear weapons program of the early 1960s.

In every phase of that program, the on-site manager assumed clear responsibility for his organization and operated through a short, unambiguous chain of command to leaders in Beijing. Nie Rongzhen had overall program responsibility; he reported directly and frequently to bodies such as the Fifteen-Member Special Commission and the Central Military Commission. When the political struggle muddied Nie's authority and weakened his chain of command, the program suffered, but on balance the Politburo knew that undermining Nie would be tantamount to sabotaging the program itself. Time and again Premier Zhou Enlai reaffirmed Nie's charge and strengthened his position.

As we have seen, the nuclear program did not escape the societal chaos of the 1950s and 1960s. Nevertheless, as we noted in our discussion of the coalition of politics and technology, the leadership mustered all its political will and economic resources to sustain at least a minimal stability in the program. Nie fought for constancy in program performance standards, scheduling, and, to the best of our knowledge, cost estimates. He kept reassuring his colleagues and subordinates of the Politburo's solid backing, and the Party and military leaders responded with slogans that advocated steady, solid progress. When one compares the program's calming slogans and steadfastness with the hysterical headlines of the Great Leap Forward or the Cultural Revolution, the contrasts are quite startling. Had Mao and his colleagues appreciated the importance of stability for all China's modernization plans, the country's history since 1957 would bear far fewer scars.

On the positive side, the tenor of the mass political movements helped minimize bureaucratic bumbledom and excessive reporting requirements in the nuclear program, as elsewhere. The antibureaucratic spirit of the late 1950s kept cadre organizations from stifling local managers and specialists. The popular movements also can claim credit for mining the uranium that fueled the first atomic bomb. Thus, as far as we can tell, the program's management staffs remained small even as the influx of newly recruited scientists and engineers raised the quality of decision making and program oversight.

We know least about the communications-with-users feature. Chinese military and political officials had to contend with strict security regulations surrounding Project 02, and the army postponed until the mid-1960s the creation of the Second Artillery Corps, the unit that would control China's strategic weaponry. Mao's revolutionary codes downgraded all military hardware, and his enduring authority delayed any serious strategic rethinking that would identify and empower potential users. Senior military officers acted as managers or ideologues, not potential users. In comparing China's program with the successful nuclear programs of other nations, we are struck by the virtual absence of military-strategic planners in the early phases of all of them. We do know that the Chinese moved missile and aircraft engineers to selected bureaus of Song Renqiong's Second Ministry of Machine Building, and that Nie's Defense Science and Technology Commission maintained close communications with all the components inside the overall strategic program, including the Fifth Academy's missile design units, throughout the years 1956-65. While secrecy may have inhibited the direct involvement of the ultimate military users in the early years of the nuclear quest, that involvement would become increasingly critical as the program matured and shifted from test devices to deliverable weapons. The Chinese did work to strengthen workable user-designer links as time went on.

Finally, we have witnessed the Chinese preoccupation with careful prototyping and testing even under the most primitive conditions. The fears of a fiasco and the memory of Soviet advisers produced a medium for devil's advocates in the Chinese program, and ensured a penchant for rigorously testing unproven technologies, concepts, and methods. Furthermore, because the program enjoyed an absolute priority, Nie Rongzhen was able to order a high level of redundancy in procurement. That priority and redundancy permitted a systematic and time-consuming empirical approach when direct knowledge or good theories proved unavailable. On balance, the exhaustive testing of components in a trial-and-error process inhibited the escalation of program costs. Moreover, the consequent exhaustiveness of experiments conducted during the search for viable subsystem prototypes may explain why China showed little subsequent interest in conducting full-scale nuclear tests on a grand scale.

In seeking an explanation for the success of China's nuclear program, we find that the program incorporated some or all of the critical management features to an important degree; the record of other weapons programs remains undeterminable on this point. Yet perhaps equally important for explaining the program's capacity to foster positive commitment and performance, the political system handed Chinese science and technology a historic challenge and, despite major lapses and mistakes, gave them the priorities and tools needed to face that challenge. As exemplified by Deng Xiaoping, who said he would assume the blame for any setbacks and assign others the credit for success, the politicians let Chinese scientists and engineers participate as equals in the most exciting technological revolution in modern history. The experts understood not only the task but also the honor.

We have seen that pockets of scientists, soldiers, technicians, and political cadres worked together in a spirit of inventiveness and accomplishment, and took pride in doing so. The historic undertaking for China coincided with their own chance to prove their scientific competence and their enduring worth to society. It was Nie Rongzhen's organizational genius to recognize and build on that coincidence of interests. The extreme secrecy and isolation of most installations in the nuclear program probably added to the atmosphere of high adventure and the promise of unparalleled rewards.

Powerful personal motives undoubtedly undergirded the pervasive commitment of individual participants in this high-stakes gamble. For one thing, each could identify with the national appeal of the program's goals and the universality of the standards for making decisions and measuring achievement. The participants could believe that attaining those goals would be widely recognized and yield lasting benefits for the nation and themselves. They had a special conviction that every talented person could make an essential contribution and would be allowed to do so. The idea spread, moreover, that the lessons and methods drawn from the program could be transferred, and that passing them on would make the participants valuable in other careers over the long term. Further, most of the people in the program obviously succumbed to the aura of mystery and adventure, and liked their priority access to senior leaders and scarce resources. They had a sense of being personally empowered by the nuclear program even when physically endangered by its hazards. In our judgment, in producing the sustained level of involvement and performance, all these factors outweighed the fear of foreign threats and of personal failure. Perhaps what Chinese scientists and leaders had most feared was that the presence and authority of Soviet scientists would lessen their own recognition and authority. Independence had a very powerful meaning for individual Chinese in the nuclear weapons program.

Yet Mao Zedong little understood the centrality of Nie's achievement in mobilizing and organizing committed and talented participants. Mao certainly did not know how to sustain motivation; indeed, in the Cultural Revolution, he came close to wrecking it. In this sense, we would conclude that political systems undertake large-scale technological programs at their peril. The reasons for success are complex and controversial and are seldom appreciated by the participants, especially by those at the top.

Nie Rongzhen's Assessment of the Strategic Program

The Chinese leader of the nuclear weapons development has assessed its positive and negative aspects. He has done so to "sum up" his country's experiences and to pass on the major lessons. Summing-up (*zongjie*) has a special place in the Chinese political process; it is a form of decision making and consensus building.[20] Nie Rongzhen's summing-up of the nuclear program deserves the last word in our study.

Nie's assessment weighs the program's correct decisions against the mistakes the Party made in carrying out the strategic program. He holds that from 1956 to 1966, the Party "really committed several serious errors." The first of these came with the Great Leap Forward, when the cadres exhibited "a proneness to boasting and exaggeration and overestimated their achievements in scientific research." In this same period, chaotic and decentralized management of scientific research resulted from the "inconsistency of knowledge." Nie conceived and sponsored the reorganization of 1958, as we have seen, to correct this mismanagement. The Great Leap years of unconstrained mass movements resulted "in the violation of the laws of scientific work." Instability within the scientific system resulted from endless irrelevant political debates and "the frequent disruption of the Party's policies toward intellectuals." Nie concludes: "These errors and deficiencies prevented us from achieving greater accomplishments in scientific research work. This [lesson] is still worth paying attention to today."[21]

He begins his evaluation of the reasons for the program's success with praise for its victory in mobilizing and motivating China's best young scientists. He recalls the order to the State Planning Commission and the Ministry of Education to take charge of the job assignments of college graduates soon after the program was initiated. Thereafter, the Central Committee approved Nie's proposal to make the State Scientific and Technological Commission, the State Planning Commission, and the Ministry of Education jointly responsible for the assignments of these graduates but to give his State Scientific and Technological Commission sole authority for the assignments of postgraduates and returned students. Thus, he remarks, "it became possible for China's scientific front to claim precedence over all others to mobilize a large number of excellent graduates every year."[22] Earlier, scholars of the rank of Peng Huanwu, Zhu Guangya, and Deng Jiaxian had organized training programs in nuclear physics and other fields related to the bomb project. Others of the coming generation had been sent to the Soviet Union and Eastern Europe. Nie's commission maintained control so that the top young graduates thus trained would belong to the strategic program.

Nie credits the Central Committee's unstinting support as a second major reason for the program's success. At a broader level, Mao's anti-Rightist campaign against intellectuals and his Great Leap Forward seriously injured Project 02. Nie asserts that the Central Committee's countermeasures rescued and promoted the program in hard times. He states: "It usually gave priority to these sophisticated [nuclear and missile] projects in the supply of materials. Although China was faced with a financial crisis [during the early 1960s], the Party Central Committee allocated special funds annually to research on new types of raw materials related to these projects."[23] The committee broke a rule, he adds, and allowed the Defense Science and Technology Commission to "give direct orders to any department to produce any important parts for these projects." The Central Military Commission supplied special official letters requiring the recipients' full compliance with the commission's orders.

Nie then states that the guiding principle of "vigorously carrying out coordination and cooperation" supported the "smooth development of the atomic bomb and guided missile programs." The unusual political, military, scientific, and technical partnership did not emerge by accident, he implies. Painstakingly formed, it necessitated the development of new attitudes, organizations, and practices. Nie fostered this process and kept it going despite constant challenges and setbacks. The environment of coordination and cooperation, moreover, "gave impetus to the establishment and development of several new industrial departments and branches of learning, to the development of China's production of new types of raw materials, precision instruments, and heavy-duty industrial equipment."

In the final analysis, the leader of the nuclear weapons program sums up the strategic weapons achievement in terms of self-reliance, the disciplined setting of priorities, and national pride. The policy of relying on China's own efforts, he declares, "was not only correct, but also vital." The alternative of depending solely on foreign aid or pur-chased foreign technologies would have lengthened the development process and "made us vulnerable to outside manipulation."[24] More-over, as he pointed out to his subordinates in the 1950s, it was simply "impossible for anyone to give his most advanced things to others." This fact was a "fundamental principle" for future policy making in defense.[25]

On priority setting, Nie applauds the defense program's ability to acquire "new raw materials, precision instruments and meters, and large equipment," the equivalent of a family's "seven daily necessities." He commends his colleagues for having achieved a consensus and or-dered their priorities on these defense necessities. Otherwise, the stra-tegic program "would never have gotten off the ground."[26]

Years later, Nie recalled the emotion he felt on October 27, 1966, after the launching of the nuclear-tipped test missile. In that moment of great elation as the rocket headed west, he remembers: "I was proud of our country, which had long been backward but now had its own sophisticated weapons."[27] Nie was speaking for all in the program—indeed for all his countrymen.

Appendixes

Statement of the Government of the People's Republic of China, October 16, 1964

The text is from Break the Nuclear Monopoly, Eliminate Nuclear Weapons *(Beijing, 1965), pp. 1-5. We have eliminated the bold type used in some sections and made trivial changes in the punctuation.*

China exploded an atomic bomb at 15:00 hours on October 16, 1964, thereby successfully carrying out its first nuclear test. This is a major achievement of the Chinese people in their struggle to strengthen their national defence and oppose the U.S. imperialist policy of nuclear blackmail and nuclear threats.

To defend oneself is the inalienable right of every sovereign state. To safeguard world peace is the common task of all peace-loving countries. China cannot remain idle in the face of the ever-increasing nuclear threats from the United States. China is conducting nuclear tests and developing nuclear weapons under compulsion.

The Chinese Government has consistently advocated the complete prohibition and thorough destruction of nuclear weapons. If this had been achieved, China need not have developed nuclear weapons. But our proposal has met with stubborn resistance from the U.S. imperialists. The Chinese Government pointed out long ago that the treaty on the partial halting of nuclear tests signed in Moscow in July 1963 by the United States, Britain, and the Soviet Union was a big fraud to fool the people of the world, that it was an attempt to consolidate the nuclear monopoly of the three nuclear powers and tie the hands of all peace-loving countries, and that it had increased, and not decreased, the nuclear threat of U.S. imperialism against the people of China and of the whole world. Even at that time, the U.S. Government openly declared that the conclusion of this treaty did not in the least mean that the United States would not conduct underground tests or that it would not use, manufacture, stockpile, export or spread nuclear weapons. Facts over the past year and more have fully proved this point.

During this period, the United States had not stopped manufacturing various nuclear weapons on the basis of nuclear tests it has already conducted. Seeking ever-greater perfection, the United States has, moreover, during this same period conducted several dozen underground nuclear tests to improve further the nuclear weapons it manufactures. In stationing nuclear submarines in Japan, the United States is posing a direct threat to the Japanese people, the Chinese people, and the peoples of all other Asian countries. Through the so-called multilateral nuclear force, the United States is now trying to put nuclear weapons into the hands of the West German revanchists, thereby threatening the security of the German Democratic Republic and the other socialist countries in Eastern Europe. U.S. submarines carrying Polaris missiles with nuclear warheads are prowling the Taiwan Straits, the Bac Bo Gulf (Tonkin Gulf), the Mediterranean Sea, the Pacific Ocean, the Indian Ocean, and the Atlantic Ocean, everywhere threatening peace-loving countries and all the peoples who are fighting against imperialism, colonialism, and neo-colonialism. Under these circumstances, how can it be considered that U.S. nuclear blackmail and nuclear threats against the people of the world have ceased to exist just because of the false impression created by the temporary halting of atmospheric tests by the United States?

The atomic bomb is a paper tiger. This famous statement by Chairman Mao Tse-tung is known to all. This was our view in the past and this is still our view at present. China is developing nuclear weapons not because it believes in their omnipotence nor because it plans to use them. On the contrary, in developing nuclear weapons, China's aim is to break the nuclear monopoly of the nuclear powers and to eliminate nuclear weapons.

The Chinese Government is loyal to Marxism-Leninism and proletarian internationalism. We believe in the people. It is the people, and not any weapons, that decide the outcome of a war. The destiny of China is decided by the Chinese people, while the destiny of the world is decided by the people of the world, and not by nuclear weapons. China is developing nuclear weapons for defence and for protecting the Chinese people from U.S. threats to launch a nuclear war.

The Chinese Government hereby solemnly declares that China will never at any time or under any circumstances be the first to use nuclear weapons.

The Chinese people resolutely support all the oppressed nations and peoples in their struggles for liberation. We firmly believe that, by relying on their own struggles and by helping one another, the people of the world are bound to triumph. China's success in making nuclear weapons is a great encouragement to the revolutionary people of the world in their struggles and a great contribution to the cause of defending world peace. On the question of nuclear weapons, China will not commit the error of adventurism or the error of capitulationism. The Chinese people can be trusted.

The Chinese Government fully understands the good intentions of peace-loving countries and peoples in demanding an end to all nuclear tests. But more and more countries are coming to realize that the more exclusive the monopoly of nuclear weapons held by the U.S. imperialists and their partners, the greater the danger of a nuclear war. They are very arrogant when they have those weapons while you haven't. But when those who oppose them also have such weapons, they will not be so haughty, their policy of nuclear blackmail and nuclear threats will not be so effective, and the possibility of complete prohibition and thorough destruction of nuclear weapons will increase. We sincerely hope that a nuclear war will never break out. We are deeply convinced that, so long as all peace-loving countries and peoples make joint efforts and persist in the struggle, nuclear war can be prevented.

The Chinese Government hereby solemnly proposes to the governments of the world that a summit conference of all the countries of the world be convened to discuss the question of the complete prohibition and thorough destruction of nuclear weapons, and that as the first step, the summit conference conclude an agreement to the effect that the nuclear powers and those countries which may soon become nuclear powers undertake not to use nuclear weapons either against non-nuclear countries and nuclear-free zones or against each other.

If those countries in possession of large numbers of nuclear weapons are not even willing to undertake not to use them, how can they expect countries not yet in possession of such weapons to believe in their sincerity for peace and to refrain from taking defensive measures that are necessary and within their capabilities?

The Chinese Government will, as always, exert every effort to promote, through international consultations, the realization of the lofty aim of complete prohibition and thorough destruction of nuclear weapons. Until that day comes, the Chinese Government and people will firmly and unswervingly follow their own path to strengthen their national defence, defend their motherland, and safeguard world peace.

We are convinced that man, who creates nuclear weapons, will certainly be able to eliminate them.

China's Nuclear Weapons Tests, 1964-1978

Test	Date	Yield	Location and delivery system	Other information
1	10/16/64	20 KT	Ground (tower-mounted; 120 m)	Fission, U^{235}
2	5/14/65	20-40 KT	Air (TU-16/Hong 6 bomber)	Fission, U^{235}
3	5/9/66	200-300 KT	Air (Hong 6 bomber)	Fission, U^{235} and some thermonuclear material (lithium 6)
4	10/27/66	20-25 KT	Med.-range ballistic missile (CSS-1/ Dongfeng 2) >800 km	Fission, U^{235}; launched from Shuangchengzi to Lop Nur
5	12/28/66	300-500 KT	Ground (tower-mounted)	Fission, U^{235} and some thermonuclear material (lithium 6)
6	6/17/67	3 MT	Air (Hong 6 bomber)	Thermonuclear device, fission-fusion-fission type using U^{235}, U^{238}, heavy hydrogen, lithium 6
7	12/24/67	15-25 KT	Air (Hong 6 bomber)	U^{235} and U^{238} plus lithium 6; only a fission cycle completed
8	12/27/68	Ca. 3 MT	Air (Hong 6 bomber)	Thermonuclear device using U^{235} with lithium nucleus plus some plutonium
9	9/23/69	20-25 KT	Underground	Fission
10	9/29/69	Ca. 3 MT	Air (Hong 6 bomber)	Thermonuclear
11	10/14/70	Ca. 3 MT	Air (Hong 6 bomber)	Thermonuclear

Test	Date	Yield	Location and delivery system	Other information
12	11/18/71	Ca. 20 KT	Ground (tower-mounted)	Fission, possibly containing plutonium
13	1/7/72	<20 KT	Air (Qiang 5 bomber)	Fission, possibly containing plutonium
14	3/18/72	100-200 KT	Air (Hong 6 bomber)	Possibly a trigger device, containing plutonium, for a thermonuclear warhead
15	6/27/73	>2 MT	Air (Hong 6 bomber)	Thermonuclear
16	6/17/74	200 KT-1 MT (probably near 1 MT)	Atmospheric	Thermonuclear
17	10/27/75	<10 KT	Underground	Fission
18	1/23/76	<20 KT	Atmospheric	Fission
19	9/26/76	200 KT	Atmospheric	Fission; partial failure of fusion; "special weapon"
20	10/17/76	10-20 KT	Underground	Fission
21	11/17/76	4 MT	Air (Hong 6 bomber)	Thermonuclear; largest Chinese test to date
22	9/17/77	<20 KT	Atmospheric	Fission
23	3/15/78	<20 KT (perhaps as low as 6 KT)	Atmospheric	Fission
24	10/14/78	<20 KT	Underground	Fission
25	12/14/78	<20 KT	Atmospheric	Fission

SOURCES: The specific figures for these tests are taken from a number of unclassified U.S. government sources; where these sources conflict, those used here are from United States, Department of Energy, *Announced Foreign Nuclear Detonations—Through December 31, 1978* (Las Vegas: Nevada Operations Office, n.d.). Where the dates differ, the ones used here are based on Chinese-announced dates given in Li Jue et al., *Dangdai Zhongguo*, pp. 569-76.

NOTE: KT = kiloton; MT = megaton.

Key Figures in China's Nuclear Weapons Program, 1954-1967

This list covers the principal participants in China's nuclear program mentioned in the book. Chinese interviewed in 1986 provided their own list of key scientists, most of whom are discussed. One scientist on their list, Zhang Xingqian, a "professor," was not further identified and is thus not included here. We have added two scientists, Lu Fuyan and Wu Zhengkai, who did not appear on the Chinese list.

Political Figures

CHEN YI 陈毅

Member of the Politburo, minister of Foreign Affairs, and vice-premier. In order to improve China's international status, he repeatedly appealed to the Politburo to give priority to the strategic weapons program.

HE LONG 贺龙

Member of the Politburo, vice-chairman of the Central Military Commission (1959-67), director of the National Defense Industrial Commission (1959-63), vice-premier, and member of the Fifteen-Member Special Commission. He was in charge of the production of weaponry and military equipment.

LUO RUIQING 罗瑞卿

Member of the Central Secretariat (1962-66), chief of the General Staff (1959-66), secretary general of the Central Military Commission (1959-66), director of the National Defense Industry Office (1961-66), and member of the Fifteen-Member Special Commission. He was in charge of coordinating the research on weaponry and military equipment with production.

MAO ZEDONG 毛泽东

Chairman of the Chinese Communist Party's Central Committee and Central Military Commission. He had final authority in all strategic weapons decisions.

NIE RONGZHEN 聂荣臻

Vice-chairman of the Central Military Commission (from 1959 on) and, from 1958 on, director of the Defense Science and Technology Commission and of the State Science and Technology Commission, vice-premier, and member of the Fifteen-Member Special Commission. After 1958, he headed the overall strategic weapons program.

ZHOU ENLAI 周恩来

Member of the Standing Committee of the Politburo and premier. Participated in original decision to develop nuclear weapons. In his capacity as director of the Fifteen-Member Special Commission after November 1962, he vigorously carried out coordination among strategic organizations so as to speed up the research and development of nuclear weapons.

Scientists

CAO BENXI 曹本熹

Chemist and chief engineer of the Second Ministry of Machine Building's Fuel Production Bureau and, concurrently, deputy director of the Beijing Nuclear Engineering Research and Design Academy. He made special contributions to the production of uranium hexafluoride and the chemical separation of plutonium.

CHEN NENGKUAN 陈能宽

Physicist with an American Ph.D. After 1960, under Wang Ganchang, he supervised the development of the explosive assembly and headed the team that carried out over 1,000 experiments to discover the principles for detonating an atomic bomb.

DENG JIAXIAN 邓稼先

Physicist with an American Ph.D. After 1960, he headed the theoretical design work in the Northwest Nuclear Weapons Research and Design Academy (Ninth Academy). He made great contributions to the theoretical design of both the atomic bomb and the hydrogen bomb.

GUO YONGHUAI 郭永怀

Engineer with an American Ph.D. Deputy director of the Ninth Academy and director of one of its technological committees. In the early 1960s, he and the engineer Long Wenguang made great contributions to the design of the configuration of the atomic bomb, the environmental tests, and the flight tests. Died in a plane crash in 1968.

JIANG SHENGJIE 姜圣阶

Nuclear chemist. Chief engineer and first deputy director of the Jiuquan Atomic Energy Complex under the Second Ministry. Between late 1963 and the spring of 1964, he helped the Nuclear Component Manufacturing Plant solve technical problems relating to the smelting and casting of the nuclear component.

LONG WENGUANG 龙文光

Engineer and principal assistant to Guo Yonghuai. Worked in the Ninth Academy's Design Department and helped engineer the configuration of the bomb. He later became head of the department.

LU FUYAN 禄福延

In charge of production technology in the Second Ministry's Sixth Institute. In 1960, he headed the test production of both uranium oxide and uranium tetrafluoride. He was subsequently transferred to Plant 414 to supervise the mass production of uranium oxide.

PENG HUANWU 彭桓武

Theoretical physicist with a British Ph.D. Deputy director of the Ninth Academy and director of one of its technological committees. In the 1960s, he was assigned to take overall charge of the theoretical design of both the atomic and the hydrogen bomb projects.

QIAN JIN 钱晋

Associate professor who refined the techniques for manufacturing the high explosives and the electric spark detonators of the first bomb. Persecuted and died in the Cultural Revolution.

WANG GANCHANG 王淦昌

Physicist with a German doctorate. Deputy director of the Ninth Academy and director of one of its technological committees. He was in overall charge of developing and testing the explosive assembly and initiator for the atomic bomb.

WU ZHENGKAI 吴征铠

Chemist. Head of the Chemical Department of Fudan University. In 1960, he was appointed head of Research Department 615 of the Institute of Atomic Energy. Under him, Section 615A did theoretical calculations for the Lanzhou Gaseous Diffusion Plant and Section 615B did the trial production of uranium hexafluoride.

YU DAGUANG 俞大光

Professor who helped design the overall multiple-line synchronous firing mechanisms for the explosive assembly.

YU MIN 于敏

Theoretical physicist. Served as deputy head under Deng Jiaxian of the Ninth Academy's Theoretical Department and made special contributions to the theoretical design of China's hydrogen bomb.

ZHOU GUANGZHAO 周光召

Theoretical physicist trained by Peng Huanwu and at Dubna in the Soviet Union. After 1960, first deputy head of the Ninth Academy's Theoretical Department. On several occasions he verified the theoretical calculations for the design of the atomic bomb. He headed a team to calculate the mechanical design of the bomb.

ZHU GUANGYA 朱光亚

Physicist with an American Ph.D. After transfer to the Ninth Academy, he was assigned to supervise the organization of scientific research on the atomic bomb and later served as academy deputy director. He also served as deputy director of the Defense Science and Technology Commission.

Administrative Officials

GUO YINGHUI 郭英会

One of the initial leading cadres of the Ninth Academy. He served as assistant to Li Jue and headed the academy's administrative wing.

LI JUE 李觉

Head of the Second Ministry's Ninth Bureau and director of the Ninth Academy. Under the leadership of the Second Ministry, he was assigned to take overall charge of the research and development of the atomic bomb project.

LIU JIE 刘杰

Between the mid-1950s and the mid-1960s, as vice-minister and minister (1960-66) of the Second Ministry, he oversaw the day-to-day operations of the ministry. Member of the Fifteen-Member Special Commission.

LIU XIYAO 刘西尧

Vice-minister and later minister of the Second Ministry and deputy commander of the First Atomic Bomb Test On-Site Headquarters. He was assigned to take charge of the administrative management of the research and development of the atomic bomb project. Under his leadership, the hydrogen bomb project developed rapidly.

QIAN SANQIANG 钱三强

Physicist with French doctorate. Director of the Institute of Atomic Energy and vice-minister of the Second Ministry. He oversaw scientific research on the atomic bomb project.

SONG RENQIONG 宋任穷

The first minister (1956-60) in charge of China's nuclear industry, first called the Third Ministry of Machine Building (1956-58) and then the Second Ministry.

WANG JIEFU 王介福

Director of the Lanzhou Gaseous Diffusion Plant.

WU JILIN 吴际霖

Deputy director of the Ninth Bureau and director of one of the Ninth Academy's technological committees. He was Li Jue's most important assistant. Persecuted and died in the Cultural Revolution.

YUAN CHENGLONG 袁成隆

Vice-minister of the Second Ministry. He was in charge of all ministry business related to nuclear weapons production. In 1963 and 1964, he stayed at the Lanzhou Gaseous Diffusion Plant and the Nuclear Component Manufacturing Plant to help solve key technical problems.

ZHANG AIPING 张爱萍

Deputy chief of the General Staff, deputy director of the Defense Science and Technology Commission, deputy director of the National Defense Industry Office, director of the First Atomic Bomb Test Commission, commander of the First Atomic Bomb Test On-Site Headquarters, and member of the Fifteen-Member Special Commission.

ZHANG YUNYU 张蕴钰

First commander of the Lop Nur Nuclear Weapons Test Base. He was in overall charge of preparations for various nuclear tests.

ZHAO ERLU 赵尔陆

Senior minister in charge of the research and development of weaponry and military equipment. Deputy director of the National Defense Industry Office and member of the Fifteen-Member Special Commission.

ZHU LINFANG 祝麟芳

Deputy chief of the Nuclear Component Manufacturing Workshop of the Nuclear Component Manufacturing Plant in Jiuquan Prefecture. He led the team that solved technical problems related to the smelting, casting, and machining of the uranium core for the first bomb.

Notes

Notes

For complete authors' names, titles, and publishing data on works cited in short form, see the References Cited section, pp. 293-309. In these Notes, we have not translated the names of four frequently cited publications: *Renmin Ribao* [People's Daily], *Xinhua* [New China News Agency], *Xin Hua Banyuekan* [New China Semimonthly], and *Xin Hua Yuebao* [New China Monthly]. The following abbreviations are used:

FBIS   *Foreign Broadcast Information Service*
FRUS   U.S. Department of State, *Foreign Relations of the United States* (Washington, D.C., vols. for 1949-57)
JPRS   Joint Publications Research Service
NRH   Nie Rongzhen, *Nie Rongzhen Huiyilu* [Memoirs of Nie Rongzhen] (Beijing, 1984)
NYT   *New York Times*

Chapter One

1. Text as in *Break the Nuclear Monopoly,* pp. 1-5; this text is reproduced in Appendix A.

2. Statement of Sept. 29, 1964, in NYT, Sept. 30, 1964.

3. Text of Oct. 18, 1964, statement in NYT, Oct. 19, 1964.

4. For Zhou's statement, see *Break the Nuclear Monopoly,* pp. 9-10; and on the partial test ban treaty the documents are in *People of the World.*

5. Interview with a senior Chinese military officer, 1983.

6. *Nie Rongzhen Huiyilu; Mimi Licheng.* The Magical Sword Branch of the Nuclear Industry Ministry, which edited *Mimi Licheng,* is associated with the Magical Sword Literary and Art Society, inaugurated in August 1983. The society seeks to "promote mass literary and art work on the national defense science and technology and national defense industry fronts." *Xinhua,* Aug. 11, 1983. The society publishes the journal *Shenjian* [Magical Sword].

7. Li Jue et al., chief eds., *Dangdai Zhongguo de He Gongye* [Contemporary China's Nuclear Industry]. This book, part of the Contemporary China series of more than 200 volumes, is the most complete Chinese history to date

of the nuclear weapons program. The four chief editors are Li Jue, Lei Rong-tian, Li Yi, and Li Yingxiang.

8. Hewlett and Anderson (U.S.); Rhodes (U.S.); Gowing, *Independence* (U.K.); Goldschmidt (international); Irving, *Virus House* (Germany); a study of the Soviet program, now in preparation by the political scientist David Holloway.

9. In this section, we have profited from reading George, *Presidential Decisionmaking*, especially pp. 1-12.

10. Mao [5], pp. 295-309. For a discussion, see J. Lewis, *Leadership*, Chap. 2.

11. Mao [8], p. 422.

12. For the most serious scholarly assessments of this evidence, see Borg and Heinrichs. The essays in this book by no means reach a consensus on the evidence of the Chinese Communist Party's pre-1950 policy options toward the United States or what conclusions can be reached from that evidence.

13. Mao [8], p. 416.   14. Mao [14], p. 87.
15. Mao [19], pp. 97-101.   16. *Ibid.*, p. 100.
17. Mao [6], p. 181.   18. Mao [8], pp. 415, 416.
19. For the major documents, see *Sino-Soviet Treaty*.

20. Secretary of State to U.S. embassy, Paris, Feb. 11, 1950, in FRUS *1950*, 6: 309.

21. Rusk to Acheson ("Memorandum by the Assistant Secretary"), April 26, 1950, in *ibid.*, p. 335.

22. Rusk to Acheson ("Extract from a Draft Memorandum"), May 30, 1950, in *ibid.*, p. 349.

23. For a discussion, see Whiting, *China Crosses the Yalu*; and George and Smoke, Chap. 7.

24. See "Effects of Operations in Korea," NIE-32 (July 10, 1951), in FRUS *1951*, 7.2: 1737-43, at p. 1742.

25. Quoted in Peng Dehuai, p. 472.

26. *Ibid.*, pp. 473-74; Li Jukui, p. 44.

27. Peng Dehuai, p. 473.

28. Mao [10], p. 43; "Why We Must Participate."

29. "Effects of Operations in Korea," in FRUS *1951*, 7.2: 1741.
30. *Ibid.*, pp. 1741-42.   31. *Seven Letters*, p. 25.
32. Mao [11], pp. 116-18.   33. "Two Different Lines."
34. George, *Chinese Communist Army*, p. 199.
35. *Ibid.*, p. 200; Whitson, p. 95.
36. George, *Chinese Communist Army*, pp. 171-75.

37. Whitson, p. 95. Several sources have assumed that Lin Biao served as the first CPV commander and was replaced by Peng in 1951. This seems not to be the case, as indicated by Peng's memoirs. For a discussion of this issue, see Farrar-Hockley, especially p. 292.

38. Whitson, pp. 95, 98. The Military Academy (Junshi xueyuan) was established on January 15, 1951.

39. "Courses of Action Relative to Communist China and Korea," March 14, 1951, in FRUS *1951*, 7.2: 1598-1605; "Vulnerabilities of Communist China," May 22, 1951, in *ibid.*, pp. 1673-82.

40. Kennan to Acheson, June 20, 1951, in *ibid.*, 7.1: 537.

Chapter Two

1. Chu Chi-hsin, p. 6.
2. Interviews with Chinese specialists, 1986.
3. For a brief discussion, see Blacker and Duffy, p. 158.
4. See Eisenhower, *White House Years*, pp. 72-73, 93-97.
5. NYT, Dec. 6, 1952.
6. NYT, Dec. 24, 1952.
7. NYT, Dec. 15, 1952.
8. For a scholarly examination of what Eisenhower actually said and did at the time, see Keefer, pp. 267-89.
9. Eisenhower, *White House Years*, Chap. 7. White House document NSC 147, dated April 2, 1953, gave the most complete review of options toward Korea. Entitled "Analysis of Possible Courses of Action in Korea," it examined, among other things, the pros and cons of using atomic weapons to end the war. The full text is in FRUS *1952-54*, 15.1: 839-57. See especially pp. 845-46. See also "Communist Capabilities and Probable Courses of Action in Korea," NIE-80 (April 3, 1953), in *ibid.*, pp. 865-77.
10. Eisenhower, *White House Years*, p. 181.
11. Adams, p. 48.
12. For a discussion of the sources on this strange event, see Keefer, especially pp. 280-81.
13. *Guangming Ribao* [Bright Daily], Jan. 23, 1953.
14. The principal source for the Chinese views in this paragraph is Jiang Zhenghao, "Some Aspects." Unless otherwise noted, the quotations in this and the next two paragraphs are from this paper, which is quoted with Mr. Jiang's permission. Mr. Jiang served on the Chinese delegation at the Panmunjom armistice talks and later in various diplomatic posts.
15. For a careful review of the controversy over the repatriation of prisoners, see Bernstein.
16. Conversations with Chinese specialists, 1984-85.
17. In addition to Jiang Zhenghao, "Some Aspects," this paragraph is based on conversations with Chinese specialists, 1984-85. Mao Zedong's review of how "we have won a great victory in the war to resist U.S. aggression and aid Korea" is in Mao [11], pp. 115-20.
18. See "Communist Capabilities," in FRUS *1952-54*, 15.1: 865, which notes: "A highly organized, well-integrated defensive zone extends possibly 15 to 20 miles to the rear of present battle positions. Many fortified areas have been constructed in rear of this zone and are being improved and expanded."
19. Eisenhower, *White House Years*, p. 181. For a text of the letter of Feb. 22, 1953, see FRUS *1952-54*, 15.1: 788-89. In a letter to the Department of the Army, General Mark Clark noted that he had a "serious doubt that the Communists would agree to any such proposal." *Ibid.*, p. 789.
20. See "Marshal Kim Il Sung," p. 7. On March 30, 1953, the New China News Agency broadcast a statement by Premier Zhou Enlai proposing that negotiations should begin immediately on the exchange of sick and wounded, followed by an overall settlement of the prisoner-of-war question. For the text, see *People's China*, No. 8 (April 16, 1953), pp. 5-7, and Department of State *Bulletin*, 28.720 (April 13, 1953), pp. 526-27.

21. American views of the Korean Armistice negotiations have been fully documented in many sources. See especially FRUS *1952-54*, 15.1 and 15.2: 938-1445; Eisenhower, *White House Years*, Chap. 7; Clark, especially Chaps. 16-18; Goodman; Hermes; Joy; Rees; and Ridgway, *Korean War*.

22. From 1949 on, the Chinese prepared two daily compilations of translated Western news articles, *Cankao Ziliao* [Reference Materials] and *Cankao Xiaoxi* [Reference News]. The first of these is distributed to senior cadres and the second to both senior and junior cadres. Both are widely available. For a review of some issues of *Cankao Xiaoxi*, see Schwarz.

23. See FRUS *1952-54*, 2.1: *passim*.

24. For the principal sources on the New Look, see Eisenhower, *White House Years*, Chap. 18; FRUS *1952-54*, 2.1; Jurika, especially pp. 319-27; and Kinnard, Chap. 1.

25. NYT, May 1, 1953. At this time, presidential press conferences did not allow direct quotations, so this quotation, like others from such meetings in this period, is a paraphrase. The transcripts of Eisenhower's press conferences have now been published; we have chosen to use the contemporary newspaper paraphrases because these are what the Chinese would have read.

26. Eisenhower publicly commented on the new policy about the time the new basic strategy was approved. In October 1953, amid speculation about Soviet thermonuclear weapons ("super bombs"), the president emphasized the strength of the American arsenal and called for a major buildup of U.S. air forces. NYT, Oct. 9, 1953. Military analyst Hanson Baldwin reviewed the New Look two days after the policy was adopted and a week later discussed the growing nuclear arms race and the increased stress on the use of atomic weapons. NYT, Nov. 1 and 8, 1953. Brief references to changes in American strategic thinking continued to appear in the American and European press throughout the first nine months of 1954.

27. The documents leading up to and including NSC 162/2 are found in FRUS *1952-54*, 2.1. The text of NSC 162/2 is at pp. 577-97.

28. *Ibid.*, p. 593. This sentence was the source of differences of interpretation between officials at the departments of State and Defense. See "Memorandum by the Undersecretary," Dec. 3, 1953, in *ibid.*, pp. 607-8.

29. Text in *ibid.*, p. 597.

30. Memorandum of NSC discussion, Aug. 27, 1953, in *ibid.*, p. 445.

31. *Ibid.*, p. 447.

32. NYT, Dec. 15, 1953.

33. NYT, Jan. 8, 1954.

34. For the text of the speech, see NYT, Jan. 13, 1954. Dulles subsequently discussed his "strategy to deter aggression." See Dulles, "Policy," pp. 357-59.

35. "Eisenhower Clamors for Preparations." See also "Dulles Has the Nerve"; Wu Quan, "New Look"; and "United States Is Afraid."

36. This is based on Rosenberg, "Origins." For typical press comments on these new deployments, see NYT, Sept. 6, 1953.

37. Rosenberg, "Origins," pp. 27-28.

38. *Ibid.*, p. 31.

39. In addition to the sources cited in note 35, see Wu Quan, "Comprehensive Foreign Policy"; and Jiang Nan, "Indian Prime Minister." Jiang Nan paid special attention to Nehru's statement that the U.S. policy of massive retaliation "included possible attack on the China mainland."

40. See, for example, Eisenhower's comments in NYT, Jan. 6, 1954, and his commitment in the State of the Union message to "continue military and economic aid to the Nationalist Government of China" (Taiwan); NYT, Jan. 8, 1954. On April 5, Dulles issued a direct warning to Beijing against further assistance to the Vietminh in Vietnam and suggested the Chinese threat might extend beyond Southeast Asia to engulf Australia and New Zealand. NYT, April 6, 1954.

41. Rosenberg, "Smoking," p. 27.

42. "Summary Statement," Oct. 11, 1954, in FRUS *1952-54*, 2.1: 750.

43. NYT, March 30, 1954.

44. Text in FRUS *1952-54*, 14.1: 278-306.

45. Documents of the Korean phase of the Geneva Conference are in FRUS *1952-54*, 16: 3-394. For a discussion of the Indochina phase of the conference, see Kahin and Lewis, Chap. 3.

46. These messages are reproduced in FRUS *1952-54*, 16: 14-142.

47. At the request of the Korean government, Rhee's letter of March 11 was not declassified and has not been released. *Ibid.*, pp. 14, 35-36.

48. The text of the letter is in *ibid.*, p. 44.

49. See, for example, "Communique of the Delegations," p. 7; "Welcome the Delegation"; and Jiang Nan, "New China News Agency."

50. U.S. ambassador, Seoul, to Department of State, May 11, 1954, in FRUS *1952-54*, 16: 245.

51. NYT, Dec. 27, 1953. This reflected a top-secret memo of Nov. 11, 1953; see FRUS *1952-54*, 2.1: 597.

52. NYT, Dec. 15, 1953, Jan. 13, 1954.

53. NYT, Dec. 26, 1953.

54. NYT, Feb. 24, 1954.

55. NYT, Feb. 8, 1955; FRUS *1952-54*, 14.1: 157 (March 19, 1953), 14.1: 333 (Nov. 18, 1953). Citing unpublished memoirs by Wellington Koo, Taiwan's ambassador to the United States, Thomas E. Stolper says the idea of a mutual defense pact was first suggested to Washington in March 1953. On the treaty's background, see Stolper, pp. 21-26.

56. Rankin, pp. 189, 190. These quotes appear in an entire chapter devoted to the treaty, pp. 171-214.

57. For these early PRC reactions, see "Thoroughgoing Betrayal." According to this source, Beijing first picked up hints of discussions of a U.S.-Taiwan mutual security pact from Nixon's visit to Taiwan in late 1953. This is consistent with Dulles's later testimony.

58. For examples, see FRUS *1952-54*, 14.1: 344, 367-70, and 399-401. Most public discussion of these recommendations focused on a Southeast Asia treaty. See NYT, April 6 and 11, May 13 and 14, and June 10, 1954. By June, the American press was referring to a U.S.-Taiwan mutual security treaty as essentially a fait accompli. This was based on a report from a Taipei newspaper that the two governments had agreed tentatively on such a pact. NYT, June 30, 1954.

59. See Jurika, Chap. 27, especially p. 422.

60. Rankin, p. 194.

61. FRUS *1952-54*, 14.1: 345-401 *passim*.

62. Rankin, pp. 194-95; NYT, June 20, 1954; Stolper, pp. 25-26.

63. Rankin, pp. 197-98.

64. Jurika, pp. 425-27.

65. *Ibid.*; "Memorandum by the Counselor," April 7, 1954, in FRUS *1952-54*, 13: 1271, in which the admiral advised that "three tactical A-weapons, properly employed, would be sufficient to smash the Viet effort there [at Dienbienphu]." On China, see FRUS *1952-54*, 12.1: 512-13, 521-26, and 556.

66. See, for example, NYT, Feb. 24, June 30, and July 22, 1954.

67. For an excellent discussion of this crisis and its aftermath, see Stolper, Chaps. 3 and 4. Because our interpretation of the crisis emphasizes military questions as they bear on the subsequent Chinese nuclear decision, it differs slightly from Stolper's.

68. Eisenhower, *White House Years*, p. 459. NYT, Sept. 4, 1954, carries the first news of this attack. The administration's first public reaction, which was reserved, came on September 5 (NYT, Sept. 6, 1954). On September 8, Washington announced that the PRC intended to attack Quemoy (NYT, Sept. 9, 1954), and the next day, Dulles met with Chiang Kai-shek in Taipei and promised the Nationalists they did not "stand alone" (NYT, Sept. 10, 1954).

69. One of the authors was an officer in the U.S. Pacific Fleet at this time. For public statements on the alert, see NYT, Sept. 5 and 6, 1954.

70. Eisenhower, *White House Years*, pp. 462, 459.

71. U.S. Senate, p. 313.

72. "United States Objectives," NSC 146/2 (Nov. 6, 1953), in FRUS *1952-54*, 14.1: 307.

73. Chinese news stories on these attacks are in *Renmin Ribao*, Dec. 24, 1953, and May 20, 1954. See also *Xin Hua Yuebao*, Dec. 28, 1954, p. 116; and *Renmin Ribao*, July 16, 1954.

74. *Xin Hua Yuebao*, Aug. 28, 1954, p. 93.

75. *Renmin Ribao*, July 16, 1954. See also Ho Cheng, who details the attacks on merchant shipping by Taiwan's forces.

76. NYT, July 24-28, 1954; *Xin Hua Yuebao*, Aug. 28, 1954, p. 51.

77. The two *Renmin Ribao* articles appeared on successive days—July 23 and 24, 1954. See also Zhu De.

78. NYT, Aug. 4, 1954.

79. NYT, Aug. 10, 1954.

80. Zhou Enlai, "Report on Foreign Affairs," especially pp. 121-26.

81. NYT, Aug. 18, 1954.

82. Khrushchev 1974, p. 246. The trip is covered in pp. 245-50.

83. Text in *People's China*, No. 21 (Nov. 1, 1954), supplement, p. 5.

84. Khrushchev 1970, p. 466.

85. *Xin Hua Yuebao*, Dec. 28, 1954, pp. 9-13.

86. For a discussion of deterrence in respect to the Taiwan Strait crisis of 1954-55, see George and Smoke, Chap. 9.

87. Interviews with Chinese specialists associated with the crisis, 1984. For a published assessment by a senior Chinese military commander involved in the crisis, see Nie Fengzhi et al., pp. 39-57 *passim*. See also He Di, "The Evolution of the People's Republic of China's Policy Toward the Offshore Islands (Quemoy, Matsu)," Sept. 1987; cited by permission of the author.

88. Gittings, p. 197.

89. Jiang Zhenghao, "Sovereignty." This paper is cited with Mr. Jiang's permission.

90. U.S. delegation, Geneva, to Department of State, June 26, 1954, in FRUS *1952-54*, 16: 1251; Sixth Plenary Session, June 9, 1954, in *ibid.*, p. 1090.

91. For the text of Zhou's statement, see *People's China*, No. 15 (Aug. 1, 1954), supplement, p. 6.

92. Zhou Enlai, "Report on Foreign Affairs," pp. 110-11, 118. Zhou added that his "proposition" did "not envisage the exclusion of any country" (p. 118).

93. *Ibid.*, pp. 121, 126.

94. Mao [7], p. 56.

95. Zhou Enlai, "Report on Foreign Affairs," pp. 111, 125-26.

96. Rankin, p. 205.

97. Ridgway, *Soldier*, pp. 278-79.

98. Eisenhower, *White House Years*, pp. 462, 465, 466; U.S. Senate, p. 164.

99. SNIE-100-4-54 (Sept. 4, 1954), in FRUS *1952-54*, 14.1: 563-71.

100. NIE-43-54 (Sept. 14, 1954), in *ibid.*, especially pp. 627-45.

101. Eisenhower, *White House Years*, p. 465.

102. *Ibid.*, pp. 466, 467. One of the authors served on a destroyer that assisted in the evacuation of the Dachen Islands. In briefings by senior officers of the task force, ship captains (as consistent with SNIE-100-4-54 cited in note 99) were told that the Dachens were critical to the early warning defense against air raids coming from China's airfields to northwest of Taiwan. They felt that Quemoy and Matsu had no such value because of their location in relation to major Chinese airfields. For a similar judgment by authorities on Taiwan, see "Memorandum of a Conversation," Jan. 19, 1955, in FRUS *1955-57*, 2: 40. In this conversation, for example, Taiwan's foreign minister called the Dachens "extremely useful for radar tracking and intelligence operations." For a Chinese view of this action, see *Disaster Strikes the Tachens*; and *Xin Hua Yuebao*, May 28, 1955, p. 60.

103. We have not given much attention to the actual negotiation of the Mutual Defense Treaty between the United States and the Republic of China; the text is in *American Foreign Policy, 1950-1955*, pp. 945-47. For a lucid discussion of the treaty and relevant subsequent events, see Stolper, Chaps. 4 and 5.

104. "Summary Statement," in FRUS *1952-54*, 2.1: 743, 747.

105. "Basic National Security Policy," Nov. 15, 1954, in FRUS *1952-54*, 2.1: 772. This document is a revision of NSC 162/2.

106. "Basic National Security Policy," Dec. 14, 1954, in *ibid.*, p. 811.

107. NYT, Jan. 3, 1955.

108. Guo Moruo, "Strengthen the Peace Forces," p. 5.

109. Jiang Nan, "People."

110. Eisenhower, *White House Years*, p. 467. At an NSC meeting on Jan. 20, 1955, the president said, "It was not that any of these offshore islands was going to be easy to defend, but that the psychological consequences of abandoning these islands were so serious. . . . We must now be concerned with the morale of those soldiers who might well be called upon to defend Formosa." "Memorandum," in FRUS *1955-57*, 2: 79-80.

111. The text of the Formosa Resolution ("Joint Resolution by the Con-

gress," Jan. 29, 1955) is in FRUS *1955-57*, 2: 162-63; and NYT, Jan. 29, 1955.

112. See Halperin and Tsou, p. 125. For the relevant documents on the defense of Quemoy and Matsu, see FRUS *1955-57*, 2: 46-48, 50-52, 75-77, 79, 101, 145, 166-68, 174-75, and 181-82. The president's statement on not being "hooked" is on p. 175. On January 31, however, Ambassador Rankin in Taipei reported that Taiwan's foreign minister had mentioned "several times and quite definitely a firm agreement that [the] two governments would issue coordinated statements including specific reference to US protection for Kinmen [Quemoy] and Matsu" (p. 181).

113. Eisenhower, *White House Years*, p. 467. For a more general discussion of this point, see George and Smoke, pp. 286-88.

114. U.S. Senate, pp. 16, 68.

115. *Ibid.*, pp. 68, 74-75, 130. See also p. 71.

116. *Ibid.*, pp. 149, 162-65, 312.

117. Halperin and Tsou, p. 137.

Chapter Three

1. This assessment is based on J. Lewis, "China's Military Doctrines," pp. 148-50.

2. *People of the World*, pp. 22, 85.

3. This paragraph is based on Li Jue et al., pp. 4, 9. Mao's quotation of June 21, 1958, is from Su Fangxue, p. 4; and Su Kuoshan. Mao echoed his 1958 quote in 1960 when he said: "We should pay attention to the policy of strength and the position of strength." All states, he insisted, "give the highest priority to building their strength." Liu Suinian, p. 23.

4. Mao [3], pp. 152, 153.

5. See Eisenhower, *White House Years*, Chap. 19. The quote is from U.S. Senate, p. 65.

6. For the text of the Formosa Resolution, see NYT, Jan. 29, 1955.

7. Tan Wenrui; *Xinhua*, Jan. 24, 1955, in *Survey of China Mainland Press*, No. 974 (Jan. 22-26, 1955), p. 2.

8. "Resolutely Oppose."

9. This paragraph is based on Li Jue et al., p. 13; and Qian Sanqiang, "Cherish the Memory."

10. Li Jue et al., p. 13; Zhang Jiong, p. 108.

11. Mao [4], pp. 23, 24, 36. Italics in the original are omitted.

12. Cited in Schram, p. 229. For an analysis of Sakata's ideas and Mao's views on contradictions, see Wakeman, p. 227. Sakata's article is listed in the References.

13. Quoted in Schram, p. 251.

14. This and the next two paragraphs are based on Li Jue et al., pp. 14, 21; and Qian Sanqiang, "Cherish the Memory."

15. Mao [16], p. 168.

16. The pioneer work on China's nuclear strategy in this period is Hsieh, *Communist China's Strategy*; see especially Chap. 2. Hsieh writes (p. 26): "In contrast to 1954, ... there was, for several months after January 1955, an unprecedented volume of comment on nuclear matters." On Soviet doctrines, see Holloway, *Soviet Union*, pp. 35-39. See also Dinerstein, Chap. 2.

17. Eisenhower, *White House Years*, p. 476. This statement to the president was not made public. See NYT, March 8, 1955.

18. See Dulles, "Report," pp. 459-60 (where he says he threatened the use of "new and powerful weapons of precision which can utterly destroy military targets without endangering unrelated civilian centers"); and Eisenhower, *White House Years*, p. 477.

19. Eisenhower, *White House Years*, p. 477. For the Chinese reaction, see "Eisenhower Advocates Use." On March 29, *Renmin Ribao* stated that the United States was threatening China with massive retaliation, and that China needed to be ready to cope with sudden emergencies. See "Make Great Efforts."

20. Guo Moruo, "Ban Atomic Weapons!," p. 3. Many pamphlets appeared in 1955 to educate the populace on the dangers of nuclear weapons and to mobilize them against nuclear war. See, for example, Wang Zhiliang.

21. Interviews with Chinese specialists, 1986.

22. "Statement of Soviet Government," p. 53. The text of the Chinese State Council resolution on this offer is in *People's China*, No. 4, Feb. 16, 1955, supplement.

23. See Liu Wei. The reactor is a deuterium-moderated heavy-water reactor; its original power of 7 megawatts was upgraded in the 1960s to 10 megawatts; and it uses 1.2% enriched fuel rods. A later, swimming-pool reactor designed by the Institute of Atomic Energy used 10% enriched material and had a total U^{235} inventory of 5 kg in the mid-1970s. The cyclotron of 1.2-m pole diameter was also later upgraded to a sector-focused variable-energy cyclotron. See Panofsky, pp. 40-43; and Tao Cun.

24. "Resolution of State Council," p. 53.

25. This information on the Soviet role in China's search for and use of uranium is based on Li Jue et al., pp. 20, 138, 168.

26. Lindbeck, p. 10.

27. Guo Moruo, "Strengthen the Peace Forces," p. 5.

28. Li Jue et al., p. 5; Qian Sanqiang, "Peaceful Utilization." See also Qian Weichang. Among the group of physicists in 1949 were Wu Youxun, Peng Huanwu, Wang Ganchang, Qian Sanqiang, He Zehui, Li Shounan, and Zhao Zhongyao.

29. The returned Chinese included Li Siguang (an eminent geologist), Hua Luogeng (an outstanding mathematician), Zhao Zhongyao (who had done major work on radiation), and many others who were to become prominent in the academy.

30. "Survey of the Chinese Academy of Sciences," p. 923; Li Jue et al., p. 6.

31. In 1950, Premier Zhou Enlai created the "Experts Work Group," headed by Wu Xiuquan, and approved a document on methods for strengthening the work of these foreign experts. Yang and Wu.

32. The Institute of Modern Physics was renamed three times. On Oct. 6, 1953, it became the Institute of Physics; on July 1, 1958, the Institute of Atomic Energy; and at the end of 1984, the Chinese Academy of Atomic Energy Science. Li Jue et al., pp. 363, 559. Typical annual reports on the work of the academy are Guo Moruo, "Summary," pp. 184-85; and Guo Moruo, "Report on the Present Status," p. 197. One source lists the nuclear scientists assigned to the institute that "later became the research center of China's nuclear physics." They were (in the order given) Peng Huanwu, Wang Ganchang,

Zhao Zhongyao, Yang Chengzhong, Yang Chengzong, Zhu Hongyuan, Xiao Jian, Deng Jiaxian, Dai Zhuanzeng, Jin Xingnan, Li Shounan, Xin Xianjie, Huang Zuqia, Lu Zuyin, Yu Min, Xu Jianming, Ye Minghan, Zhu Guangya, and Hu Ning. Zhang Jiong, p. 106.

33. "Report by Delegation," p. 209.

34. Qian Sanqiang, "Survey."

35. Unless otherwise noted, this history is based on Qian Weichang, "Physics in China," *Renmin Ribao*, Aug. 13, 1949; Li Jue et al., pp. 5, 6, 8; and T. Y. Wu, pp. 631-43.

36. Feng and Chen.

37. For a short biography of Qian Sanqiang, see Klein and Clark, 1: 188-90.

38. For background on Zhou Guangzhao, see Gu Mainan, "One Out of a Hundred Thousand." A capsule history of his career is given in Appendix C.

39. "To Study Soviet Theory." Materials for the study of Soviet Communist Party history, documents of the Nineteenth Congress of the Soviet Party, and a volume on the Soviet socialist economy were prepared for cadre study. For a sample, see *Xuexi* [Study], 1953, Nos. 1-12. A discussion of cadre study campaigns is in J. Lewis, *Leadership*, pp. 145-56.

40. "Directive," p. 200.

41. Guo Moruo, "Report on the Academy's Basic Situation."

42. "Devote Major Efforts." The agreement was signed on April 29.

43. Qian Weichang was a senior rocket engineer who had completed advanced study in the United States. See *Renmin Ribao*, May 17, 1957; "Rightist Qian Weichang"; Theodore Chen, p. 209; and "All-China Scientific Association."

44. Zhou Enlai, *Report on Intellectuals*, especially pp. 34-35.

45. Theodore Chen, p. 161.

46. NRH, p. 777.

47. Information in this and the following paragraph is from Li Yingxiang et al.

48. Liu Jie; Li Jue et al., pp. 15-16, 561.

49. For an important discussion of Chen Yun's life and views (though not his role in the nuclear program), see Bachman.

50. NRH, pp. 711, 715. Nie's memoirs do not mention the Three-Member Group.

51. For Bo Yibo's biography, see Klein and Clark, 2: 738-42.

52. Li Jue et al., pp. 15-16; Liu Jie. On the creation of the Third Office, see "Order of the State Council."

53. Most of the information in this paragraph is from Li Jue et al., pp. 15-16, 475; and Liu Wei.

54. "Li Peng Makes a Speech"; Li Jue et al., pp. 20-21.

55. Some sources state that the Third Ministry was established in April 1955 and abolished in May 1956. However, the Third Ministry of that period was responsible not for the nuclear industry but for the electrical industry; when formally abolished, it was replaced by the Ministry of Electrical Machine Industry. The new Third Ministry in charge of the nuclear industry was created on Nov. 16, 1956 (see *Xinhua*, Nov. 17, 1956; Li Jue et al., p. 16.) The State Council appointed Song Renqiong the minister and the following vice-ministers: Liu Jie, Yuan Chenglong, Liu Wei, Lei Rongtian, and Qian

Sanqiang. Some information in this and the following paragraph is from interviews with Chinese specialists, 1985-86.

56. *Xin Hua Yuebao*, Sept. 25, 1952, p. 14. The original Second Ministry, created on Aug. 7, 1952, followed a tortuous institutional path to the present Ministry of Ordnance Industry, so named in May 1982. The State Council merged it and the Ministry of Electrical Machine Building with the First Ministry of Machine Building in 1958, and then reestablished it as the Third Ministry in September 1960.

57. Duan Junyi et al.

58. Song Renqiong's biography is in Klein and Clark, 2: 787-90. The way Song's network of contacts functioned will be discussed in Chap. 5.

59. Zhou Enlai, *Report on Intellectuals*, p. 38.

60. Guo Moruo, "Comprehensive Plan," p. 139.

61. Nie Rongzhen, "Development," pp. 334-35.

62. Specifically, the Central Military Commission ordered Deputy Chief of the General Staff Zhang Aiping to head the Equipment Planning Department of the General Staff and to assume leadership of planning scientific research for conventional weapons. Zhang's biography is in Klein and Clark, 1: 9-11. A corps-level commander at the end of the revolutionary war, Zhang rose to commander and political commissar of the East China Sea Fleet. He became deputy chief of the General Staff in 1954. In the 1960s, as we shall see, he commanded various aspects of the nuclear test program.

63. The origins and development of the missile program will be the subject of our subsequent study. References to this program will thus be minimal here.

64. Zhang Aiping, "Several Questions," p. 85.

65. Nie Rongzhen, "Congratulatory Letter"; "Ten Prominent Scientists," p. 28.

66. At this time, Huang Kecheng served as head of the PLA's General Logistics Department.

67. We have no adequate biography of An Dong, who served as Nie's administrator in a number of organizations; in the early 1950s, he had been office director of the General Staff while Nie was its acting chief. In his memoirs, Nie has high praise for An Dong (see, for example, NRH, p. 713), who was "cool-headed" and could deal with problems "in a systematic way." In the mid-1960s, An was implicated in the case against the then-chief of the General Staff, Luo Ruiqing; he committed suicide after Luo was purged.

68. NRH, p. 762; Dong Kegong et al. At the Eighth Party Congress in September 1956, Deng's title was changed from secretary general to general secretary.

69. NRH, p. 770.

70. "Appointments of Nie Rongzhen and Bo Yibo"; NRH, pp. 762-63; "Report by Chinese Academy." On May 12, 1957, the State Council approved the formal establishment of the Scientific Planning Commission and confirmed Nie Rongzhen's appointment as director. On May 23, Guo Moruo, president of the Chinese Academy of Sciences and a commission deputy director, told an academy meeting that the commission had been made a "standing organization" of the State Council and would coordinate the activities of "numerous departments relating to scientific programs." *Renmin Ribao*, May 31, 1957.

71. NRH, pp. 770, 773-74, 793-95; Li Jue et al., p. 16. Nie's efforts to

train and mobilize a force of defense scientists and engineers are discussed in Chap. 5, below.

72. In May 1958, the Central Military Commission established the Fifth Department, which quickly acquired de facto control of the Aviation Industrial Commission and the Division for Scientific Research of the General Staff's Equipment Planning Department. NRH, p. 783.

73. NRH, p. 774.

74. Nie Rongzhen, "Development," p. 337.

75. *Ibid.*, p. 339.

76. Interview with Chinese specialists, 1985.

77. When Song Renqiong was appointed to head the ministry in November 1956, Mao specifically charged him with ensuring coordination and cooperation among the various organizations for the development of nuclear weapons. NRH, pp. 788-89.

78. Li Yingxiang et al.

79. Dai and Zhao; Li Jue et al., p. 189.

80. Most of the information in this and the next paragraph is from NRH, p. 777.

81. Interviews with a Chinese specialist, 1984-85. We shall examine the fate of the military aircraft industry in our later study of the strategic missile program.

82. NRH, p. 782.

83. NRH, p. 783. Not to be confused with the Fifth Academy, the Fifth Department administered the Aviation Industrial Commission and key aspects of the conventional military equipment program.

84. For the biographies of Chen Geng, Liu Yalou, and Wan Yi, respectively, see Klein and Clark, 1: 113-16, 1: 632-34, and 2: 886-87. An outstanding military leader in the revolution, Chen Geng served as deputy commander of the Chinese People's Volunteers under Peng Dehuai in Korea. In the mid-1950s, he became deputy chief of the General Staff and was, for a time (1956-57), acting chief. When he joined the new commission, Chen also became president of the Harbin Military Engineering Institute, which Nie ordered to train research and design personnel for military equipment, "including guided missiles and atomic weapons" (NRH, p. 796). Chen died in 1961. Liu Yalou was a veteran revolutionary political commissar and military commander who had served as Lin Biao's chief of staff. He commanded the PLA air force from 1949 to his death in 1965. Wan Yi also brought to the commission a background as a military commander under Lin Biao in the revolutionary period. He had served as deputy minister of the Second Ministry of Machine Building under Zhao Erlu and, in 1957, became a key member of the Scientific Planning Commission. In 1958, he also headed the PLA's Equipment Planning Department. On An Dong, see note 67, above.

85. The Central Military Commission formally established this academy in October 1956. It was reorganized into the Seventh Ministry of Machine Building in January 1965.

86. The information on the organs listed in Figs. 1 and 2 is based largely on interviews with Chinese specialists, 1986; and Li Jue et al., pp. 26, 33, 41, 257, 418, 436-37, 442.

87. These massive proceedings, published in 33 volumes, are U.N. docu-

ment A/CONF.15/1. They cover technical details of raw material resources, the production of nuclear material, basic metallurgy, reactor technology, and basic and applied nuclear physics.

88. Khrushchev 1970, Chap. 18; Khrushchev 1974, Chap. 2.

89. Khrushchev 1970, pp. 462-63.

90. *Ibid.*, pp. 465, 466.

91. *Ibid.*, pp. 467, 470. In an interview in 1946, Mao said: "The atom bomb is a paper tiger which the U.S. reactionaries use to scare people. It looks terrible, but in fact it isn't. Of course, the atom bomb is a weapon of mass slaughter, but the outcome of a war is decided by the people, not by one or two new types of weapon. All reactionaries are paper tigers." Mao [19], p. 100. These ideas were resurrected at the time of the 1958 Taiwan Strait crisis. For a relevant collection of documents, see *Imperialism and All Reactionaries Are Paper Tigers*, especially pp. 3-33.

92. "Two Different Lines."

93. Khrushchev 1974, pp. 268-69.

94. Conversation with a senior Soviet specialist, 1986.

95. Gu Mainan, "Deng Jiaxian, Veteran Scientist"; information from Chinese specialists, 1986; Li Jue et al., p. 32.

96. Khrushchev 1974, p. 269.

97. This and the next paragraph are based on Li Jue et al., pp. 20-22; and NRH, pp. 800-801. The unofficial Soviet pledge of full-scale assistance was given by a delegation of Soviet atomic energy scientists visiting China.

98. NRH, p. 803.

99. The information on the uranium part of the agreement is from an interview with a Chinese specialist, 1986; and Li Jue et al., p. 43. On the nuclear submarine, see Li Jue et al., p. 32.

100. NRH, pp. 803-4.

101. Ford, pp. 160-73.

102. See especially Borisov, pp. 45-50. "Borisov" is the pseudonym of Oleg Borisovich Rakhmanin, a senior staff official on the Central Committee of the Soviet Communist Party.

103. Khrushchev 1974, p. 258. The Soviet government first made this request to build a long-wave radio station in April 1958; He Xiaolu, p. 171. He Xiaolu notes that at the time the Chinese believed that with such a station the Soviets would "control our intelligence information and secret communications."

104. Khrushchev 1974, pp. 261-63. For a Western interpretation of the effect of the Taiwan Strait crisis on the dispute, see Ford, pp. 168-71.

105. Interview with a Chinese specialist, 1985.

106. NRH, p. 804. Nie does not identify the types of specialists. The information on the two nuclear weapons specialists is from an interview with a Chinese specialist, 1986. On the arrival of the technical specialists and the August 1958 accord, see Li Jue et al., pp. 22, 218.

107. NRH, p. 804. Soviet authors assert that Mao himself first raised the proposal for a joint fleet as "an act of provocation." See Borisov, pp. 69-70.

108. Communication from Nie Rongzhen to the authors, Feb. 15, 1986.

109. Whiting, "Quemoy," pp. 263-70. Mao's quotations in this paragraph are from Whiting's article.

110. He Xiaolu, p. 162.

111. In July 1963, the Chinese denounced the limited test ban treaty signed by the United States, Great Britain, and the Soviet Union. See *People of the World*, especially pp. 1-6.

112. Unless otherwise noted, the information in this paragraph is from NRH, pp. 804-5; and "Riddle of Research and Development," p. 5. For a general idea of the state of negotiations on a test ban in these years, see Blacker and Duffy, pp. 102-9, 126-29.

113. "Statement by the Spokesman of the Chinese Government" (Aug. 15, 1963), in *People of the World*, pp. 28-29. The following month, the Chinese repeated the charge in "Origin and Development," adding that the gift to Eisenhower was intended to "create the so-called 'spirit of Camp David.'"

114. Li Jue et al., p. 565.

115. Mao [1], pp. 517-18; Mao [21], pp. 308-11.

116. Mao [16], p. 168.

117. In articles in the 1960s, this line about strategy and tactics was linked to Mao's ideas on people's war. In December 1936, he had said: "Our strategy is 'pit one against ten' and our tactics are 'pit ten against one.'" Mao [12], p. 135. On the linkage of this statement to the new strategy, see Li Tso-peng.

118. Mao [21], p. 310.

119. Holloway, *Soviet Union*, especially pp. 29-43; the quote is from p. 29.

120. Dinerstein, p. 10.

121. Quoted in Holloway, *Soviet Union*, p. 32.

122. Khrushchev 1974, p. 255.

123. Mao speech, Moscow, November 1957, quoted in "Statement by the Spokesman of the Chinese Government" (Sept. 1, 1963), in *People of the World*, p. 42.

124. The account of the Mao-Khrushchev conversation in this paragraph and the two that follow are based on Khrushchev 1974, pp. 256-57.

125. Mao [2], pp. 494-95.

126. Quoted in "Comrade Mao Zedong."

127. Mao [13], pp. 108-9.

128. "Greet the Upsurge," p. 15. In a speech on September 5, Mao said that if the enemy "is determined to fight, it is they who will strike first, and it is they who will strike with atomic bombs. . . . If we have to fight, then we'll fight, and after we've fought, we'll rebuild. For this reason we must now build up the militia. The militia must be developed in every people's commune." Mao [17], p. 88. Mao's stress on the militia accorded with his general views on the validity of people's war to defeat all enemies, however well armed.

129. Mao [13], p. 108.

130. Mao [20], pp. 136-37. Some years later, the Chinese Institute of Literature issued *Stories About Not Being Afraid of Ghosts* (Beijing, 1961) to taunt the Soviets for their alleged ghosts of nuclear war, but the theme began much earlier.

131. NRH, p. 804.

132. "Origin and Development."

133. NRH, p. 804.

134. On the eve of the meeting, the commander of the air force wrote: "China's working class and scientists will certainly be able to make the most

up-to-date aircraft and atomic bombs in the not-distant future. . . . We can use atomic weapons and rockets made by the workers, engineers, and scientists of our country in coping with the enemies who dare to invade our country and undermine world peace. By that time, another new turning point will probably be reached in the international situation." Liu Yalou, pp. 9-10.

135. A Taiwan publication has given the text of these guidelines. See Guo Hualun, p. 13. This source does not state the exact date the Central Military Commission issued the eight guidelines.

136. *Jiefangjun Bao* [Liberation Army Daily], Jan. 16, 1958, in *Survey of China Mainland Press*, No. 1786 (June 6, 1958), p. 6. For an article that provides some context and a different perspective on the application of foreign experience, see Ford, especially pp. 160-65.

137. Mao [15], pp. 16-20 *passim.*

138. Quoted in Gittings, p. 231. In the announcement of the first hydrogen bomb test in June 1967, the Chinese heralded Mao's 1958 prediction: "Chairman Mao Tse-tung pointed out as far back as June 1958: I think it is entirely possible for some atom bombs and hydrogen bombs to be made in 10 years' time." *Xinhua* release, June 17, 1967, in FBIS: *Communist China*, June 19, 1967, p. CCC1.

139. *Renmin Ribao*, Aug. 1, 1958. This discussion is based on Ford.

140. *Renmin Ribao*, Aug. 2, 1958.

141. On Sept. 9, 1959, the Soviet news agency issued a statement on the border dispute. See "Origin and Development."

142. NRH, p. 805.

143. He Xiaolu, p. 166.

144. The information in this and the preceding paragraph is from NRH, pp. 805-6; and Li Jue et al., p. 32. The Soviet scientist Mikhail Klochko has recorded his observations about Chinese science, which suggest an alternative explanation: namely, Moscow considered China so backward that there was no way to help the PRC develop advanced weapons on its own. In 1964, Klochko wrote (p. 208): "We may conclude therefore that it will be a long time before China joins the nuclear club." Though in error in asserting that "the question would never arise of Khrushchev's handing Soviet nuclear weapons over directly to Mao," he does confirm that "all too often, the Soviet Union sent to China only middle-echelon specialists, especially when it came to matters of possible military application."

145. By July 16, the Soviet Union had delivered a note announcing its intention to recall 1,390 experts from China between July 28 and September 1. He Xiaolu, p. 172; "Riddle of Research and Development," p. 5; Li Jue et al., pp. 33-34.

Chapter Four

1. For a popular treatment of the scientific quest to understand and manipulate the uranium atom, see Bickel, *Deadly Element.*

2. In the discussion of uranium geology, we have used two works published during the 1950s: Clegg and Foley, *Uranium Ore Processing*; and Holden, *Physical Metallurgy of Uranium.*

3. Clegg and Foley, p. 4.

4. *Ibid.*

5. *Ibid.*, p. 6.

6. The scintillation counter is regarded as being "far superior to a Geiger counter for aerial prospecting and generally . . . best for surface prospecting." *Ibid.*, p. 25.

7. Most of the information in this paragraph is from *ibid.*, pp. 27-33.

8. This section is based principally on Li Jue et al., pp. 11-13, 23, 26, 102-6, 108, 111-15, 134-35, 138.

9. Li Yingxiang et al.

10. Wang Aimin et al., pp. 80, 83.

11. Zhou Jinhan.

12. Zhou Enlai, *Report on Intellectuals*, p. 11.

13. One widely cited article in this regard is Di Zhongheng, "Chinese Communist Nuclear Forces." See especially Part 1: 3-11, and Part 2: 17. This 1976 Hong Kong article states that the nuclear physicists Qian Sanqiang and He Zehui, Qian's wife, organized a Sino-Soviet prospecting team to investigate uranium ore resources in Xinjiang as early as 1949. This team purportedly functioned prior to the establishment of the People's Republic on October 1. Di also states that two uranium mines opened near the provincial capital, Ürümqi, the following March. Soviet help allegedly included blueprints, machines, and experts, as well as the building and support of a uranium ore refinery. Four more mines were said to have been added in 1951 (two outside Xinjiang). Despite the substantial detail given in this article, we believe it is unreliable.

14. Whitaker and Shinn, p. 487. The Soviet writer I. N. Golovin states (p. 65): "The Government handed this assignment of graphite and uranium production to the Commissariat of non-ferrous metallurgy." Though the term "non-ferrous" includes uranium and thorium, and is sometimes used in the Soviet Union and China when the reference is to uranium, it does not necessarily mean uranium in either the Chinese or the Soviet usage.

15. A U.S. intelligence estimate of December 1960 stated: "The exploitation of native uranium resources has been under way with Soviet assistance since 1950. Over 10 deposits are now being worked, and we believe that ore with a uranium metal equivalent of several hundred tons is being mined annually and retained in China." U.S., CIA, "Chinese Communist Atomic Energy," p. 2 (declassified January 1986).

16. By the 1980s, uranium mines were operating at 26 locations in 24 provinces and autonomous regions. After 1984, nearly 60,000 people (including 14,000 technical personnel) served in 51 prospecting teams and 7 research institutes under the ministry's Geological Bureau; another 6,500 people worked in uranium geology teams under the Ministry of Geology and Mineral Resources and the Ministry of Metallurgical Industry. Li Jue et al., p. 103, and the map opposite p. 238. For an early discussion on the subject, see Gourievidis.

17. According to Wu Xiuquan (p. 25), the Soviet side provided the company with machinery, operational equipment, and technical know-how. Wu, a former vice-minister of foreign affairs and deputy chief of the General Staff, states that the Xinjiang region "abounds in copper, lead, aluminum, molybdenum, manganese, tin and other metals," but he does not mention uranium.

By 1985, when Wu's book was published, the Chinese were making no secret of later Soviet assistance in uranium exploration. As early as 1964, the Chinese had claimed that they had significantly aided Soviet military programs. In a famous letter to Moscow of February 29, the Chinese Party Central Committee said: "Up to the end of 1962 . . . China furnished the Soviet Union with more than 1,400 million new roubles' worth of mineral products and metals. . . . Many of these . . . are indispensable for the development of the most advanced branches of science and for the manufacture of rockets and nuclear weapons." Lithium, beryllium, and a number of other weapons-related metals are mentioned, but, again, uranium is not among them. *Seven Letters*, p. 25.

18. Zhao Zixing.

19. For a comparison of various countries' knowledge at that time, see Stead, pp. 714-21.

20. The Third Ministry became the Second Ministry of Machine Building in February 1958 and the Ministry of Nuclear Industry in May 1982. For a discussion of this ministry, see Chap. 3. After 1956, four more prospecting teams—182 (in Shanxi Province), 209 (Sichuan), 406 (Liaoning), and 608 (Jiangxi)—began operating. Li Jue et al., p. 23.

21. See Wang Yanting et al., p. 1031.

22. This paragraph is based on Wang Aimin et al., pp. 79, 80, 83.

23. Some sources refer to the mine as the Chenzhou Uranium Mine. Chenzhou is the historical name for Chenxian.

24. See Wang Jian et al., p. 6. Other information in this and the preceding paragraph is from interviews with Chinese specialists, 1986; and Li Jue et al., pages cited in note 8.

25. Unless otherwise cited, the discussion of Luo Pengfei's find is based on Zhao Fuxin and Zheng Shurong, pp. 58-61, 78. These authors discuss the search in the Nanling mountains but do not name the precise location; our assumption is that the locale is Lianxian.

26. *Ibid.*, p. 59. The character "lao" means old and is used as an affectionate title.

27. *Xinhua*, Dec. 2, 1979, in FBIS: *People's Republic of China*, Dec. 4, 1979, p. L15. This article does not identify the region, but from this report and others that appeared at the time, it is clear the mining region being discussed is in the Nanling mountains.

28. Zhao Fuxin and Zheng Shurong, p. 63.

29. *Ibid.*, pp. 65, 66, 74.

30. *Wenhui Bao* [Cultural Exchange News] (Hong Kong), Aug. 3, 1980, in FBIS: *People's Republic of China*, Aug. 5, 1980, p. U4.

31. See *Dagong Bao* [Impartial News] (Hong Kong), April 17, 1980, in *ibid.*, April 18, 1980, p. U1. According to this source, the Guangdong geological group number 705, working in the Nanling mountains, had made uranium discoveries almost every year since early 1964. See also Guangdong Radio, Dec. 26, 1979, in *ibid.*, Jan. 18, 1980, p. P2.

32. Zhao Fuxin and Zheng Shurong, p. 76. Picture 37, following Li Jue et al., p. 238, shows an external view of part of the Chenxian mine.

33. Wang Jian et al., p. 12.

34. For an authoritative discussion of shrinkage stoping, see R. Lewis and Clark, pp. 453-69.

35. *Ibid.*, p. 448.

36. *Ibid.*, p. 453; Wang Jian et al., p. 10.

37. R. Lewis and Clark, p. 453.

38. Wang Jian et al., p. 10; Wang Jingtang, p. 53.

39. Wang Jian et al., p. 8.

40. *Ibid.*, p. 12. For a technical discussion of Soviet leaching techniques that would have been available to the Chinese, see Galkin et al., pp. 39-43.

41. The discussion in this and the following paragraph is based principally on Wang Aimin et al., especially p. 81.

42. A general picture of how the military fared in these years is given in General Political Department, People's Liberation Army, *Gongzuo Tongxun* [Bulletin of Activities]. For a translation, see J. C. Cheng. A discussion of the relevant articles is found in J. Lewis, "China's Secret Papers," pp. 73-77.

43. By this time, the nuclear ministry had been renamed the Second Ministry of Machine Building, and Liu Jie had become the minister. On an inspection trip to Chenxian in 1963, Liu discovered that the drillers, contrary to ministry safety instructions, were working without access to water. The result was an increase in the density of dust in the tunnels far above the allowable 2 mg per cubic meter. Minister Liu personally ordered a Party cadre to fix the problem in three days. Wang Aimin et al., p. 84, gives the date for this event as March 1959; our sources indicate the date should be 1963.

44. The first national attention publicly given to protection against radon poisoning came in 1981. At the first national academic conference on radioactive environments in mines, participants exchanged information on occupational hazards in mines and proposed ways to reduce such harmful substances as radon. See China News Agency (Hong Kong), July 14, 1981, in *China Report: Science and Technology*, 136 (JPRS 79453; Nov. 16, 1981), p. 19. For a general idea of state-of-the-art mining safety, see Rock, pp. 111-21.

45. The term "ultra-Leftist" is Chinese. For a scholarly review of how Chinese evaluate ultra-Leftism during and after the Great Leap Forward, see Joseph, Chap. 4.

46. This and the next two paragraphs are based on Song Erlian; Wang Hanfu; Li Jue et al., pp. 30-31; and information from a Chinese specialist, 1986. The ministry also issued the slogan "Everyone should engage in the science of atomic energy."

47. Galkin et al., Chaps. 5-7, summarizes the state of Soviet knowledge (as of 1958) of the treatment of uranium at this stage in the process and lists many of the sources that would have been made available to the Chinese by Soviet advisers; see especially pp. 18, 34, 117-20, 140-41, and 154-55. Somewhat later, Galkin and others authored a textbook for the use of engineers specializing in the "technology of natural radioactive elements"; see Galkin and Sudarikov.

48. Wang Aimin et al., pp. 89-90.

49. Zhao Fuxin and Zheng Shurong, p. 76. Picture 25, opposite Li Jue et al., p. 15, shows one of the early uranium ore processing facilities like the one at Lianxian.

50. Xiang Jun, p. 120. The information on the ministry's collection system is based on interviews with Chinese specialists, 1986.

51. Galkin et al., Chap. 9, describes the textbook methods for producing uranium tetrafluoride.

52. Xiang Jun, pp. 118, 119. In accordance with their agreement of August 1956, the Soviets had designed the Sixth Institute. Li Jue et al., p. 26.

53. The remainder of this section is based principally on Xiang Jun, pp. 123-31; and Li Jue et al., pp. 44, 149. Lermontov is a small town near the city of Pyatigorsk in the Caucasus. The area of the Pyatigorsk Plateau, a section of the northern foothills of the Greater Caucasus, contains a number of mountain peaks of igneous origin. Some of these are exposed volcanic cores and are of a type in which uranium is sometimes found. Some sources have identified a uranium mine in the region; Xiang Jun appears to confirm that uranium oxide is produced at Lermontov.

54. Xiang Jun, p. 130.

55. *Ibid.*, p. 131.

56. This entire section is based on Jiang Ji, pp. 100-117; and Li Jue et al., pp. 26, 131, 148, 150, 155, 157, 163. The quote here is from Jiang Ji, p. 101. Hengyang was a city of about 240,000 in 1959. An important rail junction and market center, the city manufactures heavy equipment and machinery as well as chemicals and farm implements. It was about 70 miles by rail from the Chenxian Uranium Mine. The precise location of Plant 414 can be deduced from statements in Jiang Ji, pp. 116-17. Picture 41, following Li Jue et al., p. 238, shows the interior of Plant 414, described in this section.

57. Quoted in Jiang Ji, p. 102.

58. The story of Plant 202 is based on Li Jue et al., pp. 189-91. Pictures 46 and 48, following *ibid.*, p. 238, show the interior and exterior of Plant 202.

59. NRH, pp. 814, 816.

60. NRH, pp. 806, 814.

61. Niu Zhanhua, p. 133; Li Jue et al., pp. 43, 185.

62. See Guo Jian, pp. K7-K8; and Li Jue et al., p. 363. By the end of 1956, the institute had 584 staff members; 258 were scientific research personnel and 99 engineering technical personnel. According to Li Jinqi, the Institute of Atomic Energy had more than 1,300 engineers in 1986. He states that it "contributed to the successful trial explosion of China's first atom bomb and first hydrogen bomb" and conducts "basic research for national defense purposes." For a general description of the institute in 1979, see Bromley and Perrolle, pp. 87-100; and Bloembergen, pp. 92-93.

63. Guo Jian and Shuang Yin, pp. 1-3; Ren Gu, p. 1104. The first source gives an incorrect month for the beginning of the reactor's operation.

64. U.S. sources first identified Li Yi as a deputy director of the Institute of Atomic Energy in May 1965; U.S., CIA, *Directory* (1966), p. 454. By the 1980s, Li had moved to the new High Energy Physics Institute as deputy director. In October 1980, he became the first president of the China Particle Accelerator Society; *ibid.* (1981), pp. 21, 119.

65. Li Jue et al., p. 25; Niu Zhanhua, pp. 135, 148.

66. Niu Zhanhua, p. 142.

67. Information in this and the next paragraph is from *ibid.*, p. 134; and Li Jue et al., pp. 185, 420.

68. Monel is the trademark name for a nickel-copper alloy developed in 1905. Containing about 66% nickel and 31.5% copper (plus small amounts of other metals), Monel has strong corrosion-resistance properties.

69. Niu Zhanhua, p. 135.

70. This was the Zhangdian Medical Apparatus and Instruments Factory in Zibo, Shandong.

71. Li Jue et al., p. 185; Niu Zhanhua, p. 138. Huang Changqing's reactor was 2 m high and 0.5 m in diameter. Niu Zhanhua, p. 148. A possible model of the Soviet reaction chamber described by Niu Zhanhua is in Galkin et al., p. 151; however, the chamber pictured is designed to produce uranium tetrafluoride.

72. Li Jue et al., pp. 44, 185; Niu Zhanhua, p. 137. It should be noted that Li Jue et al. gives two different months for this test, October and December.

73. Niu Zhanhua, p. 139. "Xiao" means small and is a friendly title often used by a speaker addressing someone younger than he.

74. *Ibid.*, pp. 140-41.

75. *Ibid.*, pp. 142-43.

76. *Ibid.*, p. 146.

77. The information in this and the next paragraph is from *ibid.*, pp. 145, 147; and Li Jue et al., p. 185.

Chapter Five

1. U.S., Department of Energy, p. 1. For a discussion of the development of Soviet nuclear weapons, see Holloway, *Soviet Union*, pp. 20-27.

2. Liu Shuqing and Zhang Jifu, pp. 34-35; interviews with a Chinese specialist, 1986; Li Jue et al., p. 42.

3. "Sino-Soviet Joint Communique," p. 34. The total amount of credits was 520 million rubles. "Communique on Sino-Soviet Talks," p. 31.

4. "Statement of Soviet Government," p. 53.

5. "Conclusion of the Sino-Soviet Accord."

6. At the inaugural meeting of the four departments of the Chinese Academy of Sciences in June, the academy's president, Guo Moruo, and the physicist Wu Youxun stressed the importance of nuclear physics. Guo appointed key defense scientists to the Physics, Mathematics, and Chemistry Department. See *Xin Hua Yuebao*, Sept. 28, 1955, pp. 227, 242, 243, and 246.

7. NRH, p. 795.

8. The major survey of Soviet assistance to China is Filatov. On the establishment of the Dubna institute, see *Xin Hua Banyuekan*, April 21, 1956, pp. 62-63; and Li Jue et al., p. 519. According to Moscow Radio, Feb. 1, 1981, "For several years, the number of Chinese scholars and experts working at that institute totaled 140." In FBIS: *Soviet Union*, Feb. 4, 1981, p. B2. For a full text of the Dubna agreement, see *Xin Hua Banyuekan*, Sept. 6, 1956, pp. 89-91.

9. NRH, p. 794.

10. The information in this and the following three paragraphs is from Qian Sanqiang, "Cherish the Memory"; and Li Jue et al., pp. 17, 491-92, 494-95.

11. Li Jue et al., p. 51; interviews with Chinese specialists, 1986. For similar scalings of the program, see Liu Jie; and "Riddle of Research and Development," p. 6. In 1986, Gu Mainan and Gu Wenfu reported (p. 3) that 300,000 staff and workers served on the "nuclear industrial front." Of these

300,000, about 60,000-70,000 were engineering and technical personnel and over 1,000 were senior scientific research personnel. Xue Jianhua.

12. Li Jue et al., pp. 168, 204, 232.

13. Mao [9], p. 288.

14. This and the following three paragraphs are based on interviews with Chinese specialists, 1986; and Li Jue et al., pp. 28-29. See also Wang Xianjin et al.

15. Zhou Yongkang; Xu Zhucheng. The figure of 10.7 billion yuan in 1957 is based on International Monetary Fund and World Bank figures for changes in the retail price indexes. IMF, pp. 274-75; World Bank, p. 334. The dollar conversion is based on a rate of 2.617 yuan per U.S. dollar in 1957, given in "Conversion Rates," p. 29.

16. These figures are from State Statistical Bureau, pp. 23-24. For a compilation of state expenditures from 1950 through 1959, see N. R. Chen, pp. 446-47. Defense spending for 1958 totaled 5 billion yuan.

17. Dai Yaping, "Nuclear City," pp. K19-K20; Li Jue et al., p. 42.

18. For a general introduction to plutonium, see William Miner. Miner is one of the authors of a more technical discussion of plutonium in *The Rare Metals Handbook*, 2d ed. (New York, 1961), Chap. 18.

19. *Encyclopaedia Britannica: Macropaedia*, 13: 318.

20. On nuclear reactor designs, see Nero, especially Chap. 1.

21. Hewlett and Anderson, p. 182.

22. U.S., CIA, "Chinese Communist Atomic Energy," p. 2.

23. Dai Yaping, "Nuclear City," pp. K19-K20. We have placed the location of the reactor site (part of the Jiuquan Atomic Energy Complex) near the Subei Mongolian Autonomous County on the basis of our analysis of information provided in Dai Yaping, "Nuclear Plant." The nuclear ministry approved the Subei site on Jan. 30, 1958. Li Jue et al., p. 204.

24. Dai Yaping and Zhao Jin. The 1958 population figure of 50,000 is from Ullman, p. 35.

25. The discussion of the three plants devoted to the production of plutonium metal and plutonium weapons is in Li Jue et al., pp. 68-69, 209, 418, and 430; and Dai Yaping, "Nuclear Plant," p. K5. Technical aspects of the reactor are described in Li Jue et al., pp. 204-15. On the technical aspects of China's chemical separation program, see *ibid.*, pp. 216-39. Information on the Plutonium Processing Plant, built in 1963, is in *ibid.*, p. 458. Pictures 49-52 and 54, following *ibid.*, p. 238, show several of these plutonium facilities.

26. Interviews with Chinese specialists, 1986. These codes were (1) "Provisional Regulations Concerning Health and Protection in Radioactive Work"; (2) "Standards for Gauging the Maximum Allowable Amounts of Ionized Radiation"; and (3) "Detailed Regulations Concerning Health and Protection in Work with Radioactive Isotopes." These codes divided factories into three zones: (a) workshops and equipment (*shebei*) that used radioactive materials; (b) workshops and maintenance facilities that had no radioactive equipment but might be radioactive; and (c) clean spaces that were not radioactive. The codes strictly controlled zone (a). This dangerous (*zang*) zone was isolated from the clean zone, and people had to pass through a safety zone to enter it. Monitors tested the radioactive film badges on the workers' belts and

the dosimeters in their upper left pockets when they left the zone. Everyone entering a dangerous zone had to be registered, to wear a mask, special gloves, and special rubber boots in the radioactive areas, and to shower and change clothes afterward.

27. Dai Yaping, "Nuclear Plant," p. K5; Li Jue et al., p. 233.

28. Most of the information in this and the next paragraph is from an interview with a Chinese specialist, 1985; and Li Jue et al., pp. 205, 421.

29. "Our Country."

30. Michael Minor, p. 576; Di Zongheng, 2: 22. Both of these articles assume the main reactor is in Baotou. According to Chinese interviewed, this is incorrect. The article cited in the preceding note indicates that the Chinese had 10 reactors in 1985. The following year, Jiang Shengjie (former chief engineer as well as first deputy director of the Jiuquan Atomic Energy Complex) claimed that there were more than that. Jiang, "Over Ten Reactors."

31. Whitaker and Shinn, p. 490.

32. *Ibid.*, p. 491; Liu Qijun, p. 40 (for the 1960s reports); Di Zongheng, 2: 22 (for the 1970s).

33. Michael Minor, p. 572; Liu Qijun, p. 41.

34. The information in this paragraph and the one that follows is based on Li Jue et al., pp. 565-66.

35. China's first thermonuclear explosion, on June 17, 1967, used a uranium trigger. See Appendix B for a list of China's nuclear tests through 1978.

36. Krass et al., p. 5.

37. See *ibid.*, Chaps. 1 and 2. For a more technical study, see Patton et al., especially Chap. 6: "Plant Design." The United States used electromagnetic separators for the enrichment of fuel for its first uranium bombs but abandoned this method in favor of gaseous diffusion. The Chinese Institute of Atomic Energy purchased two of these separators from the Soviet Union and experimented with this type of separation process for uranium. These separators now produce isotopes of elements other than uranium.

38. Arkin and Fieldhouse, p. 262. The United States also built gaseous diffusion plants in Paducah, Ky., and Piketon, Ohio.

39. See Hewlett and Anderson, pp. 136-41.

40. Although the case is not truly comparable, during the period 1943-60, the United States spent over two billion dollars on the construction and improvement of its gaseous diffusion plants. Patton et al., p. 1.

41. Hewlett and Anderson, p. 140.   42. Ullman, p. 35.

43. Xu Honglie.   44. Dai Yaping, "First Visit," p. K24.

45. The information in this paragraph is based on Dai and Zhao; and Chen Honggeng, p. 151. We have noted this decision of mid-October 1957 in Chap. 3. On Oct. 15 and Oct. 17, 1957, the Third Ministry approved the preliminary specifications for designing the Lanzhou plant and the Nuclear Fuel Component Plant in Baotou as the first ones in the series of mass-production facilities. Li Jue et al., pp. 168, 189.

46. Chen Honggeng, p. 151. Most of the information on the site selection is from Li Jue et al., p. 168.

47. For a general consideration of political networks in China, see J. Lewis, *Political Networks*.

48. For a thorough discussion of relationships within this army, see Whitson, Chap. 3.

49. *Ibid.*, p. 164.

50. One of these leaders, Peng Dehuai, writes (pp. 444-45): "Slanderers of the Hundred Regiments Campaign: You have gone over to the side of Japanese imperialism and the Chiang Kai-shek clique. Please read the [congratulatory] telegram Chairman Mao sent me. Why is your view so different from Chairman Mao's?"

51. See Whitson, pp. 177-86; and Clubb.

52. The information on Song Renqiong and Wang Jiefu is based on Chen Honggeng, p. 151; and Klein and Clark, 2: 787-90.

53. The biographical information on Zhang Pixu, Liu Zhe, and Wang Zhongfan is from Chen Honggeng, pp. 151-53.

54. *Ibid.*, pp. 153-54; on Deng Xiaoping's decision, see Li Jue et al., p. 27.

55. The information in this paragraph and the next is based principally on Liu Xiyao; Liu Shuqing and Zhang Jifu, pp. 34-35; and Li Jue et al., pp. 30, 168, 171-72.

56. We adopt Liu Xiyao's version of the "first write" slogan, using *xie*, instead of Liu and Zhang's *xue* (study).

57. For a discussion of the engineering and other technical aspects of building the plant, see Li Jue et al., pp. 419, 428-29. For additional information and a discussion of the security problem, see Chen Honggeng, pp. 156, 159, 160.

58. Chen Honggeng, pp. 156, 157. In a report of Jan. 6, 1961, on the "Provisional Regulations of the Chinese People's Liberation Army for the Safe-Keeping of State Military Secrets (Draft)," the General Political Department seems to have moved back to the earlier preoccupation with secrecy, especially as it related to the "super [scientific] departments" concerned with strategic weapons. See J. C. Cheng, pp. 232-38 *passim*.

59. Chen Honggeng, p. 161.

60. According to *ibid.*, the issue of cleanliness kept recurring in this area of violent winds and perpetual dust. The advisers often returned to the problem and "threatened to suspend the installation if the Chinese failed to maintain acceptable standards." The directorate's solution was to "command workers to level the land around the plant with bulldozers. . . . All electrical cables were put in trenches." The plant sent cadres south to Qinghai to find sod, which they shipped back to cover a vast area around the plant. Wang Jiefu won again, but the Soviets charged him with treating his work force brutally, a charge that the higher levels dismissed. The date of the installation is from Li Jue et al., p. 171. Picture 47, following *ibid.*, p. 238, shows the interior of the main workshop.

61. Chen Honggeng, p. 162; Li Jue et al., pp. 28, 565-66.

62. Chen Honggeng, p. 155. The information in this paragraph on the ministry's directive is from Li Jue et al., p. 28.

63. Dai Yaping, "Nuclear City," p. K19. This Politburo meeting was discussed in Chap. 3.

64. "Riddle of Research and Development," p. 5; Li Jue et al., p. 36. This decision, "The Eight-Year Program for the Atomic Energy Cause, 1960-1967," was issued on Dec. 23, 1959.

65. Su Fangxue, p. 5.
66. For additional information on the transfer of these specialists, see Dai and Zhao; Li Yingxiang et al.; Ding Houben, p. 16; and "Riddle of Research and Development," p. 5.
67. Chen Honggeng, p. 162; Li Jue et al., p. 33. Chen puts the date of departure on August 3, but Li states it was on July 8.
68. Chen Honggeng, pp. 162-63.
69. Dai Yaping, "Nuclear City," p. K19. In 1960, the state made an emergency allocation of "millions of catties" of soybeans to the nuclear plants in Lanzhou and Jiuquan Prefecture. Li Jue et al., p. 37.
70. Chen Honggeng, p. 163. Liu Jie formally replaced Song Renqiong as minister in September 1960. Liu, who had been a provincial and regional Party official in the early 1950s, served as vice-minister of geology and then vice-minister of the Second Ministry. "Appointments of Ministers."
71. Chen Honggeng, pp. 159, 164.
72. Quoted in Chen Zujia; "Riddle of Research and Development," p. 5; and Li Jue et al., p. 36.
73. Chen Zujia. Zhou Enlai first issued this oft-repeated instruction in 1959. Li Jue et al., pp. 36, 565.
74. Chen Honggeng, p. 163; Li Jinting et al., p. 18.
75. Dai and Zhao. Zhou Enlai's instruction echoed the words of a Politburo decision in June 1959, following receipt of Moscow's June 20 letter. Li Jue et al., pp. 36, 565.
76. Gu Mainan, "One Out of a Hundred Thousand."
77. Chen Honggeng, p. 163.
78. *Ibid.*, pp. 163, 164. Yuan Chenglong's background remains something of a mystery. This is one of the few mentions of his name in reports on the nuclear program, even though he served as vice-minister under both Song Renqiong and Liu Jie. The atomic bomb development program came under his direct authority until he was transferred from the ministry to become director of the Political Department of the National Defense Industry Office just before the Cultural Revolution.
79. Chen Honggeng, pp. 164, 165. This was a time for slogans and directives with numbers. See J. Lewis, *Leadership*, pp. 155-56.
80. Chen Honggeng, p. 164. A detailed discussion of the technical problems encountered in completing the Lanzhou plant is given in Li Jue et al., p. 174. The Chinese did not successfully test the Soviet-supplied barrier components with uranium hexafluoride until late 1962. Li Jue et al., pp. 175-76.
81. Chen Honggeng, p. 164.
82. *Ibid.*, p. 165.
83. This paragraph is based on Huang Fengchu and Zhu Youdi, p. 19; and Li Jue et al., p. 430.
84. Most of the information in this section is based on interviews with Chinese specialists from 1981 to 1983; and Li Jue et al., pp. 47-48, 567-68.
85. Personal communication from Nie Rongzhen, Feb. 15, 1986; Li Jue et al., p. 567.
86. Translations of 29 issues of the *Work Bulletin* are in J. C. Cheng; the quote is on p. 66.
87. *Ibid.*, pp. 66-67.

88. *Ibid.*, p. 100.

89. *Ibid.*, p. 250.

90. *Ibid.*, p. 253. For a discussion of the Chinese statements in the *Work Bulletin*, see Hsieh, "China's Secret."

91. NRH, p. 810. On Liu Shaoqi's alleged role at the Beidaihe meeting, see "Strategic Guiding Principles." Note that this article criticizing Liu predated his rehabilitation in 1980. The dismal fate of China's military aircraft industry mirrored the fate of the Soviet aircraft industry in Khrushchev's era. Since the decline of this industry in China has an important bearing on the missile program, we will deal with the industry's fate in our later study of that program.

92. NRH, p. 811.

93. NRH, pp. 811-12. The July "Resolution on Certain Questions in the Construction of the Nuclear Industry" called for strengthening all aspects of the program and tightening the transportation security system. Li Jue et al., p. 41.

94. NRH, p. 812. Chen repeated his remark on pawning his trousers to a group of Japanese newspapermen two years later. NYT, Oct. 29, 1963.

95. Quoted in Zhang Jun, p. 16.

96. NRH, pp. 812-13.

97. The report on a ground force training conference in 1961 called on PLA units at and above the regimental level to "study the principles of the use of atomic and chemical weapons and also [to] find out how to use our atomic and chemical weapons for sudden attacks." On July 13, the Central Military Commission ordered the relevant units to carry out the report's recommendations. See J. C. Cheng, pp. 684, 687.

98. *Ibid.*, pp. 732, 734.

99. Li Jue et al., p. 47; *Monan*, pp. 111-12; Liu Shuqing and Zhang Jifu, p. 25.

100. NRH, p. 819.

101. Chen Honggeng, p. 165; Li Jue et al., p. 47. The first source gives November 7 as the date of the Politburo decision.

102. The information on the commission's leadership is based on Gu Yu; and "Riddle of Research and Development," p. 6. According to Gu, both Nie Rongzhen and Luo Ruiqing were in charge of the commission's routine duties.

103. Various sources have previously identified some members of the commission. Marshal Nie Rongzhen provided the full list in a personal communication, Feb. 15, 1986. The ordering of the list is Nie's.

104. NRH, p. 819. Gu Yu adds that the commission "determined the key projects, the allocation of funds, and the coordination among various departments."

105. Li Jue et al., p. 48; "Riddle of Research and Development," p. 6. "Riddle" says that 9 meetings of the commission were held in this period, but Li Jue et al., pp. 48, 209, gives both 9 and 13 as the number.

106. NRH, pp. 785, 797, 819.

107. Wang Yougong et al.

108. Chen Honggeng, p. 166; Li Jue et al., p. 176. By this point in 1962, the construction and installation of the entire production line relevant to uranium processing and enrichment was 80% completed. "Riddle of Research and Development," p. 6.

109. This section is based on Chen Honggeng, pp. 166-71; and Li Jue et al., pp. 176-77.

Chapter Six

1. Niels Bohr and John A. Wheeler collaborated on the article "The Mechanism of Nuclear Fission" (1939). For the reflections of some of the major participants in the revolution in nuclear physics in the 1930s, see Stuewer.

2. In general, the statement concerning publications is based on Glasstone and Dolan, *Effects of Nuclear Weapons*, Chap. 1. This classic volume was first published in 1950 as *The Effects of Atomic Weapons*. The first major unclassified volume to be published on nuclear weapons is Smyth, *Atomic Energy for Military Purposes* (1945). Specialists, including those in China, usually refer to this volume as the Smyth Report.

3. Glasstone and Dolan, p. 13.   4. *Ibid.*, pp. 13-14.

5. *Ibid.*, p. 15; Smyth, p. 210.   6. Glasstone and Dolan, p. 15.

7. *Ibid.*, pp. 15-16.

8. The Information Bureau of the Second Ministry, which was discussed in Chap. 3, scoured the literature on the explosive mechanisms. Chinese interviewed in 1986 state that general writings, such as the Smyth Report, provided little help, but that they got some valuable hints from Jungk, *Brighter Than a Thousand Suns*, and Groves, *Now It Can Be Told*. All such books and reports on the Manhattan Project were translated into Chinese and carefully annotated for use by Chinese scientists.

9. The definitive treatment of the British weapons effort is Gowing, *Independence and Deterrence*, Vol. 2: *Policy Execution*. The head of the Manhattan Project, Leslie R. Groves, writes (p. 147) that a national magazine published "an article hinting at the theory of implosion." Chinese sources indicate that this statement, for example, provided clues to where information on implosion might be found.

10. Gowing, *Independence and Deterrence*, 2: 457. An excellent description of the original American implosion devices is in O'Keefe, Chap. 3.

11. Holloway, "Research Note." In his analysis of the early Soviet nuclear weapons program, Holloway has reached two conclusions (p. 196): "First, although there are clear elements of reciprocal influence in the Soviet and American nuclear weapons decisions of 1949-52, the actions that are salient on one side are not necessarily so on the other. . . . Second, Soviet decision making shows elements both of reaction to American actions and of an internal dynamic." Holloway notes how one program stimulated aspects of the other.

12. Li Jue et al., pp. 257-58; "Riddle of Research and Development," p. 5; Liu Jingzhi and Li Peicai; Liu Shuqing and Zhang Jifu, p. 4.

13. Hewlett and Anderson, Chap. 7; the quotations are from pp. 228-29. Picture 63, following Li Jue et al., p. 238, shows buildings at the Ninth Academy and mountains in the distance.

14. Liu Shuqing and Zhang Jifu, pp. 2-4.

15. *Ibid.*, pp. 3-4; Li Yingxiang et al.

16. Li Jue et al., p. 261.

17. Liu Shuqing and Zhang Jifu, p. 4.

18. For a discussion of the construction of the railroad into the Haiyan area near the Ninth Academy between 1958 and 1961, see Bai Yunshan, pp. 30-31.

19. The information in the rest of this section is principally from Liu Shuqing and Zhang Jifu, pp. 29-30; and Li Jue et al., pp. 261-62.

20. The discussion of Wu Jilin's personal history is based on Liu Shuqing and Zhang Jifu, pp. 9-10; and Li Jue et al., p. 263.

21. On the timing of Wu's moves, see Li Yingxiang et al.; and Li Jue et al., p. 270.

22. The information on Zhu Guangya is based principally on Liu Shuqing and Zhang Jifu, p. 27; and Su Fangxue, p. 5. In 1951, Zhu had written a popular book on nuclear weapons entitled *Yuanzineng he Yuanzi Wuqi* [Atomic Energy and Atomic Weapons].

23. This information on Guo Yinghui is from Li Yingxiang et al.

24. This account of Wang Ganchang's background is based on "Ten Prominent Scientists," p. 27; and on information from scientists who know him.

25. Such stories were circulating as late as 1976. See Di Zongheng, Part 2, p. 19. Wang left Germany well before the German nuclear program began in the fall and winter of 1939-40. See Irving, Chap. 2. It is true that Professor Werner Heisenberg, a major physicist in Berlin, became involved in defense-related nuclear issues while Wang was still in Germany, but Wang did not work with him or indeed even know him.

26. Liu Shuqing and Zhang Jifu, p. 6.

27. For this and other details on Wang Ganchang's career, see Ding Houben, p. 16; "Riddle of Research and Development," p. 5; and Li Jue et al., p. 565.

28. Li Jue et al., p. 270; Liu Shuqing and Zhang Jifu, p. 11.

29. Liu Shuqing and Zhang Jifu, p. 7.

30. *Ibid.*, pp. 7, 11; Feng Yuan and Chen Dong.

31. The details on Guo Yonghuai are from Liu Shuqing and Zhang Jifu, pp. 7-8, 11; and Li Jue et al., p. 270.

32. Tien, "Engineering," p. 388. In the Wade-Giles orthography, Guo's surname is Kuo.

33. After a series of ambassadorial talks in Geneva, the United States and China issued parallel unilateral statements on Sept. 10, 1955, expressing their intent "to resolve the problem of the repatriation of remaining civilians" detained by the two countries. The United States thereupon began allowing some of the 129 Chinese students and scientists it had detained to return home. Guo was among those granted a visa.

34. Interview with a Chinese specialist, 1984. On Guo's death, see Liu Shixiang, p. 20.

35. Liu Shuqing and Zhang Jifu, p. 40.

36. Liu Shixiang, p. 24; Li Jue et al., pp. 268-69.

37. Liu Shuqing and Zhang Jifu, pp. 13, 17. *Mimi Licheng* refers only to a Dr. Chen working in this area. We assume this is Chen Nengkuan because Nie Rongzhen places him, along with Qian Sanqiang, Wang Ganchang, Zhu Guangya, and Peng Huanwu, in a group of major scientists in the atomic

program; NRH, p. 788. *Renmin Ribao*, Jan. 9, 1956, identifies Chen as a returned student from the United States.

38. Liu Shuqing and Zhang Jifu, pp. 30-33.

39. This information on Deng Jiaxian is based on *ibid.*, p. 19; Gu Mainan, "Deng Jiaxian, Veteran Scientist"; Gu Mainan, "Deng Jiaxian, Man of Merit," pp. 4-8; Liu Jingzhi and Li Peicai; Su Fangxue, pp. 2-9; Zhang Aiping, "Deng Jiaxian's Illustrious Name"; and Zhang Aiping, "Memorial Speech." After Deng's death in July 1986, Chinese articles referred to him as the "father of China's atomic bomb" and the "Oppenheimer of China"; we believe these sobriquets to be exaggerations. See, for example, Gu Mainan, "Deng Jiaxian: China's Father of A-Bomb," pp. 20-22.

40. Su Fangxue, p. 4.

41. Liu Shuqing and Zhang Jifu, p. 19; Gu Mainan, "Deng Jiaxian, Veteran Scientist."

42. Liu Shuqing and Zhang Jifu, pp. 19-20; Feng Yuan and Chen Dong; Liu Jingzhi and Li Peicai.

43. Liu Shuqing and Zhang Jifu, p. 21.

44. *Ibid.*, p. 56.

45. *Ibid.*, pp. 4-5.

46. *Ibid.*, pp. 5-7.

47. *Ibid.*, p. 8. There is an oddity in this account, which has Peng Dehuai, who was dismissed from his post as defense minister in September 1959, among these senior leaders. His inclusion in the group appears to be a mistake. "Khrushchev's perfidious actions" refers to the Soviet letter of June 20, 1959, which postponed delivery of a prototype weapon. This letter is discussed in Chap. 3. The Chinese named the bomb "596" in August 1963. Li Jue et al., p. 53.

48. Except as noted, this section is based principally on Liu Shuqing and Zhang Jifu, pp. 10-18; and Li Jue et al., pp. 259-60, 264-68, 270.

49. See, for example, O'Keefe, pp. 65-67.

50. Data on the location is from "Monument."

51. The "over 30" figure comes from Ding Houben, p. 16.

52. Liu Shuqing and Zhang Jifu, p. 40. The Xi'an-based Third Institute and Plant 804, both under the ministry in charge of conventional weapons, assisted Qian Jin. Li Jue et al., p. 267.

53. Engineers at the Lanzhou Chemical Physics Institute helped develop new explosives and casting techniques. Li Jue et al., p. 268. For data on mixtures used in the U.S. nuclear program and relevant performance data, see Mader et al. Most of the available explosives were developed before or during the Second World War. Information on these would have been easily available to Chinese ordnance specialists, many of whom had worked with U.S. army specialists in China. For a description of TNT, PETN, and other military explosives, see Meyer; and Fordham, Chap. 3.

54. Scientists at the Computing Technology Institute in Beijing conducted these calculations on a Model 104 computer, then China's most advanced but primitive by Western standards. Li Jue et al., p. 267. The physicist Richard Feynman notes (p. 108) that at wartime Los Alamos the "big problem . . . was to figure out exactly what happened during the bomb's implosion, so you can figure out exactly how much energy was released and so on. [This] required

much more calculating than we were capable of. A clever fellow by the name of Stanley Frankel realized that it could possibly be done on IBM machines [used] for business purposes."

55. The date is from "Monument."

56. The number 2,000 is based on the comparable U.S. implosion device dropped on Nagasaki, which had 32 detonators fired with the simultaneous discharge of electrical capacitors. For details, see O'Keefe, pp. 77-79, 99. The detonation experiments continued until early 1964. "Monument."

57. For the English translation of a Soviet text on beryllium published in 1956, see Beus. On p. 44, the author describes the beryllium deposits found in China.

58. In the British nuclear weapons project, polonium was artificially produced at the Atomic Energy Research Establishment at Harwell. Gowing, *Independence and Deterrence*, 2: 445-46.

59. Bagnall, pp. 935, 944, 946. Bagnall writes (p. 946): "Polonium [compared to selenium and tellurium] is the most dangerous of the three elements because of its intense radioactivity; all three elements appear to be taken up by the kidneys, spleen and liver, the tissues of which undergo irreparable radiation damage in the case of polonium because of the complete absorption of the α-particle energy." The most important Western technical study of polonium, to which the Chinese would have had access, is Moyer et al., published in July 1956.

60. Most of the information in this paragraph is based on Glasstone and Dolan, pp. 16-17.

61. According to *Encyclopaedia Britannica: Macropaedia*, 12: 1072, neutrons are classified by their energy and wavelength. "It is common to describe neutrons with energies in the range from 0 to 1000 eV as slow, 1 to 500 keV as intermediate, 0.5 to 10 MeV as fast, and greater than 10 MeV as very fast."

62. Glasstone and Dolan, pp. 16-18. On p. 17, the authors note that for a complete fission of 0.1 kilotons, it "would take approximately 51 generations to produce the necessary number of neutrons" in a chain reaction initiated by only one neutron.

63. *Ibid.*, p. 17.

64. Liu Shuqing and Zhang Jifu, pp. 22-23.

65. The rest of this section is based primarily on *ibid.*, pp. 23-33; "Riddle of Research and Development," p. 6; and Li Jue et al., pp. 265, 268, 376, 569. The information on polonium is from Moyer et al., pp. 2-6.

66. This and the final three paragraphs in this section include information from Li Jue et al., pp. 53-54, 262-63, in addition to the sources given in note 65.

67. Liu Jingzhi and Li Peicai; Gu Mainan, "Deng Jiaxian, Veteran Scientist."

68. The information in this paragraph and the following one is from Su Fangxue, pp. 4-5; and interviews with a Chinese specialist, 1986.

69. Liu Shuqing and Zhang Jifu, pp. 19-20.

70. *Ibid.*, pp. 19-21; Liu Jingzhi and Li Peicai; "Riddle of Research and Development," pp. 5-6; Li Jue et al., p. 276. The next two paragraphs are based on these sources.

71. This is based on Li Jue et al., pp. 412, 418, 478.

72. See Chap. 5, note 23.

73. Liu Shuqing and Zhang Jifu, pp. 37-38.

74. Gobi is a Mongolian word meaning "a place difficult for bushes and trees to grow well." In most accounts of the nuclear weapons program, it is used generically to include not only the desert formally called Gobi in Nei Mongol, but also the deserts or wilderness in Gansu, Qinghai, and Xinjiang. This is a proper though somewhat uncommon usage. See *Ci Hai*, p. 1349.

75. Dai Yaping, "Nuclear Plant," p. K5.

76. Chen Zujia; Peng Ruoqian, pp. 172, 174, 175.

77. Peng Ruoqian, pp. 175-76.

78. Liu Shuqing and Zhang Jifu, p. 33; Peng Ruoqian, p. 176.

79. Liu Shuqing and Zhang Jifu, p. 33.

80. *Ibid.*, pp. 33-34.

81. Peng Ruoqian, pp. 183-84.

82. Liu Shuqing and Zhang Jifu, p. 34.

83. *Ibid.*, pp. 25-26; Li Jue et al., pp. 46-47. The ministry's report in September led to a series of decisions, and there is some confusion concerning which of these should be called the Two-Year Plan. See Li Jue et al., p. 568.

84. Liu Shuqing and Zhang Jifu, p. 34.

85. *Ibid.*

86. *Ibid.*, p. 35.

87. Peng Ruoqian, pp. 185-87; Liu Shuqing and Zhang Jifu, p. 37. By adopting Zhu's plan, the project reportedly saved the national treasury 21 million yuan (or about U.S.$10 million in 1963 dollars).

88. Peng Ruoqian, pp. 179-80.

89. *Ibid.*, pp. 178-79.

90. *Ibid.*, p. 180.

91. Liu Shuqing and Zhang Jifu, pp. 35-37.

92. *Ibid.*, p. 38; Peng Ruoqian, p. 185.

93. The information on Yuan Gongfu's feat is from Liu Shuqing and Zhang Jifu, pp. 38-40; Peng Ruoqian, p. 187; and "Riddle of Research and Development," p. 6. The last source states that the machining was completed in April, not on May 1.

94. Except as noted, the rest of the section is based on Liu Shuqing and Zhang Jifu, pp. 41-44. We do not know Engineer Li's full name. On the criticality experiment, see Li Jue et al., pp. 272-73.

95. Li Yingxiang et al. Liu Xiyao, who was present at the assembly stage, writes that the first bomb had been produced "by July 1964," but this may not include assembly.

96. Liu Xiyao reveals that at this same time Zhou issued a directive to "pay close attention to the combination of the atomic bomb and guided missiles."

Chapter Seven

1. In 1980, for example, archeologists from the Xinjiang Institute of Archeology completed a survey of the Loulan site, which had earlier been explored by Sven Hedin. *Xinhua*, Jan. 19 and July 1, 1980; Hedin, *Wandering Lake*. Perhaps the best-known explorer of the Lop desert was Sir Aurel Stein; see his "Third Journey"; and "Explorations."

2. The total land area of Xinjiang is 1.6 million sq. km, of which over 100,000 sq. km is devoted to the nuclear weapons test base.

3. Fa-hsien, p. 17; Huili, pp. 36-37.

4. Marsden, pp. 68, 78.

5. Schomberg, p. 318.

6. In 1915, Stein discovered that "brackish water could be reached [in the Kuruktag] by digging shallow wells in some hollows." *On Ancient Central-Asian Tracks*, p. 132.

7. Schomberg, p. 318.

8. Stein, "Explorations," pp. 18, 24.

9. Stein, "Third Journey," 2: 205.

10. Pevtsov, Chap. 7; in Chap. 8, Pevtsov provides a good description of Korla and Yanqi (Karashar), which we will summarize in our later volume on the missile program.

11. *Ibid.*, p. 312.

12. The main fishing in the region is at Bosten Hu (Bagrax Hu). For a description of this large freshwater lake east of Yanqi, see Murzayez, pp. 75-83.

13. Pevtsov, pp. 313, 328-30. In the mid-1950s, the Uygurs accounted for almost 75% of the total Xinjiang population of about 5.0 million. By the 1970s, their representation had dropped to about 55%, in a population that had grown to 8.5 million, reflecting a large influx of Han settlers from provinces to the east. For a discussion of this general phenomenon of population redistribution, see Orleans, Chaps. 4, 5.

14. Pevtsov, pp. 322, 331. There were a great many Chinese soldiers elsewhere in the province at this time as a holdover from the great Muslim rebellions and anti-foreign battles two decades before. See Wen-Djang Chu.

15. Stein, "Explorations," p. 12. For pictures of these topographical features, see *ibid.*, pp. 10, 25.

16. Hulsewe, p. 89.

17. For a discussion of this campaign and its aftermath, see Whitson, pp. 113-22.

18. This paragraph is based on Whiting and Sheng, pp. 115-18; and Wang Zhen. For a review of the complex history surrounding the Communist takeover of Xinjiang, see Forbes, Chap. 7.

19. Fu Biduo and Tian Jijin.

20. According to McMillen, p. 66, the PC Corps membership had "risen to an estimated 500,000 to 600,000" in 1966. By 1985, its ranks had swelled to some 1.0 million (or 2.25 million counting family members). See Lan Xueyi; and "PC Corps' Contributions."

21. McMillen, pp. 56-67; Whitson, pp. 114-15.

22. See Nailene Chou.

23. This section is based on Zhang Zhishan; and Su Kuoshan.

24. Su Kuoshan. October 16 is the anniversary of both the establishment of the base and the explosion of the first bomb. See *Xinhua*, Oct. 5, 1984, in FBIS: *China*, Oct. 9, 1984, pp. K13-K14.

25. Guo Diancheng and Xu Zhimin, "Real People's Heroes," p. 18.

26. *Xinhua*, Oct. 16, 1984, in FBIS: *China*, Oct. 17, 1984, p. K1; interview with a Chinese specialist, 1986. On the First Atomic Bomb Test Commission

and the First Atomic Bomb Test On-Site Headquarters, see "Celebration of the Twentieth Anniversary."

27. This paragraph is based on Guo Diancheng and Xu Zhimin, "Thunder Roars," p. 17.

28. This paragraph is based on Guo Diancheng and Xu Zhimin, "Real People's Heroes," p. 18.

29. Interviews with former PC Corps members, August 1985.

30. There was little difference between the soldier-prisoners and those confined to the famous Communist Youth League Farm in the Taklimakan Desert south of Korla. This reform-through-labor camp came under the jurisdiction of the Talimu Administrative Office and was one of the PC Corps units. Most of those confined to the farm came from China's eastern provinces.

31. These were distinguished from *zhibian liandui* (support-the-frontier companies), which were made up of volunteer youth from China's eastern metropolitan areas. The *zhibian liandui* were viewed by the local authorities as a reserve security force to be mobilized in emergencies.

32. The information in this paragraph and the next one is based principally on Guo Diancheng and Xu Zhimin, "Thunder Roars," pp. 17-18; and interviews with former PC Corps members, 1985.

33. Liu Nanchang and Liu Cheng.

34. Guo Diancheng and Xu Zhimin, "Real People's Heroes," p. 19. On the date of the Research Institute's establishment, see Nie Rongzhen, "Letter," p. 18. We believe the name of this institute is the Northwest Nuclear Technology Institute, which in 1967 was led by Cheng Kaijia. Li Jue et al., p. 291.

35. Suo Guoxin, pp. 16-19.

36. See Guo Diancheng and Xu Zhimin, "Never Forget," pp. 28-29.

37. The Chinese press frequently cites the building of the meteorological station in the high mountains of Xinjiang as a miracle of the test program. See, for example, *Xinhua*, Oct. 5, 1984, in FBIS: *China*, Oct. 9, 1984, p. K14.

38. Guo Diancheng and Xu Zhimin, "Real People's Heroes," p. 19.

39. Except as noted, the information on the delivery of the weapon is based on Liu Shuqing and Zhang Jifu, pp. 44-48.

40. The date is from "Riddle of Research and Development," p. 6.

41. On April 11, 1964, the Fifteen-Member Special Commission decided to detonate the first bomb on a tower. Li Jue et al., p. 54. The rest of this section is based principally on Liu Shuqing and Zhang Jifu, pp. 49-56; "Riddle of Research and Development," p. 6; and Li Jue et al., pp. 54-55, 274-75.

42. "Unexpected Discovery."

43. This paragraph is based on Ding Houben, p. 16, as well as Liu Shuqing and Zhang Jifu, p. 49.

44. Liu Shuqing and Zhang Jifu, pp. 51-52; "Riddle of Research and Development," p. 6; Li Jue et al., pp. 55, 275. Throughout this section, where the sources differ, we have chosen to follow "Riddle"; and Li Jue et al. For example, Liu and Zhang state that the command to insert the uranium component and initiator was given on the sixteenth, and that the command to hoist the bomb up the tower came at 6:00 that morning.

45. Interviews with Chinese specialists, 1986.

46. Li Yingxiang et al.; Guo Diancheng and Xu Zhimin, "Never Forget," pp. 28-29.

47. Xie Linhe, pp. 17-18. The fallout reached Beijing at 10:00 A.M. on October 17.

48. In interviews conducted in August 1985 with former servicemen stationed in Xinjiang, we were told that Beijing did not give the problem "serious attention" until the late 1970s. A substantial amount of data has been released in China concerning atmospheric tests conducted from 1976 to 1979. See, for example, Radiochemistry Research Laboratory, pp. 173-79; and Liang Yusheng et al., pp. 11-16. During the atmospheric test series, probably in the 1960s, Premier Zhou Enlai ordered the development of underground testing speeded up. Thereafter, the Second Ministry created what came to be named the Research Unit on Underground Nuclear Testing Phenomena; Liu Zhaolin, p. 14. In October 1967, a high-level meeting worked out plans for underground testing, but these were interrupted by the Cultural Revolution. The first underground explosion came on Sept. 22, 1969, and the Chinese basically solved the technical problems associated with underground measurement in their tests on Oct. 27, 1975, and Oct. 17, 1976. Li Jue et al., p. 291. To date, all tests subsequent to the atmospheric test on Oct. 16, 1980, have been conducted underground.

49. Ling Xiang, p. 63. This article reveals that members of expeditions into the desert, though dozens of miles from the restricted zone, have suffered from severe radiation exposure.

50. Suo Guoxin, p. 18. Suo incorrectly states that "virtually all" the scientists and technicians suffered from serious fallout radiation of "several hundred roentgens"; interviews with Chinese specialists, 1987, indicate that Suo was exaggerating.

51. Liu Shuqing and Zhang Jifu, p. 52. Picture 60, following Li Jue et al., p. 238, shows the bomb on its push cart. No. 61 shows the tower, and No. 59 an observation post and the ground zero terrain.

52. Liu Shuqing and Zhang Jifu, p. 52. August 1 is the official anniversary of the founding of the People's Liberation Army.

53. Li Yingxiang et al. See also Guo Diancheng and Xu Zhimin, "Never Forget," pp. 28-29; Su Kuoshan; and "Riddle of Research and Development," p. 6.

54. Liu Shuqing and Zhang Jifu, p. 54; "Riddle of Research and Development," p. 6.

55. Except as otherwise cited, the information in this paragraph is from Liu Shuqing and Zhang Jifu, pp. 54-55.

56. See Zheng Jixu et al.; Cai Jianwen. China did not put an advanced air-sampling system into operation until Oct. 17, 1976; Ding Houben, p. 17.

57. "Major Development."

58. This paragraph and the next one are based on Ding Houben, p. 16; and Liu Shuqing and Zhang Jifu, p. 55. The quotation is from Liu and Zhang.

59. Liu Shuqing and Zhang Jifu, pp. 1-2.

60. For the text of the official statement, see Appendix A.

Chapter Eight

1. Einstein's precise words, as quoted in Lapp, were: "The unleashed power of the atom has changed everything save our modes of thinking, and we thus drift toward unparalleled catastrophe."

2. Dinerstein, p. 7.

3. For a general discussion of this year, see Hsieh, "Sino-Soviet Nuclear Dialogue," Chap. 8.

4. "Leninism and Modern Revisionism."

5. We have used the translation of this *Hongqi* article (Nos. 3-4, March 4, 1963) in *More on the Differences Between Comrade Togliatti and Us.*

6. *Ibid.*, p. 158.

7. *Ibid.*, pp. 69-78.

8. "A Proposal Concerning the General Line of the International Communist Movement" (June 14, 1963), in *Polemic*, pp. 28-29.

9. The text of the Treaty Banning Nuclear Weapon Tests in the Atmosphere, in Outer Space and Under Water is in Blacker and Duffy, pp. 366-68.

10. "Statement of the Chinese Government Advocating the Complete, Thorough, Total and Resolute Prohibition and Destruction of Nuclear Weapons and Proposing a Conference of the Government Heads of All Countries of the World" (July 31, 1963), in *People of the World*, p. 1.

11. *Ibid.*, p. 2. For a general discussion of China's attitude toward the treaty, see Halperin, pp. 62-70.

12. "Statement by the Spokesman of the Chinese Government" (Aug. 15, 1963), in *People of the World*, p. 21.

13. *Ibid.*, p. 22.

14. *Ibid.*, pp. 27-30.

15. "Statement of the Soviet Government of August 21, 1963," in *People of the World*, p. 196.

16. "Statement by the Spokesman of the Chinese Government" (Sept. 1, 1963), in *ibid.*, pp. 38-39.

17. *Ibid.*, pp. 54-60.

18. Quoted in Hsieh, "Sino-Soviet Nuclear Dialogue," p. 164.

19. *Ibid.*, pp. 164-65.

20. "Two Different Lines," pp. 246, 255.

21. For example, see "PLA Conference"; "On Man"; and "Put Ideological Work in the Primary Position."

22. Hung Fan-ti, "Strategy of 'Flexible Response,'" especially p. 6. For a different type of critique of U.S. defense policies, see "MLF," which notes: "In order to prevent a nuclear war, the first thing to do is to oppose U.S. imperialism. . . . Nuclear weapons are created by man. They can also be destroyed by man."

23. The text of the speech is in *Xinhua*, Aug. 1, 1964, in *Survey of China Mainland Press*, No. 3273 (Aug. 6, 1964), pp. 40-45.

24. See Halperin and Lewis, pp. 58-67; and Whitson, pp. 528-57.

25. This group, led by Qian Sanqiang, was called the Neutron Physics Leading Group. Except as noted, the information in this paragraph and the rest of this section is based on "Riddle of Research and Development," pp. 6-7; and Liu Xiyao, "How China Succeeded."

26. Feng and Chen; Yu Min; Li Jue et al., pp. 61, 276.

27. Liu Jingzhi and Li Peicai.

28. Glasstone and Dolan, p. 20.

29. This paragraph and the one that follows are based on Postol. (This article, prepared for publication in *Encyclopedia Americana*, is cited by permission of the author.)

30. Liu Xiyao; "Riddle of Research and Development," pp. 6-7; Li Jue et al., pp. 62, 277-78, 367. Information in the following five paragraphs is from these sources. Liu specifically mentions using computers of the Chinese Academy of Sciences' institutes in Beijing and Shanghai. The most advanced computer used was the J-501 Model at the East China Computer Institute in Shanghai. By way of comparison, the scientists at Los Alamos in the wartime project had "Marchant computers—hand calculators with numbers. You push them, and they multiply, divide, add, and so on, but not easy like they do now. They were mechanical gadgets, failing often." Feynman, p. 108.

31. This is based partially on Yu Min; and Liu Jingzhi and Li Peicai. Yu Min had studied physics under Deng Jiaxian at Beijing University and joined the Ninth Academy in 1965. Shortly before Yu went to Shanghai, the ministry submitted its "Report on Arrangements for Making Breakthroughs on Hydrogen Bomb Technologies" (August 1965). It called for combining laboratory experiments with actual weapons tests. Li Jue et al., pp. 62, 366.

32. All the scientists reportedly were assigned to read Mao's "On Practice" and "On Contradiction," as well as political directives and tracts on "learning from Daqing," the oilfield then being touted as a model of industrial self-reliance. For a brief historical review of the early years of the Cultural Revolution, see "China, History of," *Encyclopaedia Britannica: Macropaedia*, 4: 396-400.

33. "Hold High the Great Banner."

34. In a 16-point directive of Aug. 8, 1966 ("Decision of the Central Committee of the Chinese Communist Party Concerning the Great Proletarian Cultural Revolution"), the Central Committee accepted, though in modified form, article 12, which said: "Special care should be taken of those scientists and scientific and technical personnel who have made contributions." Text in *CCP Documents*, p. 52. In January 1967, the Central Military Commission forbade assaults on "war preparation systems and security systems in the armed forces" (text in *ibid.*, p. 212), and on February 7, the central leadership issued a "Circular . . . Forbidding Exchange of Revolutionary Experience in Industrial and Mining Units, Scientific Research Organs, Design Units and Capital Construction Units Under the Industrial System for National Defense" (text in *ibid.*, p. 232).

35. Except as noted, the information in this and the following paragraph is from Liu Jingzhi and Li Peicai.

36. The information on the design unit is based on "Triumphal Song of Mao Zedong's Thought," p. ccc16.

37. For a sample of news stories and the official communique on this test, see FBIS: *Far East*, Oct. 28, 1966, pp. CCC1-CCC3. The character *xi* (double happiness) is used for weddings and other special celebrations.

38. This paragraph is based on Nie Rongzhen, "How China Develops Its Nuclear Weapons," p. 19.

39. "Mao Tse-tung's Thought Is the Victorious Banner," pp. 16-17.

40. The information on the Second Ministry is based on interviews with a Chinese specialist, 1986. Shortly after the outbreak of the Cultural Revolution, Minister Liu Jie was stripped of his power. Premier Zhou Enlai then assigned Vice-Ministers Liu Xiyao and Liu Qisen to oversee the day-to-day affairs of the ministry. They remained in power for several years.

41. The information in this and the following two paragraphs is principally from Suo Guoxin, pp. 16-18.

42. Article 8 of the "Provisional Regulations of the Chinese People's Liberation Army for the Safekeeping of State Military Secrets (Draft)" states: "Test sites for new weapons . . . must be properly classified as restricted areas." Text in J. C. Cheng, p. 239.

43. "Document of the CCP Central Committee, the State Council and the Central Military Commission," in *CCP Documents*, p. 186.

44. We shall review the events of the Cultural Revolution in our subsequent study of the missile program.

45. "Riddle of Research and Development," p. 7; Li Jue et al., p. 63. At the meeting on May 9, the commission placed the Party committee of the Lop Nur base in command of the forthcoming test. It reversed this decision a few weeks later.

46. The information in this and the following paragraph is from Ding Houben, p. 17.

47. Guo Diancheng and Xu Zhimin, "Never Forget," p. 28; Li Jue et al., p. 283. The H-bomb, dropping by parachute, is shown in picture 62, following Li Jue et al., p. 238.

48. For a compilation of documents on this test, see FBIS: *Communist China*, June 19, 1967, pp. CCC1-ccc12, and June 20, 1967, pp. ccc9-ccc20. On Nie's role, see Nie Rongzhen, "How China Develops Its Nuclear Weapons," p. 19. The Nie article incorrectly gives June 14 rather than June 17 as the date of this test.

49. This information is based on "Triumphal Song of Mao Zedong's Thought," pp. ccc17-ccc18.

50. Information in this and the next two paragraphs is from Lu Chuanzhi et al.; and Xi Qixin and Liu Jingzhi.

51. The information in the rest of this section is from "Triumphal Song of Mao Zedong's Thought," pp. ccc18-ccc20.

52. The information in this section is from Li Jue et al., pp. 49, 59-61, 63, 276, 285-91; Lu Ke, pp. 9-11; and Wang Zhuang, pp. 64-65.

53. "China's First Hydrogen Bomb," pp. 6-7.

54. *Ibid.*; "A Magnificent Victory," p. CCC4.

55. These emphases all come within what the political scientist Allen Whiting has termed "the calculus of deterrence" adopted by China in conflicts with India and with the United States in Vietnam. Whiting, *The Chinese Calculus*, especially Chap. 7.

56. NRH, p. 787.

57. Du Yuejin, p. 13.

58. Mao [9]. See especially pp. 288-89.

59. See, for example, Wang Shouyun, pp. 26-27.

60. "On Questions of Party History," p. 31.

61. As each of the four missiles became operational, the West designated them by a CSS number (CSS-1 through CSS-4). Most of the information in the rest of this section is from interviews with Chinese specialists, 1985; and from Zhang Jun, as noted.

62. Zhang Jun, p. 41.

63. *Ibid.*, pp. 562-63.

64. U.S. government reports regularly described this medium-range ballistic missile as a "single-stage, liquid propellant, transportable system developed from Soviet models. . . . It can reach targets in the Eastern USSR, peripheral nations, and some US bases in the Far East. It is an obsolescent and cumbersome missile system with slow reaction time." These reports gave an estimated range of about 600 nautical miles, or 1,100 km, which the Chinese say slightly underestimates the maximum range. See Brown, *U.S. Military Posture, FY 1976*, p. 48.

65. Zhang Jun, p. 565.

66. Jones, *U.S. Military Posture, FY 1981*, p. 76. In the late 1970s, the Chinese began to withdraw the DF-2 from the missile force. Interview with Chinese specialists, 1987.

67. Some U.S. sources state that the DF-4 (CSS-3) had a maximum range of 4,350 miles (about 7,000 km), but Chinese specialists say this is incorrect. See "Chinese Develop Missile, Satellite Launchers," p. 16.

68. This is based on Du Hua et al.; and "Outstanding Achievements."

69. U.S., Department of the Air Force, pp. 3.10-3.14.

70. The DF-5 used UDMH and an oxidizer of 100% nitrogen tetroxide.

71. Zhang Jun, p. 182. On the development of graphite control vanes, see Ley, Chap. 8. A picture of graphite vanes on the German V-2 (A-4) is in Ley, p. 195.

72. On these control systems, see U.S. Department of the Air Force, pp. 7.11-7.18. This information is from a Chinese specialist, 1986; Du Hua et al.; Zhang Jun, pp. 176, 180-82.

73. Information from a Chinese specialist, 1986.

74. Zhang Jiajun.   75. Guo Qingsheng, pp. 23-24.

76. Du Hua et al.   77. Schlesinger, p. 31.

78. Brown, *U.S. Military Posture, FY 1978*, pp. 5, 106.

79. For an examination of Chinese statements on deterrence, see J. Lewis, "China's Military Doctrines," especially pp. 151-58.

80. NRH, p. 810.

81. Wang Shouyun, pp. 26-27. Wang cites an article on medium nuclear powers by Geoffrey Kemp (*Nuclear Forces for Medium Powers*) as giving a sound explanation of China's strategy.

82. "Statement by the Government of the People's Republic of China."

83. Quoted in Yin Weixing, p. 21. Yin works in the Political Department of the Second Artillery Corps.

84. See, for example, Zhang Jianzhi. Zhang is a specialist in the Second Artillery Corps.

85. "Pay Close Attention"; "Résumé of Discussions at the Ground Force Training Conference," in J. C. Cheng, p. 684.

Chapter Nine

1. See Simpson, *Independent Nuclear State*, especially Chap. 11.

2. Interviews with Chinese specialists, 1986.

3. *Ibid.*

4. Holloway, "Innovation," p. 403.

5. Quoted in Holloway, "Entering the Nuclear Arms Race," p. 188.

6. Gowing, *Independence and Deterrence* (U.K.); Hewlett and Anderson

(U.S.); Rhodes (U.S.); Irving (Germany); Goldschmidt, pp. 121-52 (France). As of this writing, the principal sources on the Soviet program are Golovin; Kramish; Modelski; Holloway, "Entering the Nuclear Arms Race," pp. 159-97; and Holloway, "Military Technology," pp. 451-55.

7. Gowing, *Independence and Deterrence*, 2: ix. The classic treatment of Great Britain's wartime atomic program is Gowing, *Britain and Atomic Energy*.

8. Cockcroft headed the Montreal atomic energy laboratory, a joint Anglo-Canadian-French facility that worked on the chemical separation of plutonium. Penney, a mathematician and an expert on explosive effects, was a member of the British team at Los Alamos. Gowing, *Independence and Deterrence*, 2: 4-7.

9. For example, Zhao Zhongyao, a physicist who later worked at the Institute of Atomic Energy, witnessed at least one of the U.S. Bikini atoll tests in the late 1940s. However, Zhao was not a major participant in the Chinese nuclear weapons program.

10. The Soviet Union exploded a fission weapon in 1949, the British in 1952, the French in 1960, and the Chinese in 1964. The initial fusion weapons tests came in this order: United States, 1952; the Soviet Union, 1953; the United Kingdom, 1957; China, 1967; and France, 1968.

11. Gowing, *Independence and Deterrence*, 2: 497.

12. Quoted in Chap. 1, p. 1.

13. Irving, p. 267.

14. The information in this paragraph is based on Holloway, "Entering the Nuclear Arms Race," pp. 166-70.

15. Gowing, *Independence and Deterrence*, 2: 502.

16. Gowing, *Reflections on Atomic Energy History*, pp. 9-22.

17. Holloway, "Innovation," p. 394.

18. *Ibid.*, pp. 389-91; Holloway, "Military Technology," pp. 451-55.

19. United States, President's Blue Ribbon Commission, pp. 11-13.

20. For a discussion of *zongjie*, see J. Lewis, *Leadership*, pp. 160-62.

21. NRH, p. 840.

22. NRH, p. 796.

23. NRH, p. 822. The quotations in the next paragraph are from the same page.

24. Nie Rongzhen, "How China Develops Its Nuclear Weapons," pp. 15-16.

25. NRH, p. 787.

26. NRH, p. 814. In the Chinese idiom, the seven daily necessities are fuel, rice, oil, salt, soy sauce, vinegar, and tea.

27. NRH, p. 821.

References Cited

References Cited

Chinese romanizations are not provided for newspaper or journal articles. English names are given in brackets for all journals and newspapers except the two most frequently cited ones: *Renmin Ribao* [People's Daily] and *Liaowang* [Outlook]. The following abbreviations are used in this list:

FBIS   *Foreign Broadcast Information Service*
JPRS   Joint Publications Research Service
NRH   *Nie Rongzhen Huiyilu*

Adams, Sherman. *Firsthand Report*. Reprint ed. Westport, Conn., 1974.
"The All-China Scientific Association Holds an Enlarged Meeting to Criticize the Rightists Zeng Zhaolun and Qian Weichang," *Xin Hua Yuebao* [New China Monthly], Aug. 25, 1957.
American Foreign Policy, 1950-1955. Washington, D.C., 1957.
"Appointments of Ministers of Three Ministries of Machine Building by Chairman Liu Shaoqi," *Renmin Ribao*, Sept. 14, 1960.
"The Appointments of Nie Rongzhen and Bo Yibo to Be Vice Premiers by Chairman Mao Zedong," *Xinhua* [New China News Agency], Nov. 17, 1956.
Arkin, William, and Richard Fieldhouse. *Nuclear Battlefields: Global Links in the Arms Race*. Cambridge, Mass., 1985.
Bachman, David. *Chen Yun and the Chinese Political System*. Berkeley, Calif., 1985.
Bagnall, K. W. "Selenium, Tellurium and Polonium," in M. Schmidt et al., *The Chemistry of Sulphur, Selenium, Tellurium and Polonium*. Oxford, 1973.
Bai Yunshan. "Miscellaneous Recollections of the Northwest," *Libao Yuekan* [Li Pao Monthly] (Hong Kong), Oct. 9, 1980.
Bernstein, Barton J. "The Struggle Over the Korean Armistice: Prisoners of Repatriation?," in Bruce Cumings, ed., *Child of Conflict*. Seattle, Wash., 1983.
Beus, A. A. *Beryllium: Evaluation of Deposits During Prospecting and Exploratory Work*. San Francisco, 1962.
Bickel, Lennard. *The Deadly Element: The Story of Uranium*. New York, 1979.

Blacker, Coit, and Gloria Duffy, eds. *International Arms Control: Issues and Agreements.* 2d ed. Stanford, Calif., 1984.

Bloembergen, Nicolaas. "Physics," in Leo A. Orleans, ed., *Science in Contemporary China.* Stanford, Calif., 1980.

Bohr, Niels, and John A. Wheeler. "The Mechanism of Nuclear Fission," *Physical Review,* 56 (Sept. 1, 1939).

Borg, Dorothy, and Waldo Heinrichs, eds. *Uncertain Years: Chinese-American Relations, 1947-1950.* New York, 1980.

Borisov, O. B. [pseud.]. *Iz Istorii Sovetsko-Kitaiskikh Otnoshenii v 50-kh Godakh* [From the History of Soviet-Chinese Relations in the 1950s]. Moscow, 1981.

Borisov, O. B., and B. T. Koloskov [pseuds.]. *Soviet-Chinese Relations, 1945-1970.* Ed. with intro. by Vladimir Petrov. Bloomington, Ind., 1975.

Break the Nuclear Monopoly, Eliminate Nuclear Weapons. Beijing, 1965.

Bromley, Allan, and Pierre Perrolle. *Nuclear Science in China.* Washington, D.C., 1980.

Brown, George S. *United States Military Posture for FY 1976.* Washington, D.C., 1975.

———. *United States Military Posture for FY 1978.* Washington, D.C., 1977.

Cai Jianwen. "A Flying Heroine," *Renmin Ribao,* overseas ed., Feb. 28, 1987.

CCP Documents of the Great Proletarian Cultural Revolution, 1966-1967. Hong Kong, 1968.

"The Celebration of the Twentieth Anniversary of the Successful Detonation of Our Country's First Atomic Bomb," *Renmin Ribao,* Oct. 17, 1984.

Chen Honggeng. "Xinyuan" [Aspirations], in *Mimi Licheng,* listed below.

Chen, Nai-Ruenn, ed. *Chinese Economic Statistics: A Handbook for Mainland China.* Chicago, 1967.

Chen, Theodore. *Thought Reform of Chinese Intellectuals.* Hong Kong, 1960.

Chen Zujia. "Reports from the Nuclear Industrial Bases in North and Northwest China," *Renmin Ribao,* Nov. 5, 1985.

Cheng, J. Chester. *The Politics of the Chinese Red Army.* Stanford, Calif., 1966.

"China's First Hydrogen Bomb Successfully Exploded," *Peking Review,* No. 26, June 23, 1967.

"Chinese Develop Series of Missile, Satellite Launchers," *Aviation Week and Space Technology,* Aug. 25, 1980.

Chou, Nailene. "Frontier Studies and Changing Frontier Administration in Late Ching China: The Case of Sinkiang, 1759-1911." Ph.D. diss., University of Michigan, 1976.

Chu Chi-hsin. "Primary Tasks Imposed by the Five-Year Plan," *People's China,* No. 9, May 1, 1955.

Chu, Wen-Djang. *The Moslem Rebellion in Northwest China, 1862-1878: A Study of Government Minority Policy.* The Hague, 1966.

Ci Hai [A Sea of Words (Dictionary)]. Shanghai, 1980.

Clark, Mark W. *From the Danube to the Yalu.* New York, 1954.

Clegg, John W., and Dennis D. Foley. *Uranium Ore Processing.* Reading, Mass., 1958.

Clubb, O. Edmund. "Chiang Kai-shek's Waterloo: The Battle of the Huai-Hai," *Pacific Historical Review*, 25.4 (Nov. 1956).

"Communique of the Delegations of the Democratic People's Republic of Korea and the People's Republic of China to the Bilateral Talks Concerning a Political Conference," *Xin Hua Yuebao* [New China Monthly], Dec. 28, 1953.

"Communique on Sino-Soviet Talks," *Xin Hua Yuebao* [New China Monthly], Nov. 28, 1954.

"Comrade Mao Zedong on 'Imperialism and All Reactionaries Are Paper Tigers,'" *Renmin Ribao*, Oct. 27, 1958.

"The Conclusion of the Sino-Soviet Accord Relating to Soviet Aid to China's Program for the Peaceful Utilization of Atomic Energy," *Renmin Ribao*, May 1, 1955.

"Conversion Rates Between the People's Currency and Certain Foreign Currencies," *Jihua Jingji* [Planned Economy], No. 12, Dec. 9, 1957.

Dai Yaping. "A First Visit to China's 'Oak Ridge,'" *Zhongguo Xinwen She* [China News Agency], Nov. 9, 1985, in FBIS: *China*, Nov. 13, 1985.

———. "The Nuclear City in the Heart of the Grasslands," *Zhongguo Xinwen She* [China News Agency], Nov. 4, 1985, in FBIS: *China*, Nov. 8, 1985.

———. "A Nuclear Plant in the Gobi Desert," *Zhongguo Xinwen She* [China News Agency], Nov. 5, 1985, in FBIS: *China*, Nov. 7, 1985.

Dai Yaping and Zhao Jin. "Report on the Construction of Our Country's Nuclear Industrial Bases," *Jingji Ribao* [Economic Daily], Nov. 4, 1985.

Dangdai Zhongguo de He Gongye, see Li Jue et al., listed below.

Deng Xiaoping. *Selected Works of Deng Xiaoping (1975-1982)*. Beijing, 1984.

"Devote Major Efforts to Promoting the Peaceful Utilization of Atomic Energy," *Renmin Ribao*, May 2, 1955.

Di Zhongheng. "Chinese Communist Nuclear Forces," *Mingbao Yuekan* [Ming Pao Monthly], 2 parts, Aug. 1, Sept. 1, 1976.

Dinerstein, Herbert S. *War and the Soviet Union: Nuclear Weapons and the Revolution in Soviet Military and Political Thinking*. Rev. ed. Westport, Conn., 1962.

Ding Houben. "Short Stories of Professor Wang Ganchang's Activities in Nuclear Testing," *Shenjian* [Magical Sword], No. 5, Oct. 4, 1985.

"Directive of the Central Committee of the Chinese Communist Party Relating to the Cadres' Theoretical Education," *Xin Hua Yuebao* [New China Monthly], May 28, 1953.

Disaster Strikes the Tachens. Beijing, 1955.

Dong Kegong et al. "Comrade Nie Rongzhen with the Intellectuals," *Guangming Ribao* [Bright Daily], Oct. 4, 1982.

Du Hua et al. "The 'Vanguard' of the Missile Troops," *Guangming Ribao* [Bright Daily], Aug. 6, 1986.

Du Yuejin. "A Survey of the Path to Develop Simultaneously the Construction of National Defense and of the National Economy," *Liaowang*, overseas ed., Dec. 22, 1986.

Duan Junyi et al. "Comrade Zhao Erlu, a Prominent Organizer in the Construction of Our Country's Defense Industry," *Renmin Ribao*, Jan. 13, 1982.

Dulles, John Foster. "Policy for Security and Peace," *Foreign Affairs*, April 1954.

———. "Report from Asia," Department of State *Bulletin*, 32.821 (March 21, 1955).

"Dulles Has the Nerve to Argue in Favor of American Aggressive Policy," *Renmin Ribao*, Jan. 16, 1954.

Eisenhower, Dwight D. *The White House Years: Mandate for Change, 1953-1956.* Garden City, N.Y., 1963.

"Eisenhower Advocates the Use of Atomic Weapons," *Renmin Ribao*, March 18, 1955.

"Eisenhower Clamors for Long-Term War Preparations in His State of the Union Message," *Renmin Ribao*, Jan. 14, 1954.

Encyclopaedia Britannica: Macropaedia. Chicago, 1974. Vols. 4, 12, 13.

Fa-hsien. *A Record of the Buddhist Countries.* Beijing: Chinese Buddhist Association, 1957.

Farrar-Hockley, Anthony. "A Reminiscence of the Chinese People's Volunteers in the Korean War," *China Quarterly*, 98 (June 1984).

Feng Yuan and Chen Dong. "Peng Huanwu: A Famous Theoretical Physicist Who Doesn't Seek Fame," *Renmin Ribao*, June 28, 1985.

Feynman, Richard P. *Surely You're Joking, Mr. Feynman!* New York, 1985.

Filatov, L. V. *Ekonomicheskaya Otsenka Nauchno-Tekhnicheskoi Pomoshchi Sovetskovo Soyuza Kitayu, 1949-1966* [Economic Estimate of the Scientific-Technical Assistance to China, 1949-1966]. Moscow, 1980.

Forbes, Andrew D. W. *Warlords and Muslims in Chinese Central Asia: A Political History of Republican Sinkiang, 1911-1949.* Cambridge, Eng., 1986.

Ford, Harold. "Modern Weapons and the Sino-Soviet Estrangement," *China Quarterly*, 18 (April-June 1964).

Fordham, S. *High Explosives and Propellants.* Oxford, 1966.

Fu Biduo and Tian Jijin. "Powerful Industrial and Agricultural Bases Were Established in the Gobi Desert," *Jingji Ribao* [Economic Daily], Aug. 31, 1984.

Galkin, N. P., and B. N. Sudarikov, eds. *Technology of Uranium.* Jerusalem, 1966.

Galkin, N. P., A. A. Maiorov, and U. D. Veryatin. *The Technology of the Treatment of Uranium Concentrates.* New York, 1963.

George, Alexander. *The Chinese Communist Army in Action: The Korean War and Its Aftermath.* New York, 1967.

———. *Presidential Decisionmaking in Foreign Policy: The Effective Use of Information and Advice.* Boulder, Col., 1980.

George, Alexander, and Richard Smoke. *Deterrence in American Foreign Policy: Theory and Practice.* New York, 1974.

Gittings, John. *The World and China, 1922-1972.* New York, 1974.

Glasstone, Samuel, and Philip Dolan, eds. *The Effects of Nuclear Weapons.* 3d ed. Washington, D.C., 1977. [This classic volume was first published in 1950 as *The Effects of Atomic Weapons.*]

"Glorious Task of the People's Liberation Army," *Renmin Ribao*, July 24, 1954.

Goldschmidt, Bertrand. *Atomic Complex: A Worldwide Political History of Nuclear Energy.* LaGrange Park, Ill., 1982.

Golovin, I. N. *Academician Igor Kurchatov.* Moscow, 1969.

Goodman, Allan E., ed. *Negotiating While Fighting: The Diary of Admiral C. Turner Joy at the Korean Armistice Conference.* Stanford, Calif., 1978.

Gourievidis, Gerard. "Nuclear Power in China," *Revue Générale Nucléaire,* 4 (July-Aug. 1984).

Gowing, Margaret M. *Britain and Atomic Energy, 1939-1945.* New York, 1964.

———. *Independence and Deterrence: Britain and Atomic Energy, 1945-1952.* 2 vols. New York, 1974.

———. *Reflections on Atomic Energy History.* The Rede Lecture 1978. London, 1978.

"Greet the Upsurge in Forming People's Communes," *Hongqi* [Red Flag], No. 7, Sept. 1, 1958.

Groves, Leslie R. *Now It Can Be Told.* New York, 1962.

Gu Mainan [Ku Mainan]. "Deng Jiaxian: China's Father of A-Bomb," *Beijing Review,* No. 32, Aug. 11, 1986.

———. "Deng Jiaxian, a Man of Great Merit in Developing the 'Two Bombs,'" *Liaowang,* June 23, 1986.

———. "Deng Jiaxian, the Veteran Scientist Who Made Great Contributions to the Atomic Bomb and Hydrogen Bomb Development Program," *Renmin Ribao,* overseas ed., June 24, 1986.

———. "One Out of a Hundred Thousand—Zhou Guangzhao, the New Vice-President of the Chinese Academy of Sciences," *Renmin Ribao,* July 23, 1984.

Gu Mainan and Gu Wenfu. "A Strategic Diversion in the Nuclear Industry," *Liaowang,* overseas ed., Feb. 3, 1986.

Gu Yu. "How to Tackle the Organization of Key Scientific and Technical Problems," *Liaowang,* July 20, 1982.

Guo Diancheng and Xu Zhimin. Three-part series on the authors' visit to China's nuclear test base, *Liaowang,* Oct. 1, Oct. 8, Oct. 15, 1984. Part 1: "Thunder Roars Over the Boundless Gobi." Part 2: "Never Forget Premier Zhou's Teachings." Part 3: "The Real People's Heroes."

Guo Hualun. "The Study of Mao Zedong's Military Thought," in Zhou Ziqiang, ed., *Gongfei Junshi Wenti Lunji* [A Collection of Essays on Chinese Communist Military Issues]. Taipei, 1974.

Guo Jian. "The Glorious Past of Beijing's Nuclear Town in 35 Years," *Zhongguo Xinwen She* [China News Agency], Sept. 27, 1985, in FBIS: *China,* Oct. 2, 1985.

Guo Jian and Shuang Yin. "A Visit to the Institute of Atomic Energy," *Beijing Wanbao* [Beijing Evening News], in *China Report: Science and Technology,* 71 (JPRS 76994; Dec. 12, 1980).

Guo Moruo [Kuo Mo-jo]. "Ban Atomic Weapons!," *People's China,* No. 6, March 16, 1955.

———. "A Comprehensive Plan for Scientific Research and the Policy of 'Letting Diverse Schools of Thought Contend,'" in *New China Advances to Socialism.* Beijing, 1956.

———. "A Report on the Academy's Basic Situation and Tasks Ahead," *Kexue Tongbao* [Science Bulletin], Jan. 1954.

———. "A Report on the Academy's Present Status," *Xin Hua Yuebao* [New China Monthly], April 28, 1954.

———. "Strengthen the Peace Forces and Frustrate the Threats of Atomic War," *Xin Hua Yuebao* [New China Monthly], March 28, 1955.

———. "A Summary and the Main Points of the Plan for the Academy of Sciences Work in 1950 and 1951," *Xin Hua Yuebao* [New China Monthly], May 25, 1951.

Guo Qingsheng. "China Has a Nuclear Counterattack Capability—A Visit to China's Strategic Missile Units," *Liaowang*, April 22, 1985.

Halperin, Morton H. *China and the Bomb*. New York, 1965.

Halperin, Morton H., and John W. Lewis. "New Tensions in Army-Party Relations in China, 1965-1966," *China Quarterly*, 26 (April-June 1966).

Halperin, Morton H., and Tang Tsou. "United States Policy Toward the Offshore Islands," *Public Policy*, 15 (1966).

He Di. "The Evolution of the People's Republic of China's Policy Toward the Offshore Islands (Quemoy, Matsu)." Conference paper presented in September 1987.

He Xiaolu. "A Marshal and a Diplomat," *Kunlun* [Kunlun], No. 4, 1984.

Hedin, Sven. *The Wandering Lake*. London, 1940.

Hermes, Walter G. *United States Army in the Korea War: Truce Tent and Fighting Front*. Washington, D.C., 1966.

Hewlett, Richard G., and Oscar E. Anderson, Jr. *A History of the United States Atomic Energy Commission*, Vol. 1: *The New World, 1939/1946*. University Park, Pa., 1962.

Ho Cheng. "Taiwan Must Be Liberated," *People's China*, No. 17, Sept. 1, 1954.

"Hold High the Great Banner of Mao Zedong's Thought," *Jiefangjun Bao* [Liberation Army Daily], April 19, 1966.

Holden, A. N. *Physical Metallurgy of Uranium*. Reading, Mass., 1958.

Holloway, David. "Entering the Nuclear Arms Race: The Soviet Decision to Build the Atomic Bomb, 1939-45," *Social Studies of Science*, 11 (1981).

———. "Innovation in the Defense Sector," in Ronald Amann and Julian Cooper, eds., *Industrial Innovation in the Soviet Union*. New Haven, Conn., 1982.

———. "Military Technology: Battle Tanks and ICBMs," in Ronald Amann and Julian Cooper, eds., *The Technological Level of Soviet Industry*. New Haven, Conn., 1977.

———. "Research Note: Soviet Thermonuclear Development," *International Security*, 4.3 (Winter 1979-80).

———. *The Soviet Union and the Arms Race*. New Haven, Conn., 1983.

Hsieh, Alice Langley. "China's Secret Military Papers: Military Doctrine and Strategy," *China Quarterly*, 18 (April-June 1964).

———. *Communist China's Strategy in the Nuclear Era*. Englewood Cliffs, N.J., 1962.

———. "The Sino-Soviet Nuclear Dialogue: 1963," in Raymond Garthoff, ed., *Sino-Soviet Military Relations*. New York, 1966.

Huang Fengchu and Zhu Youdi, "Hou Xianglin, an Authority in Our Country's Petrochemical Engineering," *Liaowang*, March 23, 1987.
Huili, Monk, comp. *The Life of Hsuan-tsang*. Beijing, 1959.
Hulsewe, A. F. P. *China in Central Asia*. Leiden, 1979.
Hung Fan-ti. "Strategy of 'Flexible Response' and Its Contradictions," *Peking Review*, No. 15, April 10, 1964.
Imperialism and All Reactionaries Are Paper Tigers. Enlarged ed. Beijing, 1958.
Institute of Literature, Chinese Academy of Sciences. *Stories About Not Being Afraid of Ghosts*. Beijing, 1961.
International Monetary Fund. *International Financial Statistics Yearbook, 1986*. Washington, D.C., 1986.
Irving, David. *The Virus House*. London, 1967.
Jiang Ji. "Zai Xinde Shiye Li" [In a New Cause], in *Mimi Licheng*, listed below.
Jiang Nan. "Comment of the New China News Agency on the U.S. Attempt to Obstruct the Peaceful Solution of the Question of Korea [at the Geneva Conference]," *Renmin Ribao*, March 25, 1954.
——. "Comment on Indian Prime Minister Nehru's Statement on the Question of Indochina," *Renmin Ribao*, April 29, 1954.
——. "People Will Never Tolerate Those Who Play with the Flames of War," *Renmin Ribao*, Jan. 8, 1955.
Jiang Shengjie. "Over Ten Reactors Are Operating Safely in Our Country," *Renmin Ribao*, overseas ed., June 27, 1986.
Jiang Zhenghao. "Some Aspects of Zhou Enlai's Foreign Policy Thinking." Paper presented at the Center for International Security and Arms Control, Stanford University, May 16, 1985.
——. "Sovereignty and Security." Paper presented at the Center for International Security and Arms Control, Stanford University, Aug. 4, 1985.
Jones, David C. *United States Military Posture for FY 1981*. Washington, D.C., 1980.
Joseph, William A. *The Critique of Ultra-Leftism in China, 1958-1981*. Stanford, Calif., 1984.
Joy, C. Turner. *How the Communists Negotiate*. New York, 1955.
Jungk, Robert. *Brighter Than a Thousand Suns*. New York, 1958.
Jurika, Stephen, Jr., ed. *From Pearl Harbor to Vietnam: The Memoirs of Admiral Arthur W. Radford*. Stanford, Calif., 1980.
Kahin, George McT., and John W. Lewis. *The United States in Vietnam*. Rev. ed. New York, 1969.
Keefer, Edward C. "President Dwight D. Eisenhower and the End of the Korean War," *Diplomatic History*, 10.3 (Summer 1986).
Kemp, Geoffrey. *Nuclear Forces for Medium Powers*. International Institute for Strategic Studies, Adelphi Papers 106 and 107. London, 1974.
Khrushchev, Nikita. *Khrushchev Remembers*. Boston, 1970.
——. *Khrushchev Remembers: The Last Testament*. Boston, 1974.
Kinnard, Douglas. *President Eisenhower and Strategy Management: A Study in Defense Politics*. Lexington, Ky., 1977.
Klein, Donald W., and Anne B. Clark. *Biographic Dictionary of Chinese Communism, 1921-1965*. 2 vols. Cambridge, Mass., 1971.

Klochko, Mikhail A. *Soviet Scientist in Red China*. New York, 1964.

Kramish, Arnold. *Atomic Energy in the Soviet Union*. Stanford, Calif., 1959.

Krass, Allan, et al. *Uranium Enrichment and Nuclear Weapon Proliferation*. London, 1983.

Ku Mainan, *see* Gu Mainan.

Lan Xueyi. "A Picture of Prosperity in the Reclamation Areas of the Xinjiang PC Corps," *Renmin Ribao*, overseas ed., Sept. 29, 1985.

Lapp, Ralph E. "The Einstein Letter That Started It All," *New York Times Magazine*, Aug. 2, 1964.

"Leninism and Modern Revisionism," *Hongqi* [Red Flag], No. 1, Jan. 5, 1963.

Lewis, John W. "China's Military Doctrines and Force Posture," in Thomas Fingar, ed., *China's Quest for Independence: Policy Evolution in the 1970s*. Boulder, Col., 1980.

———. "China's Secret Military Papers: 'Continuities' and 'Revelations,'" *China Quarterly*, 18 (April-June 1964).

———. *Leadership in Communist China*. Ithaca, N.Y., 1963.

———. *Political Networks and the Chinese Policy Process*. Northeast Asia-United States Forum on International Policy. Stanford, Calif., 1986.

Lewis, Robert S., and George B. Clark. *Elements of Mining*. 3d ed. New York, 1964.

Ley, Willy. *Rockets, Missiles, and Men in Space*. New York, 1968.

Li Jinqi. "A Visit to the 'Atomic City' by Scientists from Hong Kong and Macao," *Renmin Ribao*, overseas ed., June 28, 1986.

Li Jinting et al. "Spaceflight Chief Engineer," *Liaowang*, Oct. 20, 1982.

Li Jue, Lei Rongtian, Li Yi, and Li Yingxiang, chief eds., *Dangdai Zhongguo de He Gongye* [Contemporary China's Nuclear Industry]. Beijing, 1987.

Li Jukui. "Memories of Logistical Work of the Chinese People's Volunteers Before the Fifth Campaign," *Xinghuo Liaoyuan* [A Single Spark Will Start a Prairie Fire], No. 5, 1985.

"Li Peng Makes a Speech to the Meeting of the Thirtieth Anniversary Celebration of the Establishment of Our Country's Nuclear Industry," *Renmin Ribao*, Nov. 5, 1985.

Li Tso-peng. "Strategically Pitting One Against Ten, . . ." *Peking Review*, No. 15, April 9, 1965.

Li Yingxiang et al. "A Description of New China's First Generation of People Who Tamed Nuclear Energy," *Beijing Ribao* [Beijing Daily], Nov. 6, 1985.

Liang Yusheng et al. "Quantitative Analysis of Distant Region Atmospheric Deposit of Debris from Nuclear Explosion [of Test 23 on March 15, 1978] Using Ge(Li) Gamma-Ray Spectra," *He Jishu* [Nuclear Technology], 6 (Dec. 1981).

Lindbeck, John M. H. "Organization and Development of Science," in Sidney Gould, ed., *Sciences in Communist China*. Washington, D.C., 1961.

Ling Xiang. "The Mystery of the Lop Nur Desert, . . ." *Wenxue Daguan* [Magnificent Spectacle of Literature], No. 4, 1986.

Liu Jie. "A Victory of the Socialist Road in the Scientific and Technological Development with Chinese Characteristics—In Commemoration of the 30th Anniversary of the Founding of the Nuclear Industry," *Renmin Ribao*, Nov. 1, 1985.

Liu Jingzhi and Li Peicai. "A Description of Deng Jiaxian, a Theoretical Physicist and a Pioneer in Our Country's Atomic and Hydrogen Bomb Effort," *Guangming Ribao* [Bright Daily], July 18, 1986.

Liu Nanchang and Liu Cheng. "The Nuclear Science City," *Renmin Ribao*, overseas ed., July 19, 1987.

Liu Qijun. "A Survey of Chinese Communist Research in Nuclear Science," *Feiqing Yanjiu* [Studies on Chinese Communism] (Taipei), Aug. 31, 1964.

Liu Shixiang. "A Green Leaf in China's Nuclear Industry," *Shenjian* [Magical Sword], No. 5, Oct. 4, 1985.

Liu Shuqing and Zhang Jifu. "Jinglei, Woguo Diyike Yuanzidan Baozha Ji" [A Loud Crash of Thunder—Report on the Detonation of Our Country's First Atomic Bomb], in *Mimi Licheng*, listed below.

Liu Suinian. "Several Conditions Concerning the Submission and Implementation of the Eight Character Guiding Principles, . . ." *Dangshi Yanjiu* [Studies of Party History], Dec. 28, 1980.

Liu Wei. "A Review of the Construction of China's First Heavy Water Reactor and First Cyclotron," *Jingji Ribao* [Economic Daily], Nov. 4, 1985.

Liu Xiyao. "How China Succeeded in Making Hydrogen Bombs with the Highest Speed," *Guangming Ribao* [Bright Daily], Nov. 3, 1985.

Liu Yalou. "Seriously Study Mao Zedong's Military Thinking," *Jiefangjun Bao* [Liberation Army Daily], May 23, 1958, in *Survey of China Mainland Press*, No. 1900, Nov. 24, 1958.

Liu Zhaolin. "A Rugged Path," *Jiefangjun Wenyi* [Literature and Art of the People's Liberation Army], No. 1, 1980.

Lu Chuanzhi et al. "He Xianjue Has Been Devoted to Defense Scientific and Technological Undertakings for Twenty Years," *Guangming Ribao* [Bright Daily], Aug. 23, 1986.

Lu Ke, "Pursue the Sun," *Kunlun* [Kunlun], No. 4, 1986.

Mader, Charles L., James N. Johnson, and Sharon L. Crane, eds. *Los Alamos Explosives Performance Data*. Berkeley, Calif., 1982.

"A Magnificent Victory for Mao Tse-tung's Thought," *Jiefangjun Bao* [Liberation Army Daily], June 18, 1967, in FBIS: *Communist China*, June 19, 1967.

"Major Development of the Weapons and Equipment of Our Armored Force," *Renmin Ribao*, overseas ed., July 2, 1987.

"Make Great Efforts to Build a Powerful People's Air Force," *Renmin Ribao*, March 29, 1955.

"Mao Tse-tung's Thought Is the Victorious Banner for Scaling the Heights of National Defence Science and Technology," *Peking Review*, No. 42, Oct. 14, 1966.

Mao Zedong [Mao Tsetung] (all works published in Beijing unless otherwise noted; the nonhyphenated form of Tsetung, adopted in Vol. 5 of the *Selected Works*, is used throughout).

    [1]. "All Reactionaries Are Paper Tigers," in *Selected Works of Mao Tsetung*, Vol. 5. 1977.

    [2]. "Be Activists in Promoting the Revolution," in *Selected Works of Mao Tsetung*, Vol. 5. 1977.

    [3]. "The Chinese People Cannot Be Cowed by the Atom Bomb," in *Selected Works of Mao Tsetung*, Vol. 5. 1977.

[4]. "On Contradiction," in *Four Essays on Philosophy*. 1966.
[5]. "On Practice," in *Selected Works of Mao Tsetung*, Vol. 1. 1964.
[6]. "On Some Important Problems of the Party's Present Policy," in *Selected Works of Mao Tsetung*, Vol. 4. 1961.
[7]. "On the Chungking Negotiations," in *Selected Works of Mao Tsetung*, Vol. 4. 1961.
[8]. "On the People's Democratic Dictatorship," in *Selected Works of Mao Tsetung*, Vol. 4. 1961.
[9]. "On the Ten Major Relationships," in *Selected Works of Mao Tsetung*, Vol. 5. 1977.
[10]. "Order to the Chinese People's Volunteers," in *Selected Works of Mao Tsetung*, Vol. 5. 1977.
[11]. "Our Great Victory in the War to Resist U.S. Aggression and Aid Korea and Our Future Tasks," in *Selected Works of Mao Tsetung*, Vol. 5. 1977.
[12]. "Problems of Strategy in China's Revolutionary War," in *Selected Military Writing of Mao Tsetung*. 1966.
[13]. "The Second Speech [at the Second Session of the Eighth Party Congress]," in *Miscellany of Mao Tsetung Thought (1949-1968)*, JPRS 61269-1 (Feb. 20, 1974).
[14]. "Some Points in Appraisal of the Present International Situation," in *Selected Works of Mao Tsetung*, Vol. 4. 1961.
[15]. "Speech at the Group Leaders Forum, . . ." *Chinese Law and Government*, 1.4 (Winter 1968-69).
[16]. "Speeches at the National Conference of the Communist Party of China," in *Selected Works of Mao Tsetung*, Vol. 5. 1977.
[17]. "Summaries of Speeches at the Supreme State Conference," *Chinese Law and Government*, 9.3 (Fall 1976).
[18]. "Talk at an Enlarged Central Work Conference," in Stuart Schram, ed., *Chairman Mao Talks to the People*. New York, 1974.
[19]. "Talk with the American Correspondent Anna Louise Strong," in *Selected Works of Mao Tsetung*, Vol. 4. 1961.
[20]. "Talks with Directors of Various Cooperative Areas," in *Miscellany of Mao Tsetung Thought (1949-1968)*, JPRS 61269-1 (Feb. 20, 1974).
[21]. "U.S. Imperialism Is a Paper Tiger," in *Selected Works of Mao Tsetung*, Vol. 5. 1977.

Marsden, William, trans. and ed. *The Travels of Marco Polo*. Garden City, N.Y., 1948.
"Marshal Kim Il Sung and General Peng Teh-huai's Letter to General Mark Clark" [March 28, 1953], *People's China*, No. 8, April 16, 1953.
McMillen, Donald. *Chinese Communist Power and Policy in Xinjiang, 1949-77*. Boulder, Col., 1979.
Meyer, Rudolf. *Explosives*. 2d rev. ed. Weinheim, West Germany, 1981.
Mimi Licheng [A Secret Course]. Ed. Magical Sword Branch, Ministry of Nuclear Industry. Beijing, 1985.
Miner, William. *Plutonium*. Oak Ridge, Tenn., 1964.

Minor, Michael S. "China's Nuclear Development Program," *Asian Survey*, 16.6 (June 1976).

"MLF [Multilateral Force], a U.S. Plan for Nuclear Threat," *Dagong Bao* [Impartial News] (Beijing), Dec. 15, 1964.

Modelski, George A. *Atomic Energy in the Communist Bloc*. Melbourne, Australia, 1959.

Monan Sui Duo Xin Wu Xia [He Was Still Loyal Although He Endured Untold Sufferings]. Beijing, 1978.

"The Monument to the Detonation Experiments of Our Country's First Atomic Bomb at the Foot of Mount Yanshan, Huailai County, Hebei Province," *Renmin Ribao*, overseas ed., Nov. 20, 1985.

More on the Differences Between Comrade Togliatti and Us. Beijing, 1963.

Moyer, Harvey V., Lloyd B. Gnagey, and Adrian J. Rogers, eds., *Polonium*. TID-5221. Oak Ridge, Tenn., 1956.

Murzayez, Ye. *Puteshestviya bez Priklyuchenii Fantastiki* [Journey Without Adventure and Fantasy]. Moscow, 1962. Translated in JPRS 25,110 (June 16, 1964).

Nero, Anthony V., Jr. *A Guidebook to Nuclear Reactors*. Berkeley, Calif., 1979.

Nie Fengzhi et al. *Sanjun Huige Zhan Donghai* [A Campaign by the Three Armed Services in the East China Sea]. Beijing, 1985.

Nie Rongzhen [Nieh Jung-chen]. "A Congratulatory Letter to the Model Workers' Meeting Held by the Ministry of Astronautics," *Jingji Ribao* [Economic Daily], Nov. 30, 1984.

———. "The Development of Science and Technology in Our Country Over the Last Ten Years," in *Ten Glorious Years*. Beijing, 1960.

———. "How China Develops Its Nuclear Weapons," *Beijing Review*, No. 17, April 29, 1985.

———. "A Letter to All Comrades of the Nuclear Test Base," *Liaowang*, Oct. 1, 1984.

———. *Nie Rongzhen Huiyilu* [Memoirs of Nie Rongzhen]. Beijing, 1984.

Niu Zhanhua. "Shanliang de Ganghuan" [A Shining Steel Ring], in *Mimi Licheng*, listed above.

O'Keefe, Bernard. *Nuclear Hostages*. Boston, 1983.

"On Man as the Primary Factor," *Hongqi* [Red Flag], No. 10, May 23, 1964.

"On Questions of Party History, . . ." *Beijing Review*, No. 27, July 6, 1981.

"An Order of the State Council of the People's Republic of China," *Guangming Ribao* [Bright Daily], Nov. 2, 1954.

"The Origin and Development of the Differences Between the Leadership of the CPSU and Ourselves," *Renmin Ribao*, Sept. 6, 1963.

Orleans, Leo A. *Every Fifth Child: The Population of China*. Stanford, Calif., 1972.

"Our Country Has Ten Nuclear Reactors," *Renmin Ribao*, overseas ed., Oct. 29, 1985.

"Outstanding Achievements in Scientific Research on Our Strategic Missiles," *Renmin Ribao*, overseas ed., Dec. 28, 1986.

Panofsky, Wolfgang. *Observations on High Energy Physics in China*. United States-China Relations Program. Stanford, Calif., 1977.

Patton, Finis S., John M. Googin, and William L. Griffith. *Enriched Uranium Processing.* New York, 1963.

"Pay Close Attention to Anti-Atomic and Anti-Chemical Training," *Jiefangjun Bao* [Liberation Army Daily], April 20, 1957.

"The PC Corps' Contributions to the Construction of Xinjiang," *Jingji Ribao* [Economic Daily], April 25, 1986.

Peng Dehuai. *Memoirs of a Chinese Marshal.* Beijing, 1984.

Peng Ruoqian. "Gebi Zhi Guang" [A Flash of Light Over the Gobi Desert], in *Mimi Licheng,* listed above.

People of the World, Unite, for the Complete, Thorough, Total and Resolute Prohibition and Destruction of Nuclear Weapons! Beijing, 1963.

Pevtsov, Mikhail. *Puteshestviye v Kashgariyu i Kun-lun'* [Journey to Kashgariya and Kunlun]. Moscow, 1949. Translated in JPRS 42,430 (Sept. 5, 1967).

Pilat, Joseph F., et al., eds. *Atoms for Peace: An Analysis After Thirty Years.* Boulder, Col., 1985.

"PLA Conference on Political Work Closes in Peking," *Xinhua* [New China News Agency], Jan. 17, 1964, in *Survey of China Mainland Press,* No. 3143, Jan. 21, 1964.

The Polemic on the General Line of the International Communist Movement. Beijing, 1965.

Postol, Theodore A. "Nuclear Warfare" (Sept. 1986), article prepared for publication in *Encyclopedia Americana.*

"Put Ideological Work in the Primary Position," *Hongqi* [Red Flag], No. 5, March 17, 1964.

Qian Sanqiang. "Cherish the Memory of Premier Zhou's Concern for Our Country's Scientific and Technological Undertakings and His Instruction to the Scientific and Technical Personnel," *Renmin Ribao,* March 10, 1979.

———. "The Peaceful Utilization of Atomic Energy in China Is Making Great Strides," *Renmin Ribao,* Oct. 11, 1959.

———. "A Survey of Modern Science in China," *Guangming Ribao* [Bright Daily], July 21, 1953.

Qian Weichang. "Physics in China," *Renmin Ribao,* Aug. 13, 1949.

Qian Xuesen. "The Significance of Coordinating the Solution of Key Problems and the Management of Problem Solving," *Kexuexue yu Kexue Jishu Guanli* [Study of Science and Management of Science and Technology], No. 6, June 12, 1984.

Radiochemistry Research Laboratory, Northwest Nuclear Technology Institute. "Determination of the Amount of Residual Plutonium[239] in Nuclear Explosion Debris," *He Huaxue yu Fangshe Huaxue* [Journal of Nuclear Chemistry and Radiochemistry], 3.3 (Aug. 1981).

Rankin, Carl Lott. *China Assignment.* Seattle, Wash., 1964.

Rees, David. *Korea: The Limited War.* New York, 1964.

Ren Gu. "Our Country's First Atomic Reactor and Cyclotron," *Yuanzineng* [Atomic Energy], Dec. 27, 1958.

"A Report by the Chinese Academy of Sciences to the Second Session of the Commission of the Academic Departments," *Kexue Tongbao* [Science Bulletin], No. 12, 1957.

"Report by the Delegation of the Academy of Sciences to the Soviet Union," *Xin Hua Yuebao* [New China Monthly], June 28, 1954.

"Resolutely Oppose the United States' Provocation of War," *Renmin Ribao,* Jan. 29, 1955.

"Resolution of the State Council of the PRC Concerning the Proposed Soviet Aid to China for Promoting Research on the Peaceful Uses of Atomic Energy," *Xin Hua Yuebao* [New China Monthly], Feb. 28, 1955.

Rhodes, Richard. *The Making of the Atomic Bomb.* New York, 1986.

"The Riddle of Research and Development on China's Atomic Bomb, Hydrogen Bomb, and Nuclear-Powered Submarine," *Liaowang,* overseas ed., June 15, 1987.

Ridgway, Matthew B. *The Korean War.* Garden City, N.Y., 1967.

——. *Soldier.* New York, 1956.

"Rightist Qian Weichang's Plot Has Been Frustrated," *Renmin Ribao,* July 6, 1957.

Rock, R. L. "Mine Engineering and Ventilation Problems Unique to the Control of Radon Daughters," in International Atomic Energy Agency, *Radon in Uranium Mining.* Vienna, 1975.

Rosenberg, David A. "The Origins of Overkill: Nuclear Weapons and American Strategy, 1945-1960," *International Security,* 7.4 (Spring 1983).

——. "'A Smoking Radiating Ruin at the End of Two Hours,'" *International Security,* 6.3 (Winter 1981-82).

Sakata, Shiyouchi. "A Dialogue Concerning New Views of Elementary Particles," *Hongqi* [Red Flag], No. 6, June 1, 1965.

Schlesinger, James R. *Report of the Secretary of Defense to the Congress on the FY 1975 Defense Budget and FY 1975-1979 Defense Program.* Washington, D.C., 1974.

Schomberg, Reginald. "The Climatic Conditions of the Tarim Basin," *The Geographical Journal,* 75.4 (April 1930).

Schram, Stuart, ed. *Chairman Mao Talks to the People.* New York, 1974.

Schwarz, Henry G. "The *Ts'an-k'ao Xiao-hsi* [*Cankao Xiaoxi*]: How Well Informed Are Chinese Officials About the Outside World?," *China Quarterly,* 27 (July-Sept. 1966).

Seven Letters Exchanged Between the Central Committees of the Communist Party of China and the Communist Party of the Soviet Union. Beijing, 1964.

Simpson, John. *The Independent Nuclear State: The United States, Britain and the Military Atom.* London, 1983.

"Sino-Soviet Joint Communique on Scientific and Technical Cooperation Accord," *Xin Hua Yuebao* [New China Monthly], Nov. 28, 1954.

The Sino-Soviet Treaty and Agreements. Beijing, 1950.

Smyth, Henry. *Atomic Energy for Military Purposes: The Official Report on the Development of the Atomic Bomb Under the Auspices of the United States Government, 1940-1945.* Princeton, N.J., 1945.

Song Erlian. "Develop the Iron and Steel Industry and the Non-Ferrous Industry at Top Speed," *Xin Hunan Bao* [New Hunan News], July 29, 1958.

State Statistical Bureau, comp. *Ten Great Years: Statistics of the Economic and Cultural Achievements of the People's Republic of China.* Beijing, 1960.

"Statement by the Government of the People's Republic of China" (Oct. 7, 1969), *Peking Review*, No. 41, Oct. 10, 1969.

"Statement of the Soviet Government Concerning Its Scientific, Technical, and Industrial Aid to Other Countries for Promoting Research on the Peaceful Uses of Atomic Energy," *Xin Hua Yuebao* [New China Monthly], Feb. 28, 1955.

Stead, F. W. "Instruments and Techniques for Measuring Radioactivity in the Field," in United Nations, *Proceedings of the International Conference on the Peaceful Uses of Atomic Energy*, Vol. 6: *Geology of Uranium and Thorium*. Document A/CONF.8/6. New York, 1956.

Stein, Aurel. "Explorations in the Lop Desert," *The Geographical Review*, 9.1 (Jan. 1920).

———. *On Ancient Central-Asian Tracks*. New York, 1964.

———. "A Third Journey of Exploration in Central Asia, 1913-16," *The Geographical Journal*, 2 parts, 48.2-48.3 (Aug., Sept. 1916).

Stolper, Thomas E. *China, Taiwan, and the Offshore Islands*. Armonk, N.Y., 1985.

"Strategic Guiding Principles for Strengthening the Build-up of National Defense, . . ." *Guangming Ribao* [Bright Daily], Jan. 20, 1977.

Stuewer, Roger H., ed. *Nuclear Physics in Retrospect: Proceedings of a Symposium on the 1930s*. Minneapolis, Minn., 1979.

Su Fangxue. "The Light of Life Is Brighter Than the Sun," *Shenjian* [Magical Sword], No. 5, Oct. 4, 1986.

Su Kuoshan. "An Interview with Zhang Yunyu, the Former Deputy Director of the Commission of Science, Technology, and Industry for National Defense," *Renmin Ribao*, overseas ed., July 29, 1987.

Suo Guoxin. "China: The Seventy-Eight Days of 1967," *Zhongguo* [China] (Changsha), No. 4, 1986.

"A Survey of the Chinese Academy of Sciences During the First Half of the Year," *Xin Hua Yuebao* [New China Monthly], Aug. 15, 1950.

Tan Wenrui. "Oppose the United States' Preparation for Atomic War," *Renmin Ribao*, Jan. 16, 1955.

Tao Cun. "A Visit to the Chinese Academy of Atomic Energy," *Jingji Ribao* [Economic Daily], Oct. 5, 1986.

"Ten Prominent Scientists," *Banyue Tan* [Semimonthly Talks], July 10, 1984.

"A Thoroughgoing Betrayal," *Renmin Ribao*, March 7, 1954.

Tien, Chang-Lin. "Engineering," in Leo A. Orleans, ed., *Science in Contemporary China*. Stanford, Calif., 1980.

"To Study Soviet Theory and Experiences Acquired in Socialist Construction Is an Important Task for All Party Cadres," *Renmin Ribao*, April 25, 1953.

"The Triumphal Song of Mao Tse-tung's Thought That Rings Ever Loud Across the Sky—On the Successful Explosion of China's First Hydrogen Bomb," *Jiefangjun Bao* [Liberation Army Daily], June 18, 1967, in FBIS: *Communist China*, June 20, 1967.

"Two Different Lines on the Question of War and Peace," in *Polemic*, listed above, pp. 221-57. Originally published in *Renmin Ribao*, Nov. 19, 1963.

Ullman, Morris B. *Cities of Mainland China: 1953 and 1958*. Washington, D.C., 1961.

"The Unexpected Discovery Before the Explosion of the Atomic Bomb," *Renmin Ribao*, overseas ed., Sept. 30, 1986.

United Nations. *Proceedings of the Second United Nations International Conference on the Peaceful Uses of Atomic Energy.* Document A/CONF.15/1. New York, 1958.

United States, Central Intelligence Agency. "The Chinese Communist Atomic Energy Program." National Intelligence Estimate 13-2-60 (Dec. 13, 1960).

——, ——. *Directory of Chinese Communist Officials.* Washington, D.C., 1966.

——, ——. *Directory of Chinese Officials: Scientific and Educational Organizations.* Washington, D.C., 1981.

——, Department of Energy. *Announced Foreign Nuclear Detonations—Through December 31, 1978.* Las Vegas, Nev., n.d.

——, Department of State. *Foreign Relations of the United States.* Washington, D.C. Vols. for the years 1949-57.

——, Department of the Air Force. *Guided Missiles Fundamentals.* AFM 52-31. Washington, D.C., 1972.

——, President's Blue Ribbon Commission on Defense Management. *A Formula for Action: A Report to the President on Defense Acquisition.* Washington, D.C., 1986.

——, Senate. *Executive Sessions of the Senate Foreign Relations Committee (Historical Series),* Vol. 7: Eighty-Fourth Congress, First Session 1955. Washington, D.C., 1978.

"The United States Is Afraid of Negotiation and Peace," *Renmin Ribao*, April 3, 1954.

Wakeman, Frederic. *History and Will: Philosophical Perspectives of Mao Tsetung's Thought.* Berkeley, Calif., 1973.

Wang Aiman et al. "Chidao de Baogao" [A Belated Report], in *Mimi Licheng*, listed above.

Wang Hanfu. "The Geological Work in Hunan Province Is Developing by Leaps and Bounds," *Xin Hunan Bao* [New Hunan News], Sept. 30, 1959.

Wang Jian et al. "Outline of Development of China's Uranium Mining Technology and Suggestions for Its Improvement," *He Kexue yu Gongcheng* [Nuclear Science and Engineering], No. 3, Sept. 1982, in *China Report: Science and Technology*, JPRS-CST-84-024 (Aug. 23, 1984).

Wang Jingtang. "Development Trend of Uranium Ore Processing Technology," *He Kexue yu Gongcheng* [Nuclear Science and Engineering], No. 3, Sept. 1984, in *China Report: Science and Technology*, JPRS-CST-85-001 (Jan. 3, 1985).

Wang Shouyun. "Defense Economics—A Realistic Question," *Baike Zhishi* [Encyclopedic Knowledge], No. 10, 1984.

Wang Xianjin et al. "The Stirring Song Composed Over the Mushroom Cloud," *Renmin Ribao*, overseas ed., July 1, 1986.

Wang Yanting et al. "The Relationship Between the Granite and Uranium Mineralization in South China," in *26ᵉ Congrès Géologique International: Résumés Abstracts*, Vol. 3. Paris, 1980.

Wang Yougong et al. "Take a Walk in the 'Science City,'" *Renmin Ribao*, Dec. 9, 1984.

Wang Zhen. "Review of the Liberation and Construction of Xinjiang," *Renmin Ribao*, Sept. 27, 1979.

Wang Zhiliang. *Yuanzineng Changshi Jianghua* [Popular Talks on Atomic Energy]. Tianjin, 1955.

Wang Zhuang. "Yu Fuhai, the Chief Bombardier of China's First Air-Exploded Atomic Bomb," *Guangjiaojing* [Wide Angle] (Hong Kong), Jan. 16, 1987.

"We Must Liberate Taiwan," *Renmin Ribao*, July 23, 1954.

"Welcome the Delegation of the Korean People to China," *Renmin Ribao*, March 15, 1954.

Whitaker, Donald P., and Rinn-Sup Shinn. *Area Handbook for the People's Republic of China*. Washington, D.C., 1972.

Whiting, Allen S. *China Crosses the Yalu: The Decision to Enter the Korean War*. New York, 1960.

——. *The Chinese Calculus of Deterrence: India and Indochina*. Ann Arbor, Mich., 1975.

——. "Quemoy 1958: Mao's Miscalculations," *China Quarterly*, 62 (June 1975).

Whiting, Allen S., and Sheng Shih-ts'ai. *Sinkiang: Pawn or Pivot?* East Lansing, Mich., 1958.

Whitson, William. *The Chinese High Command: A History of Communist Military Politics, 1927-71*. New York, 1973.

"Why We Must Participate in Korea," *Renmin Ribao*, Nov. 6, 1950.

World Bank. *China: Socialist Economic Development*. Vol. 1. Washington, D.C., 1983.

Wu Quan. "What Does Dulles' 'Comprehensive Foreign Policy' Mean?," *Renmin Ribao*, April 9, 1954.

——. "What Will Become of the U.S. 'New Look' Policy in the End?," *Renmin Ribao*, March 25, 1954.

Wu, T. Y. "Nuclear Physics," in Sidney Gould, ed., *Sciences in Communist China*. Washington, D.C., 1961.

Wu Xiuquan. *Eight Years in the Ministry of Foreign Affairs*. Beijing, 1985.

Xi Qixin and Liu Jingzhi. "From the Atomic Bomb to the Hydrogen Bomb: Two Years and Eight Months," *Guangming Ribao* [Bright Daily], July 28, 1987.

Xiang Jun. "Fengbei" [Monument], in *Mimi Licheng*, listed above.

Xie Linhe. "Premier's Zhou's Directives After the Atomic Bomb Detonation," *Shenjian* [Magical Sword], No. 4, Aug. 4, 1986.

Xu Honglie. "Lanzhou—A New Heavy Industrial Base," *Dagong Bao* [Impartial News] (Beijing), Aug. 22, 1957.

Xu Zhucheng. "Whether to Continue or Not Continue the Construction of the Baoshan Steel and Iron Complex," *Jingbao Yuekan* [The Mirror Monthly] (Hong Kong), Oct. 10, 1982.

Xue Jianhua. "China Puts Security in First Place in the Development of Nuclear Power Plants," *Renmin Ribao*, overseas ed., Aug. 27, 1986.

Yang Fangzhi and Wu Fanwu. "Reminiscences of Premier Zhou Enlai's Concern and Instruction Concerning the Foreign Experts' Work," *Renmin Ribao*, Jan. 9, 1981.

Yin Weixing. "The Weight That Shapes a Global Balance," *Zhongguo Qing-nian* [China Youth], No. 1-2, Jan. 9, 1987.

Yu Min. "Cherish the Memory of Comrade Deng Jiaxian," *Guangming Ribao* [Bright Daily], Aug. 24, 1986.

Zhang Aiping. "Deng Jiaxian's Illustrious Name Will Go Down in History," *Renmin Ribao*, overseas ed., Aug. 4, 1986.

———. "Memorial Speech Delivered at the Meeting in Honor of Comrade Deng Jiaxian," *Guangming Ribao* [Bright Daily], Aug. 4, 1986.

———. "Several Questions Concerning the Modernization of National Defense," *Hongqi* [Red Flag], No. 5, March 1, 1983.

Zhang Jiajun. "The Birth of China's Strategic Missile Troops," *Renmin Ribao*, overseas ed., May 7, 1987.

Zhang Jianzhi. "Views on Medium-Sized Nuclear Powers' Nuclear Strategy," *Jiefangjun Bao* [Liberation Army Daily], March 20, 1987.

Zhang Jiong. "March Toward a Bright China," in *Kexue de Chuntian* [Spring-time of Science]. Beijing, 1979.

Zhang Jun, chief ed. *Dangdai Zhongguo de Hangtian Shiye* [Contemporary China's Space Cause]. Beijing, 1986.

Zhang Zhishan. "How the Nuclear Bomb Base Was Built," *Renmin Ribao*, overseas ed., Aug. 22, 1985.

Zhao Fuxin and Zheng Shurong. "Xiwang Shi" [The Rock with Hope], in *Mimi Licheng*, listed above.

Zhao Zixing. "The Growth and Expansion of the Sino-Soviet Non-Ferrous and Rare Metals Corporation," *Renmin Ribao*, Jan. 2, 1955.

Zheng Jixu et al. "A Description of a Certain Air Transport Division of the Guangzhou Air Force Which Has Made Outstanding Contributions to the Modernization of National Defense," *Jiefangjun Bao* [Liberation Army Daily], June 24, 1987.

Zhou Enlai [Chou En-lai]. "Report on Foreign Affairs to the Central People's Government Council," in *Important Documents Concerning the Question of Taiwan*. Beijing, 1955.

———. *Report on the Question of Intellectuals*. Beijing, 1956.

———. *Zhou Enlai Xuanji* [Selected Works of Zhou Enlai]. Vol. 2. Beijing, 1984.

Zhou Jinhan. "The Results and Future of Scientific Research in Geophysical Prospecting in China," *Diqiu Wuli Kantan* [Geophysical Prospecting], No. 9, Sept. 22, 1959.

Zhou Yongkang. "Baoshan Iron and Steel Complex—The Newborn Iron and Steel Baby of China's Modernization," *Liaowang*, Sept. 16, 1985.

Zhu De. "Speech at the Celebration Ceremony on the 27th Anniversary of the Founding of the People's Liberation Army," *Guangming Ribao* [Bright Daily], Aug. 2, 1954.

Zhu Guangya. *Yuanzineng he Yuanzi Wuqi* [Atomic Energy and Atomic Weapons]. Shanghai, 1951.

Index

Index

In this Index an "f" after a number indicates a separate reference on the next page, and an "ff" indicates separate references on the next two pages. A continuous discussion over two or more pages is indicated by a span of page numbers, e.g., "57-59." "Passim" is used for a cluster of references in close but not consecutive sequence.

A-1 and A-2 type bombs, 152-53
Academia Sinica (Zhongyang yanjiu-yuan), Nanjing, 43f
Academy of Sciences, see Chinese Academy of Sciences
Accuracy, of missiles, 214
Acheson, Dean, 7
Adams, Sherman, 14
Administrative officials, 46-59 passim, 249-50, 263. See also Management; Organization
Aerial reconnaissance, in uranium explorations, 75
Aftertreatment (houchuli), in plutonium dissolver, 109, 112
Aircraft industry, 53, 72, 129, 234, 277
Air force, 133, 205-8 passim, 264
Alamogordo, New Mexico, bomb test at, 104, 226
An Dong, 50, 54, 263
"Analysis of Possible Courses of Action in Korea" (NSC 147), 255
Anderson, Oscar, 227
Anti-Chemical Corps, 186
Anti-Rightist campaign, 45-46, 51, 62, 141, 223, 237
Arkhipov, I. V., 63
Armies, see individual armies by name
Arms control negotiations, 12, 35-36, 64-65, 192-93; treaties from, 1, 12, 192-93, 195, 241, 266, 286
Artillery Corps, see Second Artillery Corps
Assembly, bomb, 168-69

Assistance, see Nuclear assistance, Soviet
Assured retaliation doctrine, 211
Atomic Energy Act (1946), U.S., 228
Atomic Energy Commission, U.S., 55
Atomic Energy for Military Purposes (Smyth Report), 278
August 1 Film Studio, 187
Australia, 26, 74, 257
Aviation Industrial Commission, 50f, 54, 264

Bai Wenzhi, 55, 111
Baldwin, Hanson, 256
Banian sidan plan, 211-12, 213
Baoshan Iron and Steel Complex, Shanghai, 107-8
Baotou, nuclear facilities in, 55, 97, 200n, 271, 274
"Basic National Security Policy" (NSC 162/2), 17
Beidaihe, Defense Industry Conference at, 129, 277
Beijing Iron and Steel Academy, 92
Beijing Military Region Command, 143
Beijing Nuclear Engineering Research and Design Academy, 100
Beijing Nuclear Weapons Research Institute (Beijing hewuqi yanjiusuo), 54, 140-50 passim, 159-64 passim, 196, 200, 207ff
Beijing University, 44, 106, 287
Beiping Academy (Beiping yanjiuyuan), 43, 44
Beryllium, 155-56, 269, 281

Bethe, Hans, 137, 226
Bo Yibo, 37, 47f, 95, 132
Bohr, Niels, 137, 226
Bombers, and weapons testing, 207-8
Border clashes, Sino-Soviet, 180, 216
Borisov (Oleg Borisovich Rakhmanin), 265
Born, Max, 44
Bosten Hu (Bagrax Hu), 283
Britain, *see* Great Britain
Buddhism, 171
Budget, military arsenal, 52, 107-18, 128, 234, 237, 282
Bulganin, Nikolai, 61
Bureaucratic conflict and communications, 126-34, 224, 234
Bureau of Architectural Technology (Jianzhu jishu ju), 47-48
Bureau of Geology, Hunan, 87
Bureau of Metallurgy, Hunan, 87

Cadres Bureau, Second Ministry of Machine Building, 59
California Institute of Technology, 44, 147
Camp David, "spirit of," 266
Canada, nuclear program of, 226, 290
Cankao Xiaoxi (Reference News), 256
Cankao Ziliao (Reference Materials), 256
Cao, Master Worker at Lop Nur, 187
Cao Benxi, 46, 100, 247
Central Committee, 49, 130, 145, 237; and science and technology programs, 45, 50-54 passim, 236-37; and Great Leap Forward, 51, 69, 237; and Soviet assistance, 64-65, 72, 123, 269; Ninth Academy and, 152f; and first atomic bomb test, 185; on military doctrine, 191, 211; Cultural Revolution Group under, 204; and Cultural Revolution, 204, 287. *See also* Politburo, Chinese
Central Military Commission, 45, 48f, 70-71, 117, 131, 237; and Taiwan Strait crisis, 25n-26n, 31n; and organization of nuclear program, 49f, 52, 54, 59, 233, 264; and conventional weapon research, 50, 126, 128, 263; and Lanzhou Gaseous Diffusion Plant, 115; and Soviet pullout, 123; Lin Biao and, 133f; and Lop Nur test base, 176-81 passim; and Cultural Revolution, 204; and weaponization of nuclear program, 207; and air-dropped atomic bomb, 208; and missiles, 212-16 passim

Central Secretariat, and nuclear weapons decision, 38-39
Central South Institute of Mining and Metallurgy, 95
Central South Region uranium exploration (team 309), 76, 77-78, 87
Central Special Commission, *see* Fifteen-Member Special Commission
Chagatai, 170
Chain reaction, fission, 138, 198
Changzheng-1 (CZ-1) launcher, 213
Chemical separation, for plutonium, 104, 109, 112
Chen Boda, 204
Chen Geng, 54, 63, 142, 264
Chen Jianhua, 46
Chen Nengkuan, 146-55 passim, 159f, 187f, 247, 279-80
Chen Yi, 49ff, 72, 123, 130, 144, 150, 194, 246, 277
Chen Yun, 47, 144
Cheng Kaijia, 151, 183
Chenxian Uranium Mine, 78, 81-89 passim, 269ff
Chiang Kai-shek, 11, 20f, 25, 29, 31, 63, 144, 258
China Particle Accelerator Society, 271
Chinese Academy of Atomic Energy Science, 261
Chinese Academy of Sciences, 42-47 passim, 106, 134, 223, 272, 287; Institute of Physics, 37, 47, 99, 146, 261; Institute of Modern Physics, 43, 47, 148; Institute of Atomic Energy, 47, 59, 99; Defense Science and Technology Commission and, 54; and Scientific Planning Commission, 263
Chinese Communist Party, 5-6, 70, 117, 123, 144f, 219, 223, 232; and first five-year plan, 11; anti-intellectualism of, 45-46, 51, 144, 225, 230, 236; and Great Leap Forward, 68-69, 117, 219, 236; and uranium mining and production, 86f, 95f; and Lop Nur test base, 177-78; and hydrogen bomb, 201; and Cultural Revolution, 203. *See also* Central Committee; Central Military Commission; Politburo, Chinese
Chinese Peace Committee, 40
Chinese People's Volunteers (CPV), 8f, 15, 145, 254, 264
Chinese Physics Association (Zongguo wuli xuehui), 43f
Civilian science and technology, 53
Civil war, Nationalist-Communist, 73,

116, 144, 174; Kuomintang prisoners and fugitives from, 179, 184

Clark, Mark, 255

Cleanliness, plant, 112, 119-20, 275

Climate: in mining areas, 85; of Lanzhou, 119; in Subei, 163; at Lop Nur, 181, 184-85

Cockcroft, John, 228, 290

"Cold tests," 201

"Comments on Formulating Our Country's Plan for the Atomic Energy Cause," 48

Commission of Science, Technology, and Industry for National Defense, 128n

Communes, people's, 87

Communist Party, *see* Chinese Communist Party

Communist Youth League Farm, 284

Computing systems, 154-55, 199, 200-201, 280-81, 287

Computing Technology Institute, Beijing, 280-81

Conflict: Chinese bureaucratic, 126-34, 224; Sino-Soviet, 180, 216f, 270, 276. *See also* Khrushchev; Mass movements

Congress, CCP, Eighth, 51, 68-69, 117

Congress, U.S., and Taiwan defense, 31-37 passim

Construction Bureau, Second Ministry of Machine Building, 55, 115

Conventional weapons, 49ff, 69, 114n, 126-32 passim, 263, 277. *See also* Aircraft industry

Cornell University, 147

Costs: of Chinese nuclear program, 52, 107-8, 114-15, 118, 234, 282; of U.S. nuclear program, 108, 274

Critical mass, 138

Cuban missile crisis, 193-94, 214, 216

Cultural Revolution, 46, 191, 195, 224, 232f, 236, 287; and conventional vs. strategic weapons contest, 128, 132; and hydrogen bomb, 201-6, 210; and missile program, 202-3, 214; and underground testing, 285

Cultural Revolution Group, 204

Curie Institute, Paris, 44, 156

Cyclotron, 261; Soviet supply of, 41, 47, 76, 99

CZ-1 launcher, 213

Dachen (Tachen) Islands, 26n, 27-34 passim, 259

Dai Zhuanzeng, 262

Dangdai Zhongguo de He Gongye (Con-temporary China's Nuclear Industry), 49, 253

Dapu Uranium Mine, Hunan, 78

Daqing oil field, 199, 287

Decision: on nuclear weapons program (Jan. 1955), 2, 4, 34-46 passim, 69, 76, 99, 105, 140, 142, 229f; on hydrogen bomb, 197-99

Decision making: scientific and technical, 145-46, 232. *See also* Organization

Defense Department, U.S., 213n, 215

Defense industries (general): Great Leap Forward and, 53; Soviet, 61; conventional and strategic weapons competing in, 126-31 passim; Cultural Revolution and, 287. *See also* Technology

Defense Industry Conference, Beidaihe, 129, 277

Defense policies: Chinese, *see* Military doctrine; U.S., 13-37 passim, 41, 195, 256-61 passim, 286. *See also* Nuclear policies

Defense Science and Technology Commission, 54, 59, 98-99, 102, 126-34, 203f, 223, 234, 237

Deng Jiaxian, 146, 148-49, 247, 262, 280; and nuclear component, 160-62; and thermonuclear weapons, 196, 200, 201-2; training by, 237, 287

Deng Liqun, 174

Deng Qixiu, 55

Deng Xiaoping, 49, 50-51, 116f, 123, 141, 235

Deng Zhaoming, 55

Department of Defense, U.S., 213n, 215

Dependence in nuclear program, Chinese on Soviets (1955–58), 7-12 passim, 41-42, 60-65, 71, 217-22 passim; British on U.S., 220. *See also* Nuclear assistance, Soviet

Design, 59; uranium-processing, 92; atomic bomb, 130, 137-65 passim, 225; hydrogen bomb, 199-200, 205-6; weapons, 207ff; missile, 208

Design Academy (later Bureau), Second Ministry of Machine Building, 55, 59

Deterrence: in Taiwan Strait crisis, 27-30 passim; with missiles, 198, 211, 215f; "calculus of," 288

Detonation system: atomic bomb, 139, 153-60; hydrogen bomb, 200; missile-carried thermonuclear weapon, 209

Deuterium, 200n

DF missile series, 202-3, 209-14 passim, 238, 289

V *see* Joliot-curie

Di Zhongheng, 268
Dialectics, and nuclear weapon decision, 38
Dienbienphu, 21, 258
Dinerstein, Herbert, 190
Ding Shufan, 103
Disarmament: Joliot-Curies and, 36; Chinese press on, 192; peace movement and, 195. *See also* Arms control negotiations
Dissolver, in plutonium production, 109
Dongfeng, see DF missile series
Dress codes, 59
Duan Junyi, 133
Dubna, *see* Joint Institute for Nuclear Research
Dulles, John Foster, 13, 17-26 passim, 33f, 40, 256f

East China Computer Institute, 287
Eastern Europe: Soviet nuclear aid to, 41; uranium geologists from, 73; scientists training in, 106, 237. *See also individual countries by name*
Eastern Turkestan People's Army, 174
East Zhejiang Front-Line Command, 25n, 31n
Eighth Route Army, 116, 144
Eight-Year Plan for the Development of Rocket Technology, 212
Einstein, Albert, 190, 285
Eisenhower administration, 266; and Korean War, 10-20 passim, 24; New Look policy of, 16-20 passim, 195, 256; and Taiwan policy, 17-40 passim, 257f; and Geneva Conference, 19, 21f; and arms control negotiations, 65
Eleventh Bureau, *see* Information Bureau
Engineering, of fission weapons, 139-40
Engineering Corps, PLA, 179-80, 181
Engineers, 130, 219-20, 225, 232f, 235, 271, 272-73; training abroad of, 43n, 105-6; for atomic bomb, 125, 147-61 passim, 182, 271, 280; for hydrogen bomb, 200, 271
Enrichment, of uranium, 52, 113-21, 134-36, 162, 225, 274, 277; Soviet equipment for, 41, 62, 115-20 passim, 224f; by gaseous diffusion, 62, 100, 103f, 114-21, 125, 135, 274. *See also* Lanzhou Gaseous Diffusion Plant
Environmental protection, 55
Equipment: uranium plant, 41, 62, 92f, 115-20 passim, 125, 165, 224; uranium mining, 85; costs of, 108, 114-15

Equipment Manufacturing Bureau, Second Ministry of Machine Building, 55
Equipment Planning Department, General Staff, 51, 263f
Europe: Chinese scientists in, 36, 73, 279. *See also areas and countries by name*
"Experts Work Group," 261
Explosive assembly, 153-60 passim, 164, 168, 183, 186
Explosives, 154, 280
Extract Poisonous Nails General Headquarters (Ba duding zongbu), 203

Factions and personal networks, 116-21 passim, 126-34 passim, 223
Fallout, from bomb tests, 186, 187-88, 285
Fang Qiang, 126n
Fang Zhengzhi, 146, 148, 150
Faxian, 171
Females, at Lop Nur, 181
Fermi, Enrico, 137, 226
Feynman, Richard, 280-81
Fifteen-Member Special Commission, 131-34, 198, 208f, 224, 233, 277; and hydrogen bomb, 114n, 198, 201, 204-5; and atomic bomb, 159, 164-65, 178, 182, 284; and missiles, 198, 211-12
Fifteenth Bureau, *see* Lanzhou Gaseous Diffusion Plant
Fifth Academy (Diwu yanjiuyuan), Ministry of National Defense, 50, 54, 91n, 134, 211-12, 234
Fifth Department, Ministry of National Defense, 54, 264
Fifth Institute, 91. *See also* Sixth Institute
Final Declaration of the Conference on Indochina, 28
Finance Bureau, Second Ministry of Machine Building, 59
Finances, *see* Budget; Costs
First Atomic Bomb Test Commission (Yuanzidan shouci shiyan weiyuanhui), 51, 178, 182, 184
First Atomic Bomb Test On-Site Headquarters (Yuanzidan shouci shiyan xianchang zhihuibu), 54, 178, 184
First Field Army, 174
First Ministry of Machine Building, 263
"First write" instruction, 118, 165, 275
Fission, 137-40 passim, 156-57, 197-98, 226-27, 281, 290
Fissionable material, 104-36, 153, 156,

197-98; and Soviet aid, 41, 105, 109-25 passim. *See also* Plutonium; Uranium isotopes
Fission chain reaction, 138, 198
Fission-fusion-fission bomb, 198
Five-Year Plan: First, 11, 107; Second, 107
Fleet request, Soviet, 63, 70, 265
Food: for nuclear program personnel, 86, 103, 122, 124, 143, 178-79, 276
Food crisis, national, 103, 122, 143, 178-79
Forbidden City, Zhongnanhai in, 38
Formosa, *see* Taiwan
Formosa Resolution, 32-37 passim
Fourth Research Section, Sixth Institute, 91, 93
France, 12, 21f, 26, 74, 156f; nuclear program of, 2, 202, 227, 290; Joliot-Curies in, 36, 38, 44, 137, 156
Frankel, Stanley, 281
Frenkel', Yakov, 137, 226
Frisch, Otto, 137, 226
Fuels Production Bureau, Second Ministry of Machine Building, 55, 100, 111
Fusion, 137-38, 196, 197-98, 200n, 290

Gansu Province, *see* Jiuquan Atomic Energy Complex; Lanzhou
Gao, deputy chief of Second Ministry's security department, 183
Gao Yang, 133
Gaseous diffusion, 62, 100, 103f, 114-21, 125, 135, 274; in U.S., 114f, 274. *See also* Lanzhou Gaseous Diffusion Plant
Geiger counters, 74ff, 268
General Cadres Department, Central Military Commission, 49, 117
General Staff, 51, 126, 131, 142, 164, 263f
Geneva Conference on Korea and Indochina (1954), 11, 19-22, 28
Geneva convention, on POWs, 15
Genghis Khan, 170
Geological Bureau, Second Ministry of Machine Building, 54, 81, 90, 268
George, Alexander, 9
Germany, 2, 146, 226-30 passim, 242, 279
Gittings, John, 27-28
Gobi Desert, 111, 162-63, 175f, 282
Gonganjun (Security Forces), 215
Gowing, Margaret, 227, 229, 231
Granite, uranium in, 74, 78, 80f

Great Britain, 1, 25f, 147, 192, 241, 266; nuclear weapons program of, 2, 91-92, 139, 199, 226-31 passim, 281, 290
Great Leap Forward, 53, 68-69, 94, 127, 219, 233; and science and technology, 51, 117-18, 141, 165, 236f; and uranium mining, 87-88, 233; and Lanzhou Gaseous Diffusion Plant, 117-18, 122; and interdependence, 223
Great Proletarian Cultural Revolution, *see* Cultural Revolution
Green salt, *see* Uranium tetrafluoride
Groves, Leslie R., 278
Gu Mainan, 272
Gu Wenfu, 272
Gu Yu, 277
Guam, missile targeted at, 212f
Guangdong Province, uranium programs in, 80-81, 85-89 passim, 94, 269
Guangxi Province, uranium explorations in, 75f
Guangzhou Military Region Command, 143
Guidance systems, missile, 214
"Guidelines for Developing Nuclear Weapons," 70
Gun-barrel type bomb, 104, 138, 152-53
Guo Moruo, 40, 42-43, 263, 272
Guo Yinghui, 52, 143, 145, 249
Guo Yonghuai, 46, 145-55 passim, 161, 188, 247, 279

Hahn, Otto, 137, 146, 226
Halperin, Morton, 34
Harbin Military Engineering Institute, 203-4, 264
He Dongchang, 106
He Long, 71, 126-34 passim, 246
He Xianjue, 206
He Xiaolu, 265
He Zehui, 44, 158, 200, 261, 268
Health hazards, 55, 85f, 94, 186, 187-88, 270, 273-74, 285
Hedin, Sven, 282
Heisenberg, Werner, 230, 279
Hengshan County, Hunan, 78
Hengyang, 271
Hengyang Institute of Uranium Mining and Metallurgy, 85
Hengyang Uranium Hydrometallurgy Plant (Hengyang you shuiye chang, No. 414, later 272), 85, 95-97, 271
Hengyang Uranium Mining and Metallurgy Design and Research Academy, 95
Hewlett, Richard, 227

High Energy Physics Institute, 271
Hiroshima bomb, 104, 197, 230
Ho Chi Minh, 28
Holloway, David, 66, 225, 227, 230f, 278
Hong 6 bomber, 207-8
Hongqi, 38, 69, 191-92
Hospitals, Security and Protection Bureau and, 55
Hou Qi, 86
Hou Xianglin, 125
Housing, in mining areas, 85
Hsieh, Alice Langley, 194, 260
Hu, assistant research fellow, 183
Hu Jimin, 106
Hu Ning, 262
Hua Luogeng, 261
Huai-Hai campaign, 116
Huang Changqing, 100-103 passim, 272
Huang Kecheng, 50
Huang Zuqia, 45, 196, 262
Hunan Province: uranium exploration, mining, and plants in, 47, 78-89 passim, 94f
Hundred Regiments campaign, 116f, 275
Hung Fan-ti, 195
Hungary, 61ff, 76
Hydrogen bomb: U.S., 20, 104, 197, 199, 230; Chinese, 114n, 196-206, 210, 267, 288

Ichiang (Yijiang), 26n, 31f, 37
Imitation, 226; Mao on, 71; and cost, 108; in bomb design, 140
Imperialism, U.S., and Chinese nuclear policies, 1-8 passim, 65, 68f, 191-92, 195, 241f, 286
Implosion, 104, 138-39, 152-53, 155, 161f, 278, 281
Independence: Chinese military, 35, 42, 65-72, 121-25 passim, 220-29 passim, 235, 237-38; British nuclear, 228-29
Independent Fourth Regiment, air force, 207f
India, 71-72, 127, 288
Indochina, 12, 17-22 passim, 26-35 passim, 127, 191, 224, 257, 288; armistice (July 21, 1954), 25
Information Bureau (also Information Institute; Eleventh Bureau), Second Ministry of Machine Building, 55, 278
Initiator, 139, 153-60 passim, 164, 168, 186, 225, 284
Innovation: technical, 100-103 passim,

118, 140, 155, 225; organizational, 121, 223, 225, 232
Institute of Atomic Energy (Code 601, later 401), 37, 59, 100-103, 118, 120n, 135, 146, 157-58, 261, 271; uranium hexafluoride produced by, 98-103 passim, 125, 135; reactors at, 99-103 passim, 112, 261, 272; and hydrogen bomb, 196, 200; electromagnetic separators experimented with, 274
Institute of Mechanics, 147
Institute of Modern Physics, 43, 47, 146, 148, 261
Institute of Physics, Academia Sinica, 44
Institute of Physics, Beiping Academy, 44
Institute of Physics, Chinese Academy of Sciences, 37, 47, 99, 146, 261
Intellectuals, criticism of, 45-46, 144, 225, 230, 236f. *See also* Anti-Rightist campaign; Scientists
Interdependence, nuclear: Sino-Soviet, 222-23, 228; U.S.-British, 228
Irving, David, 227, 229
Islam, 171
Isolation, of nuclear program personnel, 46, 235

Japan, 73, 104, 116, 197, 212, 281
Jet Propulsion Laboratory, California Institute of Technology, 44
Jiang Nan, 256
Jiang Shengjie, 46, 166, 247-48, 274
Jiang Tao, 55
Jiang Zhenghao, 255
Jiangxi Province, uranium mining and processing in, 75, 78, 80, 94f
Jin Xingnan, 262
Jiuquan Atomic Energy Complex (Jiuquan yuanzineng lianhe qiye, Plant 404), 54f, 103n, 107-13 passim, 121, 162-69, 182, 196, 204, 273, 276
Johnson, Lyndon B., 1-2, 229
Johnson, U. Alexis, 28
Joint Chiefs of Staff, U.S., 13, 21-22, 29, 215
Joint Institute for Nuclear Research, Dubna, 105, 123, 146, 149, 272
Joliot-Curie, Frédéric, 36, 38, 137
Joliot-Curie, Irène, 36, 38, 44, 137, 156

Kennan, George, 9-10
Kennedy, John F., 195
Khariton, Yu. B., 230

Khlopin, V. G., 229
Khrushchev, Nikita, 26, 60-72 passim, 123, 150, 217, 277, 280
Klochko, Mikhail, 267
Knowledge, in Mao's worldview, 5-6
Koo, Wellington, 257
Korean Armistice (1953), 4, 14, 15-16, 28f, 255f
Korean reunification, 19
Korean War (1950-53), 4-20 passim, 145, 211, 264; and Sino-Soviet cooperation, 7ff, 41, 127; U.S. and, 8-20 passim, 24, 27, 255; and nuclear weapons decision, 35, 73
Kuomintang, *see* Nationalists
Kurchatov, Igor, 226, 230

Laboratory No. 131, Institute of Atomic Energy, 101
Labor costs, 108
Lanzhou Chemical Physics Institute, 280
Lanzhou Gaseous Diffusion Plant (Fifteenth Bureau, Plant 504), 55, 100, 107, 113-25 passim, 130, 134-36, 162-66 passim, 274, 276; uranium hexafluoride supply for, 62, 103, 125, 135
Lapp, Ralph E., 285
"Law of the inevitability of war," 66
Leaching heap, 84-85
Leadership, 3f, 220-25 passim, 229. *See also* Administrative officials; Political leadership
Lei Rongtian, 47, 54, 76ff, 254, 262
LeMay, Curtis, 18
Lenin, V. I., 38, 66, 191, 223
Lermontov Plant, Soviet, 92, 271
Li, engineer at Jiuquan Atomic Energy Complex, 168f
Li Fuchun, 62, 132
Li Hao, 97
Li Jinqi, 271
Li Jue, 52, 54, 140-55 passim, 159f, 183-88 passim, 249, 254
Li Shounan, 261f
Li Siguang, 37f, 47, 261
Li Xiannian, 132
Li Yi, 99, 103, 254, 271
Li Yingxiang, 254
Lianxian, uranium programs in, 80-81, 87, 89, 269
Liaoning, uranium in, 75f, 87f
Liberation movements, 6, 67-68, 194, 242

Limited nuclear retaliation, 216
Limited test ban treaty, 1, 192-93, 195, 241, 266, 286
Lin Biao, 128f, 133f, 254, 264
Lithium, 155-56, 269
Lithium-6 deuteride, 200
Liu Bocheng, 9, 116f
Liu Hua, 55
Liu Jie, 37f, 47, 75, 102, 132n, 164-65, 249, 287; and food crisis, 103, 122, 143; and Lanzhou Gaseous Diffusion Plant, 121f, 136; nuclear ministry offices of, 126n, 133, 150, 262, 270, 276, 287; and atomic bomb test, 185, 188-89
Liu Jingqiu, 96-97
Liu Kuan, 78, 85f
Liu Ningyi, 195
Liu Qisen, 287
Liu Shaoqi, 46, 113, 126, 128f, 277
Liu Shuqing, 284
Liu Wei, 47f, 262
Liu Xiyao, 126n, 143, 169, 184, 188, 196-201 passim, 249, 275, 282
Liu Yalou, 54, 133, 264
Liu Yunbin, 46
Liu Zhe, 116f, 120
Long Wenguang, 147, 151, 248
Lop Nur Nuclear Weapons Test Base, 54, 149, 170-89, 203-8 passim, 232, 282-83; first atomic bomb test at, 1, 149, 182-89; first hydrogen bomb test at, 104-6, 288; missile test at, 202-3
Los Alamos, New Mexico, 139, 141, 151, 280-81, 287, 290
Loulan, 170, 282
Lu Dingyi, 132
Lu Fuyan, 91f, 93, 248
Lu Zuyin, 262
Luo Pengfei, 80-81, 269
Luo Ruiqing, 126-34 passim, 164, 246, 263, 277

Ma Zhengyi, 59
Magical Sword Literary and Art Society, 253
Magnetic separator facility, for uranium, 89
Malan, 180
Malaria, in mining areas, 85
Males, at Lop Nur, 181
Management, 3f, 220, 233-35; of Lanzhou plant, 120-21; of Ninth Academy, 150-53. *See also* Decision making

Manhattan Project, 226f, 230, 278
Manila Conference, 22-29 passim
Mao Yuanxin, 204
Mao Zedong, 96, 121, 124, 130, 174,
216, 221, 230-37 passim, 246, 287;
worldview and military doctrine of,
5-7, 9, 29, 35-42 passim, 60, 65-69,
127f, 190-91, 194, 210-18 passim,
242, 260, 265f, 290; and U.S. nuclear
threat, 6-7, 11f, 36-37, 40, 60, 64-69
passim, 217; and Soviet assistance,
6-7, 39, 42, 60, 62f, 69, 70-71, 122-
23, 221; and Korean War, 7-8, 9, 255;
and Taiwan Strait crisis, 29, 64, 265;
and nuclear weapons decision, 35-42
passim, 69, 142; and self-reliance, 42,
65-72, 123, 205, 221; and Great Leap
Forward, 68-69, 117f, 223, 237; and
Central Military Commission, 70-71,
128, 178; and uranium explorations
and production, 75, 136; and cost of
nuclear program, 107; "first write" instruction by, 118, 165; on conventional vs. strategic weapons, 127ff; and
Fifteen-Member Special Commission,
131f, 224; and Two-Year Plan, 164;
and "key link," 166, 218; and atomic
bomb assembly, 169; and detonation
of first bomb, 188; and hydrogen
bomb, 198, 205, 210, 267; and Cultural Revolution, 201-4 passim, 236;
and missile-carried thermonuclear
weapon, 209; and organization of nuclear program, 223, 264
Marco Polo, 171
Massive retaliation doctrine, U.S., 17, 20,
32, 195, 256, 261
Mass movements, 87-88, 118, 223, 233.
*See also individual movements and
campaigns by name*
Matsu, 22-34 passim, 64, 259f
Maud Report (July 1941), 227
Meitner, Lise, 44, 137, 146, 226
Metallic uranium, 90. *See also* Uranium
hexafluoride
Meteorological station, Yangpingli, Xin-
jiang, 181, 284
MiG-21 fighter plane, 72
Mikoyan, Anastas, 61
Military Academy, Nanjing, 9
Military control, in Chinese nuclear pro-
gram, 46, 48, 54, 59, 217-18, 231-32,
234
Military doctrine, 190-96, 210-18, 242;

Mao's, 6-7, 9, 29, 35-42 passim, 60,
65-69, 127f, 190-91, 194, 210-18 pas-
sim, 242, 260, 265f, 290; Soviet, 39-
40, 41, 66-68, 217. *See also* Deter-
rence; Nuclear policies
Military Science Academy, 131
Military Thought, 66
Mimi Licheng, 49, 279-80
Mines, uranium, 55, 73-89 passim, 233,
268-71 passim
Mining and Metallurgy Bureau (Twelfth
Bureau), Second Ministry of Machine
Building, 55, 85, 90
Mining techniques, uranium, 81-85
Ministry of Chemical Industry, 200n
Ministry of Coal Industry, 86
Ministry of Communications, 143
Ministry of Education, 45, 106, 236
Ministry of Electrical Machine Industry,
262f
Ministry of Foreign Affairs, 14, 106
Ministry of Geology, 37, 47, 54, 75-78
passim
Ministry of Geology and Mineral Re-
sources, 268
Ministry of Health, 55
Ministry of Heavy Industry, 144
Ministry of Machine Building, *see* Sec-
ond Ministry of Machine Building;
Third Ministry of Machine Building
Ministry of Metallurgical Industry, 200n,
268
Ministry of Metallurgy, Third Depart-
ment of, 55
Ministry of National Defense, 50, 54, 59,
91n, 134, 211-12, 234, 264
Ministry of Nuclear Industry, 2, 253, 269
Ministry of Ordnance Industry, 263
Ministry of Railways, 101
Ministry of State Farms and Land Reclamation, 175
Minority nationalities, in PC Corps,
179-80
Missile crisis, Cuban, 193-94, 214, 216
Missiles, 48, 50, 52, 129, 138, 147, 277,
288f; Soviet aid with, 41n, 72, 212;
organization of, 50, 59, 91n, 127, 134,
211-12, 234; test facilities for, 175,
202-3; and military doctrine, 191,
210-18; weaponization and, 191, 198,
202-3, 208-9; DF series of, 202-3,
209-14 passim, 238, 289; fuels for,
212, 214; targets and ranges of, 212-
14, 289

Modernization plans, 9ff, 43, 49-51, 233
Monel alloy, 101, 271
Mongol groups, in Xinjiang Uygur Autonomous Region, 172-74
"*Mozhe shitou guohe*" (In crossing the river go stone by stone), 123f, 276
Mutual retaliation, doctrine of, 67

Nagasaki bomb, 104, 281
Nanjing, 9, 43f
Nanling mountains, uranium prospecting in, 80f, 269
National Defense Industrial Commission, 59, 126-34 passim
National Defense Industry Office, 59, 126-34 passim, 276
National Defense Industry Special Commission (Guofang gongye zhuanmen weiyuanhui), *see* Fifteen-Member Special Commission
Nationalism, 4, 35, 123, 226
Nationalists, 73, 116, 144, 174, 179, 184. *See also* Chiang Kai-shek; Taiwan
National liberation struggles, 6, 67-68, 194, 242
National Security Council, U.S., 17, 24, 31f, 259
NATO, 228
NDIC, *see* National Defense Industrial Commission
NDIO, *see* National Defense Industry Office
Nehru, Jawaharlal, 14, 256
Neptunium-239, 108
Networks, personal, 116-21 passim, 126-34 passim, 223
Neutron Physics Leading Group, 286
Neutrons: in fission, 156-57, 281; classification of, 281
New China News Agency, 255
New Defense Technical Accord (Oct. 15, 1957), 41n, 62-63, 65, 69-70, 89, 115, 141
New Look policy, U.S., 16-20 passim, 195, 256
New Technological Bureau, Chinese Academy of Sciences, 134
New Zealand, 26, 257
Nickel alloys, 98
Nie Rongzhen, 2, 143, 147, 174, 180, 183, 186, 247, 279; and mass movements, 46, 53, 202ff, 223, 233, 236f; and organization, 47-54 passim, 126-34 passim, 220, 223, 230-37 passim,

263f; Defense Science and Technology Commission of, 54, 59, 98-99, 126-34, 203, 223, 234, 237; and Soviet assistance, 62-65 passim, 69-72 passim, 99; and uranium processing, 90, 99; and plutonium production, 111; and Fifteen-Member Special Commission, 132ff, 233, 277; and thermonuclear weapons, 196, 198, 202-5 passim, 209; and military doctrine, 211, 216; summing-up, 219, 236-38
"Nine-point law," 122
Ninth Academy, 121, 148, 198, 209, 287; and atomic bomb, 54, 140-64 passim, 169, 178, 182-85 passim, 189, 196; and hydrogen bomb, 196-201 passim, 205
Ninth Bureau, Second Ministry of Machine Building, 54, 140ff, 151, 207
Nixon, Richard M., 20, 257
No-first-use policies, 194, 215, 242
Non-Ferrous and Rare Metals Corporation, Sino-Soviet, 76-77
Non-Proliferation of Nuclear Weapons Treaty (July 1, 1968), 12
North Korea, 7-8, 15, 32
Northwest Nuclear Weapons Research and Design Academy (Xibei hewuqi yanjiu sheji yuan), *see* Ninth Academy
Northwest People's Liberation Army, 174
Nuclear assistance, Soviet, 2, 39-45 passim, 51ff, 60-65, 69-72 passim, 220-22, 228, 235, 265, 267; with reactors, 41, 47, 76, 99, 105, 112, 272; and prototypes, 41n, 61-65 passim, 140f, 150, 222, 280; with missiles, 41n, 72, 212; end of, 60-61, 65, 71-72, 89-100 passim, 113, 121-25, 129, 150, 160-61, 200n, 223-24, 228, 267, 270, 276; and Sixth Institute, 73, 271; and costs of nuclear program, 108, 114-15; and cleanliness, 119-20, 275; Ninth Academy and, 141, 145, 150, 152, 160-61; and nuclear component, 160-61, 163, 165
Nuclear component: design and manufacture of, 160-69; transport of, 183; at Lop Nur, 186, 284
Nuclear Component Manufacturing Plant, Jiuquan Atomic Energy Complex, 54, 111, 162-69, 204
Nuclear Education Leading Group, 106
Nuclear Fuel Component Plant (He ran-

liao yuanjian chang, Plant 202), Bao-
tou, 97, 200n, 271, 274
Nuclear Fuel Processing Plant, Jiuquan
Atomic Energy Complex, 111, 162
Nuclear ministry, *see* Ministry of Nuclear
Industry; Second Ministry of Machine
Building; Third Ministry of Machine
Building
Nuclear policies: Chinese, 1-2, 34-42
passim, 51, 65-69 passim, 190-96,
210-18, 241-43, 277, 286, 288; Soviet,
12, 39-40, 41, 51, 66ff, 105, 217, 278;
U.S., 16-17, 20, 32, 41, 195, 256, 261,
278
Nuclear threat, U.S., 6-7, 11-22 passim,
32-41 passim, 195, 229, 231, 241-42,
256, 261; Mao and, 11f, 36-37, 40,
60, 64-69 passim; and Chinese nuclear
policy, 34-41 passim, 65-69 passim,
211, 215, 217; Cuban missile crisis
and, 193-94; and thermonuclear weapons, 198
Nuclear weapons: Chinese decision for
(Jan. 1955), 2, 4, 34-46 passim, 69,
76, 99, 105, 140, 142, 229f; British
programs, 2, 91-92, 139, 199, 226-31
passim, 281, 290; Soviet programs, 2,
104, 114, 137-40 passim, 198f, 227-31
passim, 256, 278, 290; German programs, 2, 226-30 passim, 279; Project
02, 39, 76, 96, 234, 237; vs. conventional weapons, 126-32 passim, 277;
weaponization and, 130f, 207-10, 287.
See also Hydrogen bomb; Tests; United
States nuclear program
Nuclear Weapons Bureau (Ninth Bureau), 54, 140ff, 151, 207
Nuclear Weapons Research Institute, *see*
Beijing Nuclear Weapons Research
Institute
Nutrition, *see* Food

Oak Ridge (Tenn.) gaseous diffusion
plant, 114f
"On Contradiction" (Mao), 287
"One-time test, overall results" (*yici
shiyan, quanmian shouxiao*) slogan,
107
"On Practice" (Mao), 287
"On the Ten Major Relationships"
(Mao), 211
Oppenheimer, J. Robert, 85, 141
Organization, 3f, 46-59, 106-7, 126-34,
220-25 passim, 232-37 passim, 262-

64; Politburo and, 47-54 passim, 106-
7, 131ff, 164, 221-22; Nie Rongzhen
and, 47-54 passim, 126-34 passim,
220, 223, 230-37 passim, 263f; of nuclear weapons program (charts), 56-
58; innovation in, 121, 223, 225, 232;
of Ninth Academy, 150-55; multiformity in, 223. *See also* Management
"Outline Plan for Developing the Atomic
Energy Cause, 1956–1967 (Draft),"
48

Pakistan, 26
Panmunjom armistice talks, 14, 255
Particle accelerator, Soviet, 105
Party Congress, Eighth, 41, 51, 68-69,
117
PC Corps, 174f, 179, 181, 283f
Peaceful coexistence policy, Soviet, 68
"Peaceful utilization," of atomic energy,
41, 51, 99, 105
Peace movement, 195. *See also* Dis-
armament
Peierls, Rudolf E., 226
Peng Dehuai, 8f, 174, 254, 264, 275, 280
Peng Huanwu, 44-45, 46, 99, 237, 248,
261; and atomic bomb, 145-51 pas-
sim, 157f, 161, 188, 226, 279-80
Peng Zhen, 50, 123
Penney, William, 228, 290
People's Daily, see Renmin Ribao
People's Liberation Army (PLA), 59, 116,
141-42, 174-75, 275, 277, 285; in
Korean War, 9f; in Taiwan Strait crisis,
37; General Staff of, 51, 126, 131,
142, 164, 263f; Defense Science and
Technology Commission and, 54; air
force of, 133, 205-8 passim, 264; Rail-
way Corps of, 143, 175, 179, 181; PC
Corps of, 174f, 179, 181, 283f; and
Lop Nur Nuclear Weapons Test Base,
175-81; Engineering Corps of, 179-80,
181; Second Artillery Corps of, 191,
202, 207, 213, 215f, 234
Personal motives, for individual commitment, 235
Personal networks, 116-21 passim, 126-
34 passim, 223
Pevtsov, Mikhail, 172
Philippines, 22-29 passim, 212f
Physics, nuclear, 38, 43-45, 146, 272
Physics and Chemistry Engineering Acad-
emy, Tianjin, 120n
PLA, *see* People's Liberation Army

Planning Bureau, Second Ministry of Machine Building, 59
Plant Two (uranium oxide production plant), 92-96 passim
Plant Four (uranium tetrafluoride production plant), 91-98 passim, 103
Plant 172 (aircraft industry), 207
Plant 202 (nuclear fuel component plant), 97, 200n, 271
Plant 404, see Jiuquan Atomic Energy Complex
Plant 414 (Hengyang uranium hydrometallurgy plant, later Plant 272), 85, 95-97, 271
Plant Location Selection Office, 48
Plutonium: facilities for, 48, 55, 104-14 passim, 196, 273; and implosion system, 104, 139; in thermonuclear weapons, 104, 114n, 197-98, 200
Plutonium 239, 108-9, 165
Plutonium bomb explosion (Dec. 27, 1968), 113, 114n
Poland, 24f, 62, 76
Politburo, Chinese, 127, 130-31, 220, 223f, 231, 233; and first five-year plan, 11; and nuclear weapons decision, 37-45 passim, 76, 99, 105, 140; and Soviet nuclear assistance, 42, 60, 65, 72, 121, 122-23, 276; and organization of nuclear weapons program, 47-54 passim, 106-7, 131ff, 164, 221-22; Three-Member Group of, 47, 49, 95, 144; on limited test ban treaty, 192-93
Politburo, Soviet, 61f, 193
Political aspect, of military doctrine, 66f
Political leadership, 4, 116-17, 151, 220-25 passim, 229-33 passim, 246-47. See also Chinese Communist Party; Mao Zedong; State Council; Zhou Enlai
Political system: and science and technology, 45-46, 50-54 passim, 70, 232-37 passim. See also Mass movements; Personal networks; Political leadership
Polo, Marco, 171
Polonium, 156ff, 281
Population: minority, 172-74, 179-80; in Xinjiang Uygur Autonomous Region, 172-74, 283
Prevention, politics of, 66f. See also Arms control negotiations
Prisoners, in PC Corps, 179, 284
Prisoners-of-war (POWs): Korean War, 14-15, 16, 27, 255; Kuomintang, 179
"Probable Developments in Taiwan Through Mid-1956" (U.S. intelligence), 30-31
Proceedings of the Second United Nations International Conference on the Peaceful Uses of Atomic Energy, 55
Production and Construction Corps (PC Corps), 174f, 179, 181, 283f
Project 02, 39, 76, 96, 234, 237
Project 161, 100
Prototypes, 234; Soviets and, 41, 61-65 passim, 140f, 150, 222, 280
Purdue University, 148

Qian Gaoyun, 120n
Qian Jin, 146, 154, 248
Qian Sanqiang, 147, 226, 230, 249, 261, 262-63, 279-80; nuclear instruments brought from France by, 36, 156f; and nuclear weapons decision, 37-38, 43-47 passim, 99; and uranium hexafluoride production, 99, 101f; and Ninth Academy, 147, 150; and initiator, 156ff; and hydrogen bomb, 196, 200; and uranium prospecting, 268; and Neutron Physics Leading Group, 286
Qian Weichang, 44f, 262
Qian Xuesen, 50, 147
Qing era, 175
Qinghai Province: Ninth Academy in, 54, 140-55 passim, 159, 161, 182, 196, 200, 209; missile sites in, 213
Qinghua University, 44, 106
Quality, of uranium ores, 80
Quality control, in uranium production, 92, 99
Quartzite, uranium in, 81
Quemoy, 22-34 passim, 64, 258ff

R-2 and R-5 missiles, Soviet, 212
Radford, Arthur W., 13, 17-22 passim, 32, 34
Radiation counters, 74ff, 268
Radiation health hazards, 94, 186, 187-88, 273-74, 285
Radio station request, Soviet, 63f, 70, 265
Radium, 156
Radium salt, Chinese acquisition of, 36
Radon gas, radioactive, 86, 270
Rail transport, 141, 143, 181; of disassembled bomb, 182-83
Railway Corps, PLA, 143, 175, 179, 181
Rankin, Karl Lott, 20-21, 29, 260
Reaction chamber: Huang Changqing's, 101, 272; vacuum, 164

Reactor Equipment Leading Group, 114n
Reactors, 261, 274; Soviets and, 41, 47, 76, 99, 105, 112, 272; plutonium-producing, 48, 104-14 passim; at Institute of Atomic Energy, 99-103 passim, 112, 261, 272; in Jiuquan Atomic Energy Complex, 107, 111, 112-14, 273
Reconnaissance, aerial, in uranium explorations, 75
Red Flag General Headquarters (Hongqi zongbu), 203
Red Guards, 202, 203-4
Red Rebellious Corps (Hongse zaofan tuan), 203-4
"Reflectors," in bomb design, 138
Renmin Ribao (People's Daily), 1-2, 24-25, 32, 37, 40, 112, 261
"Report on Arrangements for Making Breakthroughs on Hydrogen Bomb Technologies" (nuclear ministry), 287
"Report on Problems in Speeding Up the Development of Nuclear Weapons," 208
"Report on the Development of the Atomic Energy Program Taking the Road of Self-Reliance," 164
Research, 3f, 42-45, 52, 105, 128, 130, 146; on conventional weapons, 50f, 128, 263; on uranium processing, 90-91, 95, 99-103, 118; Ninth Academy, 145, 150-51, 152, 160-62; on initiator, 157-59; missile design, 208
Research Department 615, Institute of Atomic Energy, 100-103, 118, 120n
Research Institute, at Lop Nur, 180-81
Research Unit on Underground Nuclear Testing Phenomena, 285
"Resolution on Certain Questions in the Construction of the Nuclear Industry" (Central Committee, July 16, 1961), 130, 277
Revolution Rebellion General Headquarters (Geming zaofan zongbu), 203
Revolutions, nuclear weapons and, 194
Rhee, Syngman, 19-20
Rhodes, Richard, 227
Ridgway, Matthew, 29, 31
Roads and highways, 141, 143, 179
Robertson, Walter S., 20
Rosenberg, David, 18
Rusk, Dean, 1, 7

Safety: Security and Protection Bureau, 55; in uranium mining and production, 86, 90; in nuclear energy plants, 93f, 157-58, 273-74; during tests, 186, 187-88, 203, 205
Sakata Shiyouchi, 38
Sandstone, uranium in, 80
San-gao (Three Do's), 118
Schomberg, Reginald, 171-72
Science: mass movements and, 45-46, 51, 117-18, 141, 165, 201-6, 214, 236f; political leaders and, 45-46, 50-54 passim, 70, 229-31, 236-37
Science and Technology Bureau, Second Ministry of Machine Building, 55
Science and Technology Equipment Committee, Central Military Commission, 128n
Scientific and Technical Cooperation Accord, Sino-Soviet, 105
Scientific Planning Commission, 49f, 53, 263f
Scientific Research District (Kexue yanjiu qu), 180
Scientific verification, *see* Verification, scientific
Scientists, 4-5, 36, 37-38, 42-51 passim, 107, 134, 219-37 passim, 247-49, 261-62, 272-73; training of, 36, 43n, 105-6, 145-49 passim, 236-37; and uranium programs, 38, 73-81 passim, 95-96; Cultural Revolution and, 46, 203-4, 232, 287; State Science and Technology Commission and, 54n; at Dubna institute, 105, 123, 146, 149, 272; on atomic bomb, 130, 139-40, 145-62 passim, 180-81, 186, 279-80; on hydrogen bomb, 199f, 202, 287; on missile-carried thermonuclear weapon, 209
Scintillation counters, 74, 76, 268
Second Artillery Academy, 213
Second Artillery Corps, 191, 202, 207, 215f, 234
Second Field Army, 116
Second Ministry of Machine Building, 48, 54-59 passim, 99, 121-30 passim, 234, 264, 269f, 276; original, 49, 263; and Nuclear Component Manufacturing Plant, 54, 111, 164; and Ninth Academy, 54, 121, 147ff, 160-61, 183; and uranium projects, 55, 87-103 passim, 111, 115, 118; Fuels Production Bureau of, 55, 111; and plutonium production, 55, 111-12, 114n, 196; Information Bureau of, 55, 278; Fifteen-Member Special Commission and,

133f, 205, 208; and atomic bomb test, 183-88 passim, 196; and hydrogen bomb, 196-201 passim, 205; and weaponization of nuclear program, 198, 207f; and Cultural Revolution, 203, 287

Secrecy, 48, 54, 70, 95, 151, 231, 234f, 275. *See also* Security precautions

Sections 615A and 615B, Institute of Atomic Energy Research Department, 100, 103n

Security and Protection Bureau, Second Ministry of Machine Building, 55, 112, 186

Security Forces (Gonganjun), 215

Security precautions, 48, 70, 96, 118-19, 125, 151, 162, 180-83 passim, 215, 231, 277

Self-reliance, 42, 65-72, 123-25, 199, 205, 220-21, 237-38, 287. *See also* Independence

Senate, U.S., and U.S.-Taiwan defense treaty, 32, 34

Seventh Ministry of Machine Building, 212, 264

Shandong Aluminum Factory, 144

Shanghai Baoshan Iron and Steel Complex, 107-8

Shanghai Electric Furnace Factory, 92

Shanghai Glass Machinery Factory, 92

Shangrao County, uranium mine and plant in, 78, 95

Shrinkage stoping, 81-85

Shuangchengzi, missile testing at, 209

Sichuan, gaseous diffusion plants in, 120n

Silk Road, 111, 115, 162, 170-71, 177

Simulation tests: atomic bomb, 184-85; weapons, 208

Sino-Soviet agreements, 7-12 passim, 34, 41n, 45, 60-72 passim, 89, 105, 115, 141, 220-22. *See also* Nuclear assistance, Soviet

"Situation with Respect to Certain Islands Off the Coast of Mainland China" (U.S. intelligence), 30

Sixth Bureau, Second Ministry of Machine Building, 55, 92

Sixth Institute, 90-98 passim, 271

Slavskii, Yefim Pavlovich, 61

Smyth Report, 278

Song Renqiong, 63, 115-22 passim, 135, 223, 234, 249; Third and Second Ministry headed by, 49, 52f, 115-16, 117,

142, 262, 264, 276; and Ninth Academy, 140, 142, 144, 161

Southeast Asia Collective Defense Treaty (Sept. 8, 1954), 22, 26

South Korea, 16, 19-20

Southwest Bureau, Chinese Communist Party, 117

Southwest China Associated University, 144, 148

Soviet Academy of Sciences, 229

Soviet Union, 26, 45, 229-30, 232; arms control negotiations by, 1, 35-36, 64-65, 192-93, 195, 241, 266; nuclear program of, 2, 104, 114, 137-40 passim, 198f, 227-31 passim, 256, 278, 290; Mao's worldview and, 5ff; and Korean War, 7ff, 41, 127; Chinese alliance and accords with, 7-12 passim, 34, 41n, 45, 60-72 passim, 76-78, 89, 105, 115, 141, 220-23 passim, 228; military doctrine of, 12, 39-40, 41, 66-68, 217; U.S. New Look policy and, 16, 18, 195; and Taiwan Strait crisis, 24, 26, 34, 63f; Chinese scientists in, 49, 105f, 123, 146, 149, 237, 272; Chinese debates with, 60, 66-72 passim, 191-95 passim, 217; uranium processing in, 91-92, 96, 271; Chinese lithium and beryllium supplied by, 155-56, 269; Chinese border clashes with, 180, 216; and Cuban missile crisis, 193-94, 214, 216; missile targets and, 213f; science and politics in, 229-30; aircraft industry in, 277

Spanish uranium prospectors, 74

Speer, Albert, 229f

"Spend less, get more" (*shao huaqian, duo banshi*), 107

Sputnik I, 68, 217

"Stage of flowering" (*kaihua jieguo*, 1962–64), 224-25

Stalin, Josef, 16, 40, 66, 71, 190, 230

State Construction Commission, 47

State Council, 41, 47-55 passim, 59, 76, 123, 204, 262-63

State Department, U.S., 19, 22, 31-32

State Planning Commission, 236

State Science and Technology Commission, 53, 54n

State Scientific and Technological Commission, 236-37

State Technological Commission, 53, 54n

Stein, Aurel, 172, 282f

Stolper, Thomas E., 257f

Stoping, shrinkage, 81-85
Strassman, Fritz, 137
Strategic Air Command, U.S., 18
Strong, Anna Louise, 6
Su, Deputy Chief Engineer, Second Ministry, 159
Su Hua, 55
Su Zhan, 59
Subei Mongolian Autonomous County (Subei Mongolzu zizhixian), 111, 273. *See also* Jiuquan Atomic Energy Complex
Submarines, nuclear-powered, 62
Summing-up (*zongjie*), 236
Sun, engineer with Ninth Academy, 154
Sun Yanqing, 55
Sun Zhiyuan, 47, 126n, 133
Sun Zongjie, 95-96
Suo Guoxin, 285
Supercharge, 139
Suzhou, 111
Szilard, Leo, 137, 226

Taiwan (Formosa): U.S. policy toward, 17-40 passim, 257-60 passim. *See also* Taiwan Strait crisis
Taiwan Strait crisis (1954-55), 11f, 22-40 passim, 63f, 73, 258f, 265
Taklimakan Desert, 284
Talimu Administrative Office, 284
Tamper, 139, 168
Tang Tsou, 34
Targets, Chinese missile, 212ff
Teams 309 and 519, for uranium prospecting, 76, 77-78, 87
Technicians, 4-5, 49, 106f, 130, 221-25 passim, 232, 272-73; training of, 43n, 49, 105-6; atomic bomb, 149, 159, 163-68, 180-81, 186ff; hydrogen bomb, 202, 205-6. *See also* Engineers
Technology, 45, 220, 225; and modernization plans, 9f, 43, 49-51; and political system, 45, 50-54 passim, 70, 232-37 passim; mass movements and, 45-46, 51-52, 117-18, 141, 165, 201-6, 214; in military doctrine, 66f, 211
Tenth World Conference Against Atomic and Hydrogen Bombs, 195
Test base, *see* Lop Nur Nuclear Weapons Test Base
Tests, 234, 244-45, 285; arms control negotiations and, 1, 12, 35-36, 64-65, 192-93, 195, 241, 266, 286; first atomic bomb (Oct. 16, 1964), 1-2,

182-89, 195-96, 197, 215, 229, 241-43, 283, 284-85; U.S., 18, 20, 104, 226, 290; plutonium bomb (Dec. 27, 1968), 113, 114n; atomic bomb detonation process (Nov. 20, 1963; June 6, 1964), 156-60; "cold," 201; "fundamentals of a thermonuclear explosion" (Dec. 28, 1966), 201; hydrogen bomb "booster" (May 9, 1966), 201; DF-2 missile (Oct. 27, 1966), 202-3, 209, 212, 238; hydrogen bomb (June 17, 1967), 205-6, 267, 288; weapon, 207, 290; air-dropped atomic bomb (May 14, 1965), 208; liquid-fueled missile (Nov. 1960), 212; DF-3 missile (Dec. 1966), 213; underground, 285; British, 290; French, 290; Soviet, 290
Thermonuclear weapons, 196-202; Soviet, 198, 256; missile-carried, 202-3, 209. *See also* Hydrogen bomb
Third Bureau, Ministry of Geology, 47, 54, 76, 77-78
Third Ministry of Machine Building (nuclear), 48-53 passim, 115-16, 117, 142, 262-63, 269, 274; Song Renqiong heading, 49, 52f, 115-16, 117, 142, 262, 264, 276; and uranium exploration, 77, 81; and plutonium reactor, 114n
Third Ministry of Machine Building (aircraft), 207-8
Third Office, State Council, 47
Three Do's (*san-gao*), 118
"Three hard years" (1960-62), 86, 94, 122, 127ff, 143, 178-79, 224
"Three-in-one administrative" system, 121
Three-Member Group, 47, 49, 95, 144
Three-stage bomb, 198
Tian Zhaozhong, 91
Tianjin Physics and Chemistry Engineering Academy, 120n
Timing: of decision to build bomb, 11-13; in bomb design, 138-39, 159; for completing bomb, 152, 164-65; for first test, 185
Tongxian, uranium production plants in, 91-98 passim, 103
Training: of scientists, 36, 43n, 49, 105-6, 145-49 passim, 236-37; of engineers and technicians, 43n, 49, 105-6
Transportation: development in, 143; security in, 182-83, 277
Treaties, nuclear: limited nuclear test ban, 1, 192-93, 195, 241, 266, 286;

Non-Proliferation of Nuclear Weapons (July 1, 1968), 12
Tritium, 200n
Tuoli, *see* Institute of Atomic Energy
Turkic groups, in Xinjiang Uygur Autonomous Region, 172-74
Twelfth Bureau (Mining and Metallurgy Bureau), 55, 85, 90
Twelve-Year Plan for the Development of Science and Technology, 49-51
Two-Year Plan, for atomic bomb development, 164-65, 282

Underground testing, 285
Uniforms, 59
United Kingdom, *see* Great Britain
United Nations Charter, 12
United Nations troops, in Korean War, 8, 12 15 passim
United States, 26, 74, 202; and arms control negotiations, 1, 35-36, 64-65, 192f, 195, 241, 266; and Chinese nuclear policies, 1-2, 34-41 passim, 65-69 passim, 191-92, 195, 211, 215, 217, 241f, 286; and China's first atomic bomb test, 1-2, 215, 229, 241-42; Mao's worldview and, 5, 6-7; and Korean War, 8-20 passim, 24, 27, 255; and Geneva Conference, 11, 19-22, 28; defense and nuclear policies of, 13-37 passim, 41, 195, 256-61 passim, 278, 286; and Indochina, 17-22 passim, 26, 28, 32, 127, 224, 257, 288; Taiwan policy of, 17-40 passim, 257-60 passim; Soviet military doctrine and, 40f, 67-68; Chinese scientists and students in, 43n, 73, 144-48 passim, 279-80; intelligence on Chinese nuclear industry, 77, 109-11, 213n, 268, 271, 288f; and Cuban missile crisis, 193-94, 214; Chinese missile targets and, 212ff; British alliance with, 220, 227-28. *See also* United States nuclear program
United States nuclear program, 2, 137-41 passim, 226f; policies on, 16-17, 20, 32, 41, 195, 256, 261, 278; tests in, 18, 20, 104, 226, 290; hydrogen bombs of, 20, 104, 197, 199, 230; uranium processing in, 91-92, 114f, 274; atomic bombs in, 104, 226, 281; costs of, 108, 274; plutonium production in, 109; Los Alamos, 139, 141, 151, 280-81, 287, 290; and multistage

thermonuclear bomb, 198; Manhattan Project, 226f, 230, 278
University of Berlin, 146
University of California, 146
University of Michigan, 145
Uranium, 39, 73-103, 196; geology of, 38, 73-81 passim; explorations for, 41, 47, 54, 73-81, 268-69; processing of, 55, 62, 73, 85-105 passim, 162, 224, 277; mining of, 55, 73-89 passim, 233, 268-71 passim; Soviets paid with, 61-62; quality of, 80; yellow cake, 88-92 passim; first fission experiments with, 137; radium and polonium and, 156, 158; hydrogen bomb and, 200. *See also* Enrichment, of uranium; Uranium hexafluoride; Uranium isotopes; Uranium oxides; Uranium tetrafluoride
Uranium Chemistry Technology Group, 157-58
Uranium core, *see* Nuclear component
Uranium hexafluoride, 55, 89-90, 98-105 passim, 162, 166, 276; gaseous diffusion of, 62, 100, 103f, 114f, 125, 135 (*see also* Lanzhou Gaseous Diffusion Plant); Soviets and, 62, 99, 115, 222
Uranium Hexafluoride Plant (Liufuhuayou shengchan chang), 103n
Uranium isotopes, 104, 108-9, 113f, 125, 134-39 passim, 156, 165
Uranium Mining and Metallurgical Processing Institute (Youkuang xuanye yanjiusuo), *see* Sixth Institute
Uranium Mining and Metallurgy Design and Research Academy, Hengyang, 95
Uranium Oxide Production Plant (Plant Two), 91-96 passim
Uranium oxides, 55, 90-98 passim, 105, 271
Uranium tetrafluoride, 55, 90-98 passim, 103, 162, 272
Uranium Tetrafluoride Workshop, Plant 202, 97-98
"U.S. Policy Towards Communist China" (NSC 166/I, Nov. 6, 1953), 18-19
Uygurs, 172, 283

Van Fleet, James, 8, 21, 25
Verification, scientific: for uranium enrichment, 135; of bomb design, 149, 161; before bomb test, 149, 185
Vietnam, 12, 21f, 28, 127, 257f, 288
Volcanic rock, uranium in, 80

"Walking-on-two-legs" political line, 51-52
Wan Yi, 54
Wang Aimin, 270
Wang Chengshu, 46
Wang Chengxiao, 135f
Wang Enmao, 175
Wang Fangding, 157f
Wang Ganchang, 44, 99, 145-60 passim, 182-88 passim, 205, 226, 248, 261, 279-80
Wang Heshou, 133
Wang Jiefu, 115-24 passim, 134-35, 136, 250, 275
Wang Wenyuan, 89
Wang Yalin, 55
Wang Zhen, 174f
Wang Zhongfan, 116f, 120
War: Mao on, 6, 36-40 passim, 68-69, 190-91, 216, 265f; "law of the inevitability of," 66; Central Committee on, 191-92; nuclear policy and, 211. *See also* Deterrence; Nuclear policies
Waste disposal, nuclear, 55
Weaponization, of nuclear program, 130f, 151, 191, 198, 202-3, 207-10, 287
Weapons, *see* Conventional weapons; Nuclear weapons; Weaponization
Weather station, Yangpingli, Xinjiang, 181, 284
Western Han Dynasty, 111, 170
West Germany, 242
Wheeler, John A., 137, 226
Whiting, Allen, 64, 288
Whitson, William, 9, 116
Wigner, Eugene, 137, 226
Wilson, Charles E., 21
Work Bulletin, 128
Wu Dayou, 144-45
Wu Jilin, 46, 52, 250; and atomic bomb, 143-44, 145, 150-55 passim, 159, 169, 182-88 passim
Wu Xiuquan, 261
Wu Youxun, 43, 261, 272
Wu Zhengkai, 100, 102, 248

Xia fang ("to the lower levels") movements, 94
"*Xian xie kaishu hou xie caoshu*" (First write in regular script and then write in cursive script), 118, 165, 275
Xiang Jun, 271

Xiang mountains, uranium in, 78
"Xiao," 272
Xiao Jian, 262
Xin Xianjie, 262
Xinjiang Institute of Archeology, 282
Xinjiang Military Region Command, 204
Xinjiang Province, uranium explorations in, 47, 73-78 passim, 87, 268
Xinjiang Uygur Autonomous Region, 170-89, 282-83. *See also* Lop Nur Nuclear Weapons Test Base
Xinsheng liandui (newborn companies), 284
Xu Jianming, 262
Xu Keijiang, 206
Xuanzang, 171

Yan'an, 223
Yang, technician predicting zero hour, 185
Yang Chengzhong, 262
Yang Chengzong, 36, 262
Yangpingli meteorological station, Xinjiang, 181, 284
Ye Jianying, 128-29
Ye Minghan, 262
Yellow cake, uranium, 88-92 passim
Yijiang (Ichiang, Yijiangshan Dao), 26n, 31f, 37
Yita incident, 180
Yu Daguang, 146, 154, 248
Yu Min, 146, 196, 200-201, 248, 262, 287
Yuan Chenglong, 124, 166f, 250, 262, 276
Yuan Gongfu, 167-68

Zero hour, for first test, 185
Zhang Aiping, 26n, 54, 101-2, 123, 132n, 133, 143, 208, 250, 263; and missiles, 50, 214; and first atomic bomb, 51, 54, 169, 178, 182-88 passim
Zhang Cheng, 97
Zhang Dingzhao, 75
Zhang Gengsheng, 75
Zhang Guizhi, 86
Zhang Jifu, 284
Zhang Jingfu, 134
Zhang Jun, 213n
Zhang Peilin, 166f
Zhang Pixu, 116f, 121
Zhang Tianbao, 46
Zhang Tongxing, 46

Zhang Wenyu, 46
Zhang Xianjin, 59
Zhang Xingqian, 150, 246
Zhang Yunyu, 176-77, 178, 183, 187f, 250
Zhao Erlu, 49f, 126, 132f, 211-12, 250, 264
Zhao Zhongyao, 261f, 290
Zhejiang University, 44
Zheng Hantao, 132n
Zhibian liandui (support-the-frontier companies), 284
Zhongcheng (medium-range), 213n
Zhongguancun (science quarter), Beijing, 134
Zhongguo Wuli Xuebao (Chinese Journal of Physics), 44
Zhongjincheng (medium-short-range), 213n
Zhongnanhai: nuclear decision made in, 38-39, 230; Fifteen-Member Special Commission meeting in, 164-65
Zhongshan County, uranium in, 75
Zhongyuancheng (intermediate-range), 213n
Zhou Enlai, 1, 130, 220, 223f, 230, 233, 247; and Soviet Union, 7, 63; and Taiwan Strait crisis, 25, 28-29; and Ge-neva Conference, 28; and nuclear weapons decision, 37-39, 45; and ura-nium projects, 38, 75f, 98, 135-36; and Twelve-Year Plan, 49; and tech-nical personnel, 106; slogans of, 107, 118, 123f, 276; and Fifteen-Member Special Commission, 131-32, 133, 164-65, 178, 182, 208; and Zhu Guangya, 145; and Ninth Academy, 150, 198-99; and atomic bomb, 164-65, 169, 178, 182, 186-89 passim, 208, 282; and missiles, 198, 282; and hydrogen bomb, 198-99, 205f; and weaponization, 198-99, 208f, 282; and Korean War, 255; "Experts Work Group" created by, 261; and Cultural Revolution, 287
Zhou Guangzhao, 45f, 123, 146, 149, 154, 161f, 168, 185, 249
Zhou Zhi, 168
Zhu Guangya, 99, 249, 262, 279; train-ing programs by, 106, 237; and atomic bomb, 143-46 passim, 151, 157f, 161, 184, 187f, 279-80, 282
Zhu Hongyuan, 262
Zhu Linfang, 163-68 passim, 250
Zhuguang mountains, uranium in, 78
Zhukov, G. K., 67